1 MONTH OF
FREE
READING

at

www.ForgottenBooks.com

By purchasing this book you are eligible for one month membership to ForgottenBooks.com, giving you unlimited access to our entire collection of over 1,000,000 titles via our web site and mobile apps.

To claim your free month visit:

www.forgottenbooks.com/free28849

ISBN 978-0-364-63117-1
PIBN 10028849

This book is a reproduction of an important historical work. Forgotten Books uses
state-of-the-art technology to digitally reconstruct the work, preserving the original format
whilst repairing imperfections present in the aged copy. In rare cases, an imperfection in
the original, such as a blemish or missing page, may be replicated in our edition. We do,
however, repair the vast majority of imperfections successfully; any imperfections that
remain are intentionally left to preserve the state of such historical works.

THE GAS-ENGINE.

A TREATISE ON THE

INTERNAL-COMBUSTION ENGINE

USING GAS, GASOLINE, KEROSENE, ALCOHOL,
OR OTHER HYDROCARBON AS
SOURCE OF ENERGY.

BY

FREDERICK REMSEN HUTTON, E.M., Ph.D., Sc.D.

*Emeritus Professor of Mechanical Engineering in Columbia University; Past
Secretary and President of the American Society of Mechanical Engineers.*

THIRD EDITION, REVISED.

FIRST THOUSAND.

NEW YORK:

JOHN WILEY & SONS.

London: CHAPMAN & HALL, Limited.

1908.

GENERAL

The Scientific Press
Robert Drummond and Comp
New York

WHEN a previous treatise by the author was published under the title of "The Mechanical Engineering of Power Plants." it was suggested by one of his most gifted critics that the title should be amended because the book did not cover the power-plant practice which uses gas-engines.

The point was well taken, but the omission was intentional. To have included the gas-engine would have made that book inconveniently bulky. Furthermore, the treatment of the gas-engine must be essentially different from that given to the steam-engine, and at that time the state of the art, both practically and scientifically, did not admit of the preparation of a satisfactory and exhaustive discussion. Since that time, however, there has grown up a largely increased appreciation of the fuel value of what were called the waste gases from the blast-furnace, and a wider extension of the manufacture of fuel gas in producers. The gas-engine has been extensively applied in the departments of electric lighting, and of compression, both of air and gas. It is since that time also that there has appeared the exacting demand for motors for self-propelled vehicles and for small launches, so that it has become possible to undertake that for which the time was not ripe when the criticism was made. There was, at that time, little distinctively American practice to be studied, but the principal work had been done in England, Germany, and Belgium. The introduction by Daimler of the high-speed gas-engine and the immediate development which this class of motor received in France for motor-vehicle use, has greatly stimulated the work upon this class of machine in all departments. Furthermore, the development of the carburetor and the recognition of the significance of carburation as a process in the handling of liquid fuels, enormously widened the scope and field for the internal-combustion motor. In fact, in the opinion of the writer, the development of this particular detail draws a broad line of distinction between the former and present practice which marks in effect an epoch of the development of the art.

By the term gas-engine is, therefore, meant the internal-combustion engine, whether using gas manufactured without the motor and delivered to it as combustible gas, or making its own gas by carbureting air on its way to the combustion-chamber.

The author believes that a very important field is open for development of small gas producers operating on coal in combination with the internal-combustion engine which they are to serve. It will be apparent that by either the liquid-fuel system of carburetting air, or by the producer system, the gas-engine reaps all the advantages which follow from getting rid of the boiler and its plant as details of the steam-generation system whether in stationary, marine, or motor-vehicle practice.

The plan and scope of the treatise will be apparent on inspection. The starting-point must, obviously, be the liberation of the energy resident in fuels in the form of heat and the conversion of that heat energy into mechanical energy with the physical laws and mathematical principles which are involved in such transformation. The cycle of operations which the heat medium undergoes in transforming heat energy into mechanical work next follows, and the types of motor in which these transformations occur with gas, gasoline, kerosene, and alcohol as sources of the combustible hydrocarbon. The succeeding chapters open up the details of mechanically effecting the mixture of fuel and air for the internal combustion which is desired, and the methods of carburetting, igniting, and governing. The chapter on manipulation is intended to be of service to users of engines of this class, although as a rule the unsatisfactory working of such engines is the consequence of defective ignition, carburation or mixture, already discussed, rather than the consequence of the methods of manipulation. This part concludes with a brief presentation in compact form of the results in economy and performance as determined by test.

The final chapters treat of the mathematical analysis of the laws and principles whose action has been discussed in the first part. By the mathematical form of this analysis it becomes easy to use the results of the analysis in a quantitative way for a comparison of cycles or with a view to studying the effects of varying these cycles. So far as known, this part of the work is the most complete treatment which has yet been made, and it is hoped it will leave little to be desired by those who would find a study of this sort serviceable, since it is believed to be practically exhaustive. Especial attention should be called to the formula for theoretical mean effective pressure. The development of the Otto cycle, by reason of the investment of capital for this purpose, has thrown into comparative obscurity the important possibilities offered by the other type of internal-combustion engines in which the heating takes place at constant pressure rather than at constant volume. A chapter is given to a brief treatment of this particular form of engine, but it will be apparent that this is the opening of a door at this time rather than the entering upon a full treatment. If the continuous rotative type of motor is to be developed for the direct utilization of the heat energy from combustion, it is likely to appear in the development of this class of cycle. But at this time the state of the art does not justify the giving of more space than is allotted.

The book concludes with some general statements concerning explosive mixtures, an historical summary and a brief bibliography.

The author, in conclusion, must express his obligation and indebtedness to the work of previous writers for the results and data which they have made as contributions to the arts and sciences whose applications have developed the gas-engine. References to the sources from which these data are taken will be found through the text or in the titles. He would express his particular obligation to Dr. Charles E. Lucke, who is associated with him in the work of education and research, and from the result of whose work with his permission he has made very liberal drafts. The cyclic analysis in Chapter XVII was made by Dr. Lucke while a graduate student under the author's direction, but the work is so marked by originality and industry in the prosecution of its detail that Dr. Lucke should receive the full credit which it deserves and which the author is very glad to take this occasion to ascribe. Dr. Lucke has also contributed much of signal value in the way of quantitative experimental and practical data, the results of research and tests in the laboratory. The author would express his thanks also to Prof. R. H. Fernald of St. Louis, to Mr. T. J. Foster of Scranton, and to Mr. E. P. Ingersoll, and others to whom he is indebted for permission to make use of serviceable illustrations which have already appeared elsewhere.

No attempt has been made to enter upon the field of the design of the gas-engine considered as a machine. This field of engine design has been so admirably covered by others with respect to the steam-engine and the same principles of design apply so readily to the gas-engine with the necessary modifications introduced by differences in principle, that this subject has been disregarded in the present treatment. The exception is the treatment of the cylinder volume as a derivative from the transformation of heat energy into mechanical work which may, perhaps, be found serviceable.

F. R. HUTTON.

COLUMBIA UNIVERSITY,
　　NEW YORK,
　　　September, 1903.

PREFACE TO THE THIRD EDITION.

THE practical solution of a problem involving the applications of natural law may be effectively studied from two view-points. The one is the functional view-point, involving an examination of what the machine does, and how. The second is the quantitative examination of the size it must have, to do a certain amount of industrial work in accordance with the limits set by natural laws. Both studies are necessary for successful design, but the functional idea should precede the quantitative study.

In the preparation of the first and second editions of this text; the first idea was most prominently before the author's mind, but the development of the internal combustion motor, and the numbers concerned in its design and manufacture have made it appear desirable to amplify certain parts of the treatment, so as to cover more of the quantitative requirement of the engineer and builder. - It is hoped that such features will be appreciated and will add to the reference value of the book.

Particular attention is therefore called to the Reference Table of Gaseous Fuels in connection with paragraph 29, and the availability of such table for use in determining Mean Effective Pressure. Too great emphasis cannot be put upon the importance of using the exact data of fact in the making of guarantees of efficiency and economy, rather than the unreliability of mere assertion, and this table is intended as a step towards making the heating value of the fuel the ultimate standard of these matters. The treatment of mean effective pressure, of efficiency, of the guarantee and considerable expansions of the discussion on the producer, and of the alcohols as fuels are all new features of this edition. The value of the analysis of the different cycles is also increasing as more attention is paid to these matters by both engineers and the public. The carburetor treatment has been expanded and brought more up to date.

In the previous editions the treatment of the design of engine parts (paragraph 205) has been intentionally omitted. It is also left out of this, not only because of the inconvenient bulk and cost which it would entail upon the book, but more because of the appearance of the book by a colleague and co-worker of the author, entitled "Gas-Engine Design," by Prof. Charles Edward Lucke. In this the problems of inertia effects, balancing, crank-effort and shaking forces, together with valve proportions, fly-wheels, cams, springs, and other parts are treated so exhaustively, and in so scholarly a way, that to include these topics herein would be both repetitious and superfluous. The author would express his thanks and recognition for the privilege of incorporating into this edition certain data and tabular matter compiled and computed by Professor Lucke in connection with instruction courses in which both books have been used. To Mr. R. E. Mathot of Belgium he is indebted for the originals from which the sections of foreign gas-producers have been derived.

It is the hope and desire of the author that the teachers, students, engineers, manufacturers, and others who have been kind enough to commend the previous editions, will find this new one incorporating suggested alterations even more useful than its predecessors.

F. R. HUTTON.

Columbia University,
September, 1907.

TABLE OF CONTENTS.

CHAPTER I.

INTRODUCTION.

CHAPTER II.

LIBERATION OF HEAT ENERGY. COMBUSTION.

CHAPTER III.

MECHANICAL ENERGY FROM EXPANSION OF GAS AND AIR.

CHAPTER VII.

GAS ENGINES USING GASOLINE   AUTOMOBILE ENGINES.

CHAPTER VIII.

ALCOHOL-ENGINES.

CHAPTER IX.

PROPORTIONING OF MIXTURES.

CHAPTER X.

CARBURATION AND CARBURETERS.

CHAPTER XI.

IGNITION.

CHAPTER XII.

GOVERNING.

CHAPTER XVII.

THEORETICAL ANALYSIS OF THE GAS-ENGINE.

LIST OF ILLUSTRATIONS.

THE GAS-ENGINE.

CHAPTER I.

INTRODUCTORY.

1. Sources of Motor Energy.—There are three important sources of power or energy for industrial uses. The first is known as muscular power and is that which resides in the contractile tissue of the muscles in man and animals. The second is the force by which the earth attracts all masses towards its centre, which produces weight and the acceleration due to gravity. The third is a group of forces which become manifest or are released upon chemical reactions such as occur in combustion or oxidation; the two most important manifestations of these are the forces of electricity and of heat.

2. The Limitations of Muscular Force and the Power of Gravity.—There are certain fixed limits set to the amount of energy which can be made available from any single unit, either of men or of animals.

While this limit will vary with the muscular endowment of the individual, and further with his temperament, his training, his health, his size, his race, and his species, it will be obvious that large powers can only be obtained by aggregating many such units. This is inconvenient and costly, but even more than this, a definite limit is set by the endurance of the animal unit, which must have periods of rest and recuperation. The speed of such units is also limited by the ability of the animal motor to maintain his maximum effort for any length of time. There is, finally, no considerable reserve of energy in storage to be drawn upon in case the resistance should be temporarily increased.

The power due to gravity becomes available for motor uses when a weight or mass is lifted to a higher level and is permitted to descend to a lower one. Solid weights are only serviceable when lifted by some other mechanical force; water and air are the only weights which are otherwise lifted, independent of man, to a distance farther from the centre of the earth. Water is lifted by the sun in vapor, to be deposited on the high levels of the land, whence it seeks to descend again to the tide-water levels; the winds which drive wind-motors are caused by the descent of the cold air from upper levels of the atmosphere to the lower levels by reason of its greater weight per unit of bulk which tends to displace the air which has been warmed by the earth. It is obvious, therefore, that gravity, as a motive power, is dependent upon the availability of higher levels of land at which a sufficient mass of water can be accumulated; and an adequate reservoir in any particular region or an adequate flow from a source, together with an available difference of level, are necessary conditions for the use of water-motors. With respect to windmills, it must as yet be said that while there is an abundance of energy present in the atmospheric ocean, at the bottom of which all the industries of the earth are carried on, yet the reliability, capacity, and controllability which must attach to a satisfactory industrial motor are not found in most places. The exceptions are where windmills may be used for pumping or to store some other form of energy in accumulators or otherwise, to be given out as required.

This same series of difficulties has beset the successful application of the energy stored by the winds and other disturbances in the waves of the ocean. Tide-motors depend upon the lifting of the ocean level by stellar or planetary attractions and are reliable and controllable within the limits of their capacity. They are only made of large capacity at great cost. The types of motors so far proposed to utilize either the impact of ocean waves or the lifting force of the continuous wave near the coast have not proved reliable nor permanent enough for engineers to· venture to adopt or install them as a source of continuous energy.

It will be apparent that since it is the energy of the sun which lifts the water to higher levels of land and which disturbs the equilibrium of the strata of air, there is a figurative sense in which both water-motors and windmills can be called heat-motors in the last reduction.

3. Importance of Motor Energy Derivable from Heat.— It will be at once apparent that while the energy resident in falling water is most serviceable and is destined, doubtless, to become more so as the means of transmitting energy are improved, yet there are many causes which make the form of motor utilizing the energy liberated in the form of heat to be by far the most considerable. The energy due to falling water is, with a few notable exceptions, limited in amount both by the weight available and by the height of the fall. The· weight available becomes uncertain when by reason of diminishing rainfall the amount of water received upon any watershed becomes diminished. In the case of the energy derived from the combustion of a fuel furnishing heat, there is stored an amount of available energy limited only by the supply of such combustible fuel. The energy, moreover, is in compact bulk, and in the form of compressed gas or in the liquid fuels it is of comparatively light weight with respect to the amount of energy which it can furnish. For these reasons the importance of the study of heat· motors is very great under the present conditions of industry, and the exceeding convenience which attaches to the gas-engine as a means of utilizing the energy of combustible fuel has been continually receiving increased consideration.

While it is not difficult to believe that the near future may reveal methods for generating or liberating energy directly from a fuel in the form of electromotive force and current, and this is now done where the chemical reactions in the various electrical batteries release such force and current, yet, at this writing, the importance and extent of the applications of such methods place them in the field of the physicist and the experimenter rather than in that of the engineer concerned with industrial problems.

4. Analysis of a Power-plant.—The industrial result in a power-plant is the production of something which shall have a commercial or salable value. This may be a manufactured article or it may be the transportation of persons or of goods for industrial purposes, or for pleasure, for which the community shall be willing to pay. It will be apparent, therefore, that the last link in a power-generating chain will be as extensive as the entire field of industry. The transmission to the machines or appliances which utilize the energy is also a field of wide extent which will be greatly conditioned by the purpose for which the power is to be used. For these reasons it also may be excluded from the present field of consideration and attention paid only to the problems connected with the liberation of the energy or the generation of the power in a device or appliance which is fitted to receive the energy liberated from the combustible fuel and manifested in the form of a force exerted through a space. The problem of the heat-engine, therefore, has two distinct divisions. The first is the liberation of the heat energy and its transfer to a medium capable of exerting mechanical energy. The second division is the motor or engine to receive this mechanical energy and to put it into usable form. The development in the subsequent chapters will follow the lines of these two divisions.

5. Media for Use in Heat-engines.—In the selection of a medium to receive the energy liberated from a source of heat it will be apparent that the considerations are both scientific or physical, and commercial or practical. The first and most obvious phenomenon which occurs upon the increasing of the heat energy in a body is an increase in its bulk or volume. The gases undergo the greatest change in bulk or volume for a given increase in their amount of heat energy, and would naturally be those which would be first chosen as media. While solids and liquids also undergo a similar dilatation, it is less in extent, although capable, within the range of such dilatation, of exerting a force of much greater intensity. By the use of gases which change their shape or figure very easily within a containing envelope, and which have small density or weight per cubic foot, the flow of such media through pipes and passages is more rapid and is less affected by friction or other resistance. Among the media, they will be found to differ among each other according to the ease with which different materials will pass from the gaseous state into a liquid form. Steam, which is the result of evaporating water into a gas, is the most convenient example of such media as change their state easily within the limits which are within convenient reach. Other such media are:

Ammonia (NH_3);
Acetone (C_3H_6O);
Alcohol (C_2H_6O);
Bisulphide of carbon (CS_2);
Chloride of carbon (CCl_4)·
Chloroform ($CHCl_3$):
Ether ($C_4H_{10}O$);
Naphtha and gasoline (C_6H_{14} to C_8H_{18}).

These have the advantage of making the change from liquid to vapor at temperatures lower than is necessary for steam, but are open to the serious objections on the practical side that they are costly and require to be operated in such a form of engine as shall permit that after the use in the motor proper the escaping vapor shall be condensed back to liquid so as to be used over again continuously. They are also open to objections either by reason of an offensive or pungent odor, or because they are inflammable or explosive, or produce some physiological effect on the human organism. Some are poisonous and irrespirable. These are competitors with steam as a medium rather than with air.

Air, on the other hand, as a medium is cheap, safe, odorless, innoxious, non-inflammable, and very accessible. It has the advantage, furthermore, of being able to be used with a direct contact with the flame which is a manifestation of the liberation of the heat energy, which the others are not capable of doing. The importance of the air as a medium is, therefore, sufficient to make it possible to confine the present treatment to those forms of energy which are conveniently imparted to air, and to those forms of motors which utilize the expansive force incident to such heated air. References to discussions of other media will be found in other treatises for those who may desire to pursue this particular department more fully.*

6. Sources of Heat Energy.—In seeking for a source of heat energy for utilization in a power-plant, it is apparent that the same sort of circumstances must govern. The source of heat must be conveniently accessible, cheap, and must contain large reserves of heat in a small bulk. While the oxidation of all chemical substances is accompanied with the liberation of heat, there are certain of them in which this liberation occurs with the most convenient rapidity. These substances are carbon and hydrogen in their usual natural or combined forms. These occur in nature in solid form, as coal or wood, or in manufactured coke and charcoal; in liquid form, as oils; and in gaseous form either natural or manufactured. It will be apparent in the later treatment that for many reasons the liberation of energy by the combustion or oxidation of gas is by far the most convenient, so far as the motor itself is concerned and the plant as a whole. It may be desirable, when the fuel is in solid form, to convert it artificially into gas by a gas-making process and use it in that form rather than in its natural state. For the purpose of this treatise the source of heat will be taken as a gas and its process of liberation will be the burning of this gas with the necessary proportion of air whereby the energy of the ignited gas is imparted to raise the heat energy of the air. The liquid fuels can be treated and considered as

* See Hutton's "Heat and Heat Engines," Chapters IX and XXI.

operating in a manner identical with that of gas, inasmuch as in their practical utilization the liquid fuel is injected in a finely divided state into the air and is ignited in this vaporous or atomized condition, when it behaves exactly as the gas would, so far as the effect in liberating energy is concerned.

7. Internal-combustion Method of Heating Air as a Medium. —The energy liberated from a gas, or an oil, or a solid fire, by its combustion, may be imparted to the air as a medium by three different methods:

I. The fire may be placed on one side of a metallic wall through which the heat of the fire must pass to heat the mass of gas on the other side of the wall. This plan may be called the external heating system. It is the method used in the steam-engine, so far as imparting the energy of the fire to the water and steam in the boiler is concerned, and is the method which is followed in the ordinary hot-air engine of the Ericsson or similar types.

II. The second method is that whereby a solid mass heated by the fire is afterwards removed from the fire and brought into contact with the mass of gas to which, by contact and radiation, it imparts a part of its energy. . This is even less effective than the preceding system, but may be called a combination of the external and internal heating systems.

III. The third, or internal-combustion, system is to have the fire enclosed in a vessel and maintained in activity by the mass of the gas itself, which receives directly and without an intervening mass the heat energy from that fire. In this case, obviously, the gas must be such as to furnish the necessary oxygen for this internal combustion, and of course, of all, heated air is the most convenient for this class of operation.

This last system, the internal-combustion system, is by far the most effective, since any system depending upon heating air or any gas by contact with a solid at a high temperature must necessarily be slow; it requires that the gas be in thin layers or films, and large masses of gas have to be handled with corresponding bulk or weight of the heating surface. For effective transfer the hot walls must be hotter than the receiving medium, and the difference of temperature must be so great that it is difficult to find a material for the solid which does not rapidly deteriorate from the high heat.

The heating of the medium by internal combustion has been effected either with coal, with oil, or with gas. The methods used might be tabulated as follows:

I. With the use of coal as a source of heat.

(a) Air is passed through a coal fire and, after having effected the combustion of the coal and become heated, the air passes to the working cylinder of the motor, where it exerts its expansive force. The names connected with this system are Cayley, Genty, Shaw.

(b) A coal fire is moved through an enclosed mass of air. (System of Lord.)

II. Using a liquid fuel not vaporized before entering the cylinder of the motor.

(c) The enclosed air in the motor behind the piston acts as a quiet atmosphere supporting the combustion of a jet of oil. (Diesel.)

(d) The air is caused to move past a burner and in passing it supports the piston and the heated products pass on. (Wilcox, Brayton, Nordberg & Shadall.)

(e) Oil is forced by a pump into a hot chamber vaporized therein by the heat and is then brought into contact with the air. The proportions of fuel and air are so maintained as to make the resulting gaseous mixture practically explosive, so that the combustion propagates itself through the mixture. (Hornsby, Capitaine, Mietz & Weiss.)

III. Using gas or oil which has been previously vaporized.

(f) An enclosed mass of atmospheric air supports the combustion of a quiet jet of gas-flame. (Diesel and Gibbs.)

(g) Air in motion passes a fixed gas-flame and becomes heated by it. This is the method of most of the atmospheric engines, such as Wilcox, Weiss, and the Otto Atmospheric.

(*h*) Air mixed with gas in explosive proportions is caused to pass a point where combustion is localized. (Brayton, Schmid, and Beckfeld and Reeve.)

(*i*) Air mixed with gas in such proportions that a flame will propagate itself throughout the mass is enclosed in a chamber and while at rest is inflamed by being locally ignited. To this class belong the Otto, Priestman, Nash, Westinghouse, and nearly all existing internal-combustion engines.

The present treatment will confine itself to the systems using oil or gas, by reason of the fact that where solid fuel is used the presence of corrosive products of combustion from solid fuel and the injury to the cylinder and moving parts by the dust and ashes from such fuel have removed engines of this type from competition with those using the more convenient form of the fuel. System *i* includes the engines which may properly be called "explosive" engines, since a flame at one point of the mixture is expected to propagate itself throughout it. Those in system *h* are properly "non-explosive," since the combustion is localized and the gases are in motion when heated. Engines in this latter class may be continuous non-explosive (Reeve) or intermittent non-explosive (Diesel).

7a. Analysis of the Internal Combustion Motor. — It would appear therefore that the modern type of internal combustion motor is designed to receive the energy resident in a gas or liquid fuel into its cylinder, and there, by the processes of chemistry and the laws of mechanics, to transform that heat energy into mechanical work. The change from potential energy in the fuel into actual and available energy is effected by the combustion process to be discussed in the next chapter, whereby the heat energy which is a necessary accompaniment of the chemical reaction is imparted to the air which supports it. From the cylinder the mechanical energy is transmitted by a revolving shaft to the point where work is to be done. A diagram is subjoined (Fig. 224) which will illustrate the sources of energy as input which must be supplied to such a motor when self-contained and will show also the directions of output of mechanical energy for the motor itself as well as for useful industrial work. If the motor is not self-contained, some of these sources of energy are independent of it, but must then usually be paid for in other units than those of heat and power.

The efficiency of such a motor is the fraction which expresses the relation of the output as numerator to the input as denominator. If both be expressed in heat units, the fraction is the thermal efficiency. The mechanical efficiency is the relation between the work in foot-pounds delivered to the piston and the work delivered from the revolving crank-shaft.

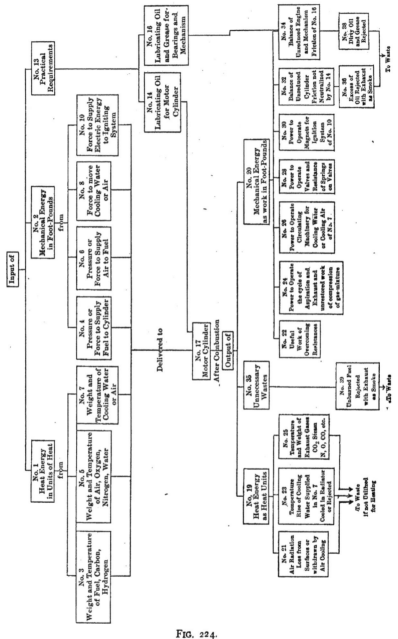

FIG. 224.

II

CHAPTER II.

LIBERATION OF HEAT ENERGY.

COMBUSTION.

8. Introductory.—It has been observed in the preceding chapter that the convenient and accessible sources for the energy due to heat compelled the engineer to have recourse to the combustion of the elements carbon and hydrogen in the oxygen of the atmospheric air. While heat appears as a transformation of mechanical energy in friction, impact, abrasion, attrition, and in overcoming electrical resistances, these sources are excluded when the object sought is heat, which may itself be transformed into mechanical energy. The hydrogen and carbon are stored in the earth as the result of processes of creation and distillation under conditions of heat and pressure in geological periods. In order that this energy may be released and made available in practice, conditions favorable to the necessary rapid oxidation must be established. This process of rapid oxidation, accompanied with the liberation of heat and light, is called combustion. While oxygen combines with many of the metals or bases or elements, as well as with carbon and hydrogen, and such processes are also properly called combustion, yet such combinations are either too costly to serve as convenient sources for heat, or else the process of oxidation is so slow that sufficient heat cannot be derived from the process in any practically short or momentary length of time.

9. Combustion. Flame. Smoke. Incomplete Combustion. —Combustion may therefore, for the purpose in hand, be defined as a combination with oxygen which takes place with sufficient rapidity to be accompanied by the phenomena of light and heat.

12

In the combustion of a solid fuel it appears to be necessary to raise the surface of the solid particles of that fuel to a temperature at which the carbon and hydrogen which it contains shall be distilled on that surface and form into a gas; in other words, the oxygen of the air cannot unite with a solid, but that solid must first be gasified before chemical combination can begin. Similarly with a liquid fuel it must either be so finely divided as to be a mist or to constitute an atmosphere loaded with a vapor of combustible material. Even then the gas must be raised to a temperature of ignition, in order that combustion may begin.

The term **fire** came into use before the combustion of gas was general or recognized, and is therefore properly restricted to the continuous process of chemical union of a combustible with oxygen by distillation from a solid such as wood or coal, or their derivatives. When there is no solid there is flame but not a fire. There may be fire without flame; but if the gas or atomized fuel supply were shut off and the flame disappeared and there were no embers or clinkers or ash left behind, which are necessarily associated with the idea of quenching a fire, then there was no fire although there was flame. The flame at the wick of an oil lamp-burner or at a gas-burner is not a fire, but the flame of a burning match or in burning shavings or waste is one. The extinction of a flame is practically an instantaneous process when the arrest of gas-flow can be: the extinction of a fire is always a gradual and progressive process. The electric spark is not a fire nor a flame, but like the flame it can start the chemical processes for which heat is requisite, and produce the same results as a true fire.

A luminous flame is a current of hot gas carrying with it solid particles at such a temperature as to glow or give out heat and light. These solid particles may be combustible or they may be mechanically suspended in the gas and glowing, but inert so far as the generation of heat is concerned. When these particles are combustible and the temperature of the current is sufficiently high they will glow and burn until they become entirely gas, and disappear. In the absence of solid particles, a current of hot gas the result of complete union with oxygen which would be called complete combustion, would be colorless and invisible. Such a flame is called a *non-luminous* one. The temperature of the current of flame is measured by the degree of incandescence of these particles. When they are intensely white they are in their hottest condition. A yellow flame is cooler, and a red flame is still cooler. A flame usually results when the supply of oxygen at the point where combustion began was not quite sufficient, or the temperature not sufficiently elevated to produce a complete combustion of the fuel at the point where such combination began. A luminous flame is much the most efficient means of heating a solid by radiation. The complete combustion in the necessary mass of air produces the highest possible temperature at the time and place of such union, and is most effective for heating a body immersed in the current of non-luminous gas.

The word incandescence strictly used refers to a condition of heat accompanied by light as an evidence of great heat energy, but without chemical action. The word incandescence, however, is often extended to include the condition in which the chemical action is relatively slow while the heat intensity is high.

Smoke is a current of hot gas carrying with it solid particles of carbon which are not hot enough to ignite and burn. or which have been cooled below the temperature required for such combustion. The term smoke is often applied to currents of gas carrying with them tarry or other matter in a finely divided state. Such a current has all the appearance of a smoke, but differs from it, inasmuch as, if it were brought up to a sufficient temperature, it would ignite and burn. A true smoke, carrying particles of lampblack, cannot be so treated, since the lampblack is only capable of ignition at temperatures considerably above those which can be brought about except in the electrical arc or with the oxyhydrogen blowpipe. These conditions make it apparent that to prevent smoke is the best that the engineer can do, and that there is no such thing as smoke consumption. It is one great advantage of the combustion of gas and finely divided oil, that the conditions of smokeless combustion are much more easily attained than with the combustion of solids. The gas does not require distillation by heat and there are no solid particles in a gas which is a true chemical mixture. In the combustion of such a gas it will happen that not infrequently with an inadequate supply of oxygen and too low a temperature the carbon will separate in the form of soot or lampblack. Thorough mixture of the gas and oxygen and high temperature will diminish this difficulty from the deposition of carbon.

Incomplete combustion is the union of carbon with oxygen to form a compound which upon combination with the necessary additional oxygen in the presence of heat will burn to the final state in which the products of such combustion are incombustible.

The usual form of this incomplete combustion is the burning of carbon to carbon monoxide (CO), which will burn to carbon dioxide or carbonic acid (CO_2) upon the supply of the necessary additional oxygen and heat. If the carbon escapes from the apparatus in use without being completely burned, there is, obviously, a loss of available energy.

10. Ignition. Explosion. Propagation of Flame. — In order that the gas distilled from the solid fuel or derived from any source should begin its combination with oxygen it is necessary that it should be set fire to. This means that the mixture of oxygen and gas, or either in the presence of the other, shall be raised to a temperature at which chemical combination shall be possible. This beginning of chemical union is called ignition and can usually be effected most conveniently by bringing a flame or an incandescent solid into contact with the mixture whereby a part of it shall be raised to the temperature required. The passage of an electrical spark through the mixture will effect this ignition also. If the mixture is sufficiently intimate and in proper proportions, the ignition, beginning at one point by flame or spark or incandescence, will propagate itself through the mass and the entire mixture will become ignited. If the mixture is not intimate and there is not the proper amount of oxygen, or if the temperature of the flame is low by reason of the poor quality of the gas, the flame may not propagate itself through the entire mass and the combustion will be incomplete.

If, on the other hand, the mixture be rich in combustible, the propagation of the chemical combination through the mass may be so rapid as to be practically instantaneous. When this occurs the expansion of the volume of the mixture due to this rapid combustion will occur with a suddenness which makes it concussive, and such an ignition or propagation of the ignition through the mass is called an explosion. The noise which is commonly attached to an explosion is a secondary phenomenon resulting from the concussive character of the expansion of the heated mass. It may either be an impact of the air, as in the case of a powerful electrical discharge among the clouds, or it may be the reaction of forces in a solid mass, as in the detonation of rock disrupted by explosives. An explosive is a solid or liquid having this property of intensely rapid propagation of its ignition coupled with a copious supply from itself of the necessary oxygen for the required chemical combination to take place without drawing that oxygen at the slow rate which would occur when the oxygen was furnished by the air.

When the problem of ignition is applied to a gaseous fuel, it will appear that there are several ways in which the gas and the oxygen may be brought together for ignition and combustion, and that each method may constitute a class.

Class I. Gas issuing from an orifice into a supporting atmosphere and where all the oxygen for combustion is derived from that atmosphere.

This first class of combustion is very imperfect, so that only low temperatures result, while large excesses of oxygen are required over what is chemically necessary. It is this very imperfection which causes the efficiency of the ordinary gas-jet as a source of light. The unequal distances travelled by molecules of gas before reaching the place where they can find and combine with the necessary oxygen gives the flame a volume, i.e., a certain portion of space is filled with the flame. In the study of combustion, as the origin of heat, this class is of no importance.

Class II. Gas mixed with oxygen insufficient in quantity for its combustion or for the formation of an explosive mixture, issuing into a supporting medium from which all necessary additional oxygen is derived.

Mixing the oxygen with the gas, previously to heating for ignition, as in Class II, is a direct aid to nature, eliminating the hunting process of Class I, or, at any rate, reducing it, and making necessary only the heating to the ignition temperature to cause combustion. This is shown in the immediate shortening of the flame over that of the previous class, and its loss of luminosity, while still retaining the volume character of the flame. It is the principle of the Bunsen burner, and the large class of apparatus which follow it for use in furnaces, heaters, cooking-stoves, and for heating water in steam-carriages.

In most of these the mixture of air with the fuel is made by causing a jet of gas to impinge on a mass of air, some of which is carried along with the air under the double influence of gas friction and the heated top of the burner, whence the mixture issues.

Combustion of Class II is characterized by the fact that there is an actual volume of flame; the flame is hotter than in Class I, which means that for a given flame volume either more gas is burned or the products of combustion are less diluted; the flame is less luminous and not of uniform color throughout its volume.

An infinite variety of details of arrangement in the exit and mixing of the air and gas may be devised with varying results for special cases, but it is true of all of them that, though the combustion be very perfect and the amount of heat generated large, yet there is always a "flame volume," indicating a struggle, as it were, on the part of the gas and air in their final combustion. The combustion, though approaching perfection in many cases, is rendered so only by the use of a large excess of the oxygen chemically required giving oxidizing products of combustion.

Class III. Gas mixed with oxygen in quantities insufficient for complete combustion, but sufficient for the formation of an explosive mixture, issuing from an orifice into a supporting atmosphere, from which all necessary additional oxygen is to be derived.

Class IV. Gas mixed with oxygen in just sufficient quantities for combustion, issuing from an orifice into any sort of atmosphere. This sort of mixture may be called a "chemical" mixture.

Class V. Gas mixed with oxygen in such quantities as to form an explosive mixture, but with insufficient oxygen for complete combustion, burned in a mass by a single explosion.

Class VI. Gas mixed with oxygen in chemical proportions, burned by a single explosion in mass.

It is only when the gas and air are previously mixed completely and uniformly in the proper chemical proportions that non-reducing, non-oxidizing products of combustion are obtained, and, since none of the heat goes to warm excesses of oxygen or of fuel, the temperature of these products must be the highest possible. Combustion of this sort is flameless, or, rather, what flame there is is without volume, having only length and breadth without thickness, and is, in fact, a surface.

Such combustion is governed by laws quite different from those under which the classes already noted operate, and it is to the combustion of chemical and other explosive mixtures that attention may for the present be mainly devoted.

Consider first the class mentioned as Class VI, in which a mass of chemical mixture—i.e., gas and its needed oxygen—is confined in a chamber. If inflammation be provoked at any point of the mass, it will, by self-propagation, finally and successively inflame the whole mass. This is the first and fundamental principle of this sort of combustion. The investigation of this propagation of inflammation by such men as Davy, Bunsen, Mallard and LeChatelier, Berthelot, and others has shown that:

(*a*) In any mixture the rate of propagation is constant for a given temperature before inflammation.

(*b*) The rate of propagation for such mixtures varies with different combustibles, being, for example, very fast for hydrogen and slow for marsh-gas.

(*c*) The rate of propagation increases with the temperature of the mixture before inflammation.

(*d*) The combustion is visible by reason of a flame-cap, or deep-blue film of flame, which travels through the mass, and which, at any instant, completely separates all the burned from the unburned mixture.

This uniformity of velocity of inflammation would indicate that in a mass where inflammation had started at a point, the flame-cap, or surface of combustion, exists at any instant on the surface of a sphere whose radius is proportional to the time elapsed.

All this has been assumed so far to take place in a large mass of gas. If, however, the enclosing vessel be given special forms, certain other characteristics are brought out. One which ls of interest is the fact that, when the enclosing vessel is a cylinder, or prism, in which the combustion surface travels with its centre on the axis, the velocity becomes affected by reduction of cross-section, and that there will always exist for every such mixture an area of cross-section so small that the self-propagation ceases. This has been explained by saying that the walls carried off heat so fast that the small flame-cap could not generate heat enough to keep itself above the temperature of ignition. Davy secured the same effect by using his screen of wire gauze, which, if interposed in the path of the combustion surface, instantly cooled the same sufficiently to prevent the ignition of the mixture on the other side, provided, of course, the temperature of the gauze itself is sufficiently low.

When a neutral diluent gas, such as N or CO_2, is added to a chemical mixture arranged for the above-discussed combustion, its effect is to reduce the rate of propagation, though not in conformity with any law yet discovered. Of course there will be a point when so much of the neutral gas is present that combustion is impossible, but no reliable data are at hand on this point, and the same conditions often give widely varying results.

While large quantities of a neutral gas may be added, without affecting the combustion except to decrease the rate of propagation, a dilution by a comparatively slight amount of oxygen will prevent it altogether. An excess of gas, it has been found, will act within certain limits like the presence of a neutral gas. By far larger amounts of fuel than of oxygen may be present in excess without arresting combustion.

In Class V, where explosive mixtures are burned in mass, the mixtures having excess of fuel, the combustion is possible within quite wide limits, with no other effect than varying the rate of propagation. In fact, a great deal of it appears to-day in gas-engines which run on the richer fuels. While, of course, in these engines the proper chemical mixture should be invariably used with only sufficient dilution to secure a proper mechanical mixture, they are seldom, if ever, constructed to maintain this properly, and, as a slight excess of oxygen beyond this relation will completely prevent inflammation, the error is always made on the other side; sooty exhausts bear testimony to this. The gas-engine also gives evidence of the fact that neutral gases decrease the rate of propagation, for in some two-cycle engines it is impossible to get a vertical combustion line on the indicator-diagram with a fixed ignition, except at very slow speeds—about 50 revolutions per minute. This is due entirely to the presence of exhaust-gases in excessive quantities as diluents to the charge.

Some of the principles above noted as belonging to masses of mixture at rest will make clearer the nature of the problem of combustion of the same mixtures when in motion issuing from an orifice.

It has been found in this latter case that a mixture of gas and air in proportions which would be explosive if it were quiescent in a chamber can be burned with perfect safety from a nozzle, provided the rate of outflow of the mixture from that nozzle is slightly in excess of the rate at which propagation of the flame would occur in that mixture. A cap of flame forms at a distance from the nozzle which will vary with the velocity of efflux. When that velocity of efflux exceeds the rate of propagation, the flame will retreat farther and farther from the nozzle until it becomes extinguished by the lowering of temperature due to the surrounding medium. The flame has blown itself out. On the other hand, if the velocity of efflux decreases, the flame will approach the nozzle, and if it is allowed to fall sufficiently, the flame will run back into the nozzle itself and thus back into the stationary mixture in the containing vessel, which will, by the propagation property which it possesses, result in an explosion. Some further facts on the treatment of this class of combustion will be presented in Chapter XIX.

SPONTANEOUS COMBUSTION is a phenomenon which has been observed where the absorption of oxygen by a body of porous character may become sufficiently rapid so that the temperature due to this chemical combination shall raise the combustible up to the point at which flaming will begin. It were better if this action were called spontaneous ignition. The conditions favorable for it are the presence of a readily oxidizable body, distributed in a finely divided state over or through some material whereby a great surface is exposed to action by oxygen. Oily rags and greasy cotton waste fill these conditions, and both are particularly liable to spontaneous ignition. If the heat of oxidation can be conducted off as fast as it is generated, spontaneous ignition is less likely to occur, but as a rule the porosity which exposes a large surface to oxidation is unfavorable to the transfer of the heat. Capillary action may also act to help the rapid oxidation process.

11. Oxygen and Air Required for Combustion. Air Required for Combustion of Carbon. — Since combustion is the chemical union of oxygen with the combustible elements, it must take place according to the laws of chemical combinations, and the weights of air for each element will be those which will furnish the oxygen weight demanded by the relations of the atomic weights in the chemical compounds which are formed.

Atmospheric air contains oxygen and nitrogen in the follow- iny proportions, at a temperature of melting ice:

	By Weight.	By Volume.
Oxygen................	0.236	0.213
Nitrogen..............	0.764	0.787
	1.000	1.000

Whence a given quantity of air weighs $\frac{1000}{236} = 4.25$ times the weight of the oxygen which it contains, and $\frac{1000}{764} = 1.31$ times the weight of the nitrogen which it contains.

By volume a given quantity of air will occupy $\frac{1000}{213} = 4.69$ times the volume of the oxygen which it contains, and $\frac{1000}{787} = 1.27$ times the volume of the nitrogen which it contains. At 62°–64° Fahr., where computations are usually made, this multiplies to replace oxygen by atmospheric air becomes 475–480, and this value is used in computing the tables.

When carbon burns to carbonic acid, which is the normal and preferred combustion process, the chemical equation for the process and result is

$$C + O_2 = CO_2,$$
$$12 + 32 = 44,$$

in which C is the symbol for one part by weight of carbon; O_2 is the symbol for the two parts of oxygen required to burn the carbon to carbonic acid, whose symbol is CO_2. The figures below each are the respective multiples of their atomic weights for combination; whence it appears that the oxygen weight needed will be given by the proportion

$$\text{Weight of oxygen required} : \text{Weight of carbon furnished} :: 32 : 12,$$

or 2.66 pounds of oxygen must be furnished to burn the one pound of carbon completely. The weight of the carbonic acid, CO_2, will be the sum of the weights of carbon and oxygen, or $1 + 2.66$ $= 3.66$ lbs.

When the combustion is effected by supplying atmospheric air, there must be supplied from the foregoing calculation concerning atmospheric air $2.66 \times 4.25 = 11.3$ lbs. of air. Add 1.0 lbs. of carbon. The products of the combustion will weigh 12.3 lbs. and will consist of carbonic acid and nitrogen.

Similarly, the volume of air in cubic feet to burn one pound of carbon can be calculated from the weight of it. At atmospheric pressure and at the temperature of melting ice a pound of air occupies 12.39 cubic feet. Hence 11.3 pounds of air will occupy $11.3 \times 12.39 = 140$ cubic feet at $32°$ F., or 152 cubic feet at $62°$ F.

When carbon (C) burns to carbonic oxide (CO) instead of to carbonic acid (CO_2),

$$C + O = CO,$$
$$12 + 16 = 28,$$

whence the oxygen is $1\frac{6}{12}$ of the unit weight of the carbon, and 1.33 pounds of oxygen or $1.33 \times 4.25 = 5.65$ pounds of air are required. The products of the combustion are 2.33 pounds of carbonic oxide. The weight of air for this combustion will be $1.33 \times 4.25 = 5.65$ pounds of air, or $5.65 \times 12.39 = 70$ cubic feet of air at $32°$ F., or 76 at $62°$ F.

If the CO burns as a combustible gas to CO_2, the additional supply of air is required as in the preceding case.

12. Air Required for Combustion of Hydrogen.—Hydrogen burns to water-vapor or steam-gas, whose chemical symbol is H_2O. The chemical equation is

$$H_2 + O = H_2O,$$
$$2 + 16 = 18,$$

whence one pound of hydrogen requires $\frac{16}{2} = 8$ pounds of oxygen, and $8 + 1 = 9$ pounds of water-vapor result as products of the combustion, if oxygen is used alone.

Eight pounds of oxygen need $8 \times 4.25 = 34$ pounds of air, making $34 + 1 = 35$ pounds of water and nitrogen as the actual weights of the products of combustion. The volume of air for hydrogen combustion is $34 \times 12.39 = 421$ cubic feet of air at $32°$ F. or 457 cubic feet of air at $62°$ F.

13. Air Required for Combustion of Compounds.—In the burning of compounds of carbon and hydrogen each acts as though the other did not exist, and the air required is the sum of the requirements of the constituents. Marsh-gas, for instance, known also as light carburetted hydrogen or methane, of composition CH_4, requires

$$C + O_2 = CO_2 \quad = 12 + 32 = 44$$
$$H_4 + O_2 = 2(H_2O) = \ 4 + 32 = 36$$
$$\text{Total} = 16 + 64 = 80$$

The added oxygen is four times the weight of the original gas, or one pound of gas gives five pounds of carbonic acid and water if no nitrogen is added. Four pounds of oxygen will be furnished by $4 \times 4.25 = 17$ pounds of air at $32°$, or $17 \times 12.39 = 208$ cubic feet of air at $32°$, and giving 18 pounds of CO_2, H_2O, and N.

The proportions of the CO_2 and H_2O were respectively $\frac{44}{80}$ of the former and $\frac{36}{80}$ of the latter; or there was one part of water to 1.32 parts of carbonic acid, since

$$36 : 44 :: 1 : 1.22.$$

Similarly, for olefiant gas, ethylene, C_2H_4, the equations will be

$$C_2 + O_4 = 2CO_2 = 24 + 64 = \ 88$$
$$H_4 + O_2 = 2H_2O = \ 4 + 32 = \ 36$$
$$\text{Total} = 28 + 96 = 124$$

That is, for a weight of gas (28) will be required a weight of oxygen (96), or 3.43 pounds for one pound of gas, making 4.43 pounds of CO_2 and H_2O, and calling for $3.43 \times 4.25 = 14.58$ pounds of air, or $14.58 \times 12.39 = 180$ cubic feet of air, at $32°$.

The products of combustion will be $14.58 + 1 = 15.58$ pounds of CO_2, H_2O, and N, and in this combustion one part of water goes to 2.44 parts of carbonic acid.

If there is sulphur enough in the fuel not to be negligible, then an additional chemical equation is required and more oxygen; S burns to SO_2, or $32 + 32 = 64$. One pound of oxygen is required for each pound of sulphur, corresponding to 4.25 pounds of air or $12.39 \times 4.25 = 52.65$ cubic feet of air at $32°$ or 57 cubic feet at $62°$ F.

Generalizing from the foregoing, it would appear possible to designate hydrocarbons by a symbol C_nH_m, in which n and m shall be the atoms of each constituent in one molecule. Accepting the generally received principles of the chemists that equal volumes of all gases contain the same number of molecules (Avogadro's law), and that each molecule is made up of two atoms, it will follow that the molecules of oxygen required for one molecule of the hydrocarbon will be

$$\frac{2n+\dfrac{m}{2}}{2} = n+\frac{m}{4},$$

or $n+\dfrac{m}{4}$ volumes of oxygen are required for the complete combustion of one volume of the hydrocarbon. The volume of air will be as before

$$v = \left(n+\frac{m}{4}\right) \times 4.75.$$

Tabulating some of these results:

Element 1 pound.	Pounds of Oxygen.	Pounds of Air.	Volume of Air at 32°.	Weight of Products of Combustion.	Composition of the Products of Combustion.
Hydrogen	8	34	421	35	Water-vapor
Carbon, C to CO_2	2.66	11.3	140	12.3	Carbon dioxide
C to CO.........	1.33	5.65	70	6.6	Carbon monoxide
CH_4............	4	17	210	18 {	Carbon dioxide / Water-vapor
C_2H_4	3.45	14.58	180	15.6 {	Carbon dioxide / Water-vapor
Sulphur.	1	4.25	53	5.25	Sulphur dioxide

It should be observed that the volume of the products of combustion must involve an assumption of a particular temperature at which the weight per cubic foot shall be known for the mixture of gases. This volume and temperature being known and called respectively v_0 and T_0, the volume v_1 at the temperature T_1 will be

$$v_1 = v_0 \frac{T_1}{T_0},$$

the pressure being supposed to be the same in both states. The following table gives the series of hydrocarbons forming the marsh-gas group, in which $m = 2n + 2$, when n is the proportion of carbon in any constituent:

1	2	3	4	5	6
Name of Gas or Liquid.	Composition by Atoms.	Per cent of C.	Per cent of H.	Boiling-Point Fahr.	Specific Gravity.
GASES.					
Methane (marsh-gas)	CH_4	75.00	25.00	....	0.559
Ethane.....................	C_2H_6	80.00	20.00		
Propane.....................	C_3H_8	81.81	18.19		
Butane.....................	C_4H_{10}	82.80	17.20	....	0.600
LIQUIDS.					
Pentane.....................	C_5H_{12}	83.33	16.67	86	0.628
Hexane.	C_6H_{14}	83.72	16.28	156	0.664
Heptane.	C_7H_{16}	84.00	16.00	208	0.669
Octane.	C_8H_{18}	84.21	15.79	257	0.703
Nonane.	C_9H_{20}	84.38	15.62	277	0.741
Decane.....................	$C_{10}H_{22}$	84.51	15.49	316	0.757
Endecane....................	$C_{11}H_{24}$	84.61	15.39	360	0.765
Dodecane.	$C_{12}H_{26}$	84.70	15.30	386	0.776
Tridecane	$C_{13}H_{28}$	84.78	15.22	420	0.792
Tetradecane.	$C_{14}H_{30}$	84.85	15.15	462	
Pentadecane.	$C_{15}H_{32}$	84.90	15.10	497	
Hekdecane.	$C_{16}H_{34}$	84.94	15.06	536	
Octodecane.	$C_{18}H_{38}$	85.04	14.96		
SOLIDS.					
Paraffin (myricyl).............	$C_{27}H_{56}$	85.26	14.74		
Paraffin (ceryl).	$C_{30}H_{62}$	85.31	14.69	699	

14. Combustion of an Analyzed Fuel. Combustion Ratio.

—The chemical analysis of a fuel gives the percentage or weight of C, H, S, and O in a pound, or the proportion in a cubic foot.

Hence the calculation for the weight or volume of air is identical with the foregoing, except by reason of the provision for satisfying the oxygen in the fuel itself. The investigations of Dulong and Despretz and others have shown the principle to hold, that when oxygen and hydrogen exist in a compound in the proper proportions to form water by union with each other, these constituents have no effect either in affecting the calorific power or the demand for outside oxygen for combustion. It is only the surplus hydrogen above that necessary to form water with the oxygen which need be considered; or instead of using the total per cent or weight of hydrogen, the latter is diminished by one-eighth of the weight of oxygen, since one part of hydrogen by weight goes to eight weights of oxygen.

Suppose, for example, in the case of a gas, its analysis gives

H	22.	per cent
CH_4	67.	" "
C_2H_4	5.	" "
C_3H_8	1.	
CO	0.6	'
CO_2	0.6	" "
N	3.0	'
O	0.8	'

$$100.0$$

Then the volume of air for its combustion per cubic foot will be found by the following calculation:

H	will require	$.22 \times 2.38$	$= 0.52360$ cu. ft.
CH_4	" "	$.67 \times 9.52$	$= 6.37840$ " "
C_2H_4	" "	$.05 \times (2 + \frac{4}{4}) \times 4.75$	$= 0.71250$ " "
C_3H_8	" "	$.01 \times (3 + \frac{8}{4}) \times 4.75$	$= 0.23750$ " "
CO	" "	$.006 \times 2.38$	$= 0.01428$ " "

Total air for imflammable gases $= 7.86628$ " "

Subtract for oxygen $.008 \times 4.75 = .03800$ " "

Total air for mixture............ 7.82828 " "

If, instead of a gas, the fuel analyzed were a liquid, the table in the preceding paragraph enables the computation to be easily made from a formula in the form

$$\text{Vol. of air} = \left(140 \times \frac{C}{100}\right) + 421\frac{H}{100},$$

which can be written without sensible error

$$V = 1.40(C + 3H$$

if the computations be made for 32° F. and the fuel has no oxygen. For example, let an oil be chosen with an analysis

Carbon.................................... 84
Hydrogen.................................. 16
                                          ———
                                          100

Then the volume of air in cubic feet at 32° will be

$$V = 1.40 (C + 3H) = 1.40(84 + 48) = 184.8.$$

At 62° it will be about 200 cubic feet. If the fuel contain oxygen and sulphur, then as before

Each per cent of C requires $140 \times C \div 100$ cu. ft. of air,
"   "   "   " H   " $241 \times H \div 100$ "   "   "
"   "   "   " S   " $52 \times S \div 100$ "   "   "

so that the above principle gives

$$\text{Volume of air} = \frac{140C + 421\left(H - \frac{O}{8}\right) + 52S}{100}.$$

By weight, for a fuel containing C and H,

$$\text{Weight of air} = 11.3C + 34\left(H - \frac{O}{8}\right).$$

This is more usually written:

$$\text{Weight of air} = 12C + 36\left(H - \frac{O}{8}\right).$$

It will be found convenient to establish a relation between the weight of the combustible and the weight of the fuel and air required for its complete combustion. If the weight of the fuel be called y and the weight of air for its combustion be called x, then the ratio between weight of fuel and weight of its products of combustion, which may be called K_r, will be denoted by

$$K_r = \frac{y}{x+y}$$

and may be called its "combustion ratio."

$$\text{For carbon,} \quad K_r = \frac{1}{12.3} = .0813$$

$$\text{" hydrogen,} \quad K_r = \frac{1}{35} = .0285$$

$$\text{" marsh-gas,} \quad K_r = \frac{1}{18} = .0555$$

$$\text{" ethylene,} \quad K_r = \frac{1}{15.6} = .0461$$

The use of this ratio of combustion will appear when computations are desired as to the increase in temperature due to combustion, and the quantity of fuel is given in pounds instead of cubic feet as in the calculations based on the data in parag. 29.

While the computations by weight made hitherto are general by reason of their independence of temperature, yet since in gas-engine problems the resulting volume from a combustion is often of prime consequence, attention must be directed to certain phenomena peculiar to this action. The law of Avogadro, that under the same conditions of pressure and temperature equal volumes of all gaseous substances whether elementary or compound contain the same number of molecules, makes it apparent that when a new substance is formed by a chemical union of atoms ($\frac{1}{2}$ of the molecule) it does not follow that the new volume is the sum of the elemental volumes. On the contrary, this relation is the exception, and experiment shows that the volume of a compound gas made up of elements which combine in relations of 1:1 are the only ones which make the compound gas twice that of the elemental ones. For example, when two volumes of hydrogen (H_2) unite with one volume of oxygen (O_1) to form water-vapor, the volume of the latter is twice that of the oxygen, and not three times that of the unit. When CO burns to CO_2 one additional volume of oxygen is required, but the resulting gas occupies only two volumes. In the case of marsh-gas (CH_4), the C requires two volumes of O, which will occupy two volumes, and the H_4 will also require two volumes O, which will occupy twice two or four volumes, making 6 in all. Hence since CH_4, being a compound, occupies 2 volumes, to secure complete combustion of both elements there must be added 4 volumes of oxygen to 2 volumes of CH_4, making also 6, and there is neither increase nor decrease.*

If several of these computations be tabulated, the following figures result:

TABLE.

Line No.	Result for 1 cu. ft. of	H	CH_4	C_2H_4	CO	C_4H_8
1	Oxygen required in cu. ft...........	0.50	2.00	3.00	0.50	6.00
2	Volume of N in air to supply O, 4.75−1.00=3.75×vol. of oxygen. .	1.87	7.50	11.25	1.87	22.50
3	Volume of air=sum of 1+2........	2.37	9.50	14.25	2.37	28.50
4	Total volume when oxygen is used, line 1+unit weight.	1.50	3.00	4.00	1.50	7.00
5	Total volume when air is used, line 2+line 4...................	3.37	10.50	15.25	3.37	29.50
6	Volume after combustion when oxygen is used.	1.00	3	4	1	8
7	Volume after combustion when air is used, line 6+line 2.	2.87	10.50	15.25	2.87	30.50
8	Change of volume when O is used, line 6−line 4.	−.50	0	0	−.50	+1
9	Change of volume when air is used, line 7−line 5.	−.50	0	0	−.50	+1

15. Calorific Power of a Fuel.—The calorific power of a fuel is the amount of heat expressed in thermal units (par. 39)

* See also the tabular data in the table of elementary gases, parag. 29.

which is liberated upon the combustion of a unit of weight of the combustible material. The calorific power of a fuel does not depend upon the rapidity of the combustion, nor on the time taken in the process of absorbing the total heat resulting from it. The temperature produced by the combustion does depend upon the rate at which the combustion takes place. Values for various calorific powers of different fuels are given in the following tables in connection with the discussion of such fuels. It should be noted that the calorific power as determined by most of the calorimeters gives a figure not directly applicable to gas-engine calculations, since the gases are discharged cold from the measuring apparatus, with the products of combustion condensed by the absorbing medium, so that the latent heat of their condensation is credited to the calorific power of the gas. In the internal-combustion engine; on the other hand, the heat is generated in the presence of hot gases which are not condensed in the apparatus itself but escape as vapors. When all the water produced by the combustion and the steam thus formed is condensed, the result of the computation is called the "higher heating or calorific value." In the computation of the "lower calorific value" the steam formed is assumed to be present as dry saturated steam, having 965 units of heat latent from 32°, or a total expenditure from 64° to 212° of 116 in the liquid and 965 in the gas, making 1081 in all. The lower value will differ from the higher by deducting this quantity of heat from the higher for every pound of water produced by the combustion of one pound of gas.* In France the higher heating value is

---

* As examples of the method to be followed in computing the difference between the high and low values the following may be used as guides:

With methane, CH_4 the atomic weight of the fuel is 16.

The atomic weight of resulting CO_2 and H_2O is . . . 80.

The water is 36 parts of this 80, or is $\frac{9}{20}$ by weight.

If the fuel weighs 16 and the products weigh 80, the fuel is $\frac{16}{80}$ of such products, or if the fuel weighed 1 pound the products would weigh 5.

Hence the water is $\frac{9}{20}$ of 5 pounds, or $2\frac{1}{4}$ pounds.

$2\frac{1}{4}$ pounds withdraw 1080 heat units per pound in cooling from steam at 212° to water at 64°, or 2428 heat units in all.

preferred, and is used by Witz and others, on the ground that it is a defect in the gas-engine that it should not utilize this latent heat as the steam-engine can, to produce external work, and the gas computation should be made in fairness to the gas on the basis of the possession of this ability. In America, Germany, and England the lower value is used because the steam cannot be condensed in practice, and heat being latent and remaining so, it does not communicate anything to the effective cycle. Experiments should always report which value of the heating power of the gas is used, since the difference will rarely be less than five per cent in actual examples, and may reach ten per cent. The following table [from Geitel*] shows three typical gas values, in meters and feet and in kilograms and pounds:

With ethylene C_2H_4 the atomic weight of the fuel is 28.

The atomic weight of resulting products of combustion is 124.

The water is 36 parts of this weight, or approximately $\frac{2}{7}$.

If the fuel weighs 28 and the products of combustion 124, one pound of fuel makes 4.4 pounds of products of which $\frac{2}{7}$ is water, $1\frac{1}{4}$ pounds of products is water. Hence these withdraw 1080 per pound or 1388 in all.

With benzene C_6H_6 the fuel weighs 72, the products 183. The water is 54 of this or practically $\frac{4}{5}$.

One pound of fuel makes a little over four pounds of products. Hence the weight of water in the products is $\frac{4}{5}$ of a pound so that $\frac{4}{5}$ of 1080 or 720 units are withdrawn.

With butylene C_4H_8 the fuel weighs 56 and the products 248 so that one pound of fuel makes $\frac{248}{56} = 4.4$ pounds of products. Of these the H_2O is $\frac{72}{248} = \frac{1}{3}$ practically or the water is $\frac{4.4}{3} = 1.4$ pounds. Hence $1.4 \times 1080 = 1512$ heat units withdrawn.

* *Das Wassergas und seine Verwendung in der Tecknik.* To transform calories per cubic meter into B.T.U. per cubic foot the factor to be used in multiplying the former approximates 8.91. An average of 9 in round figures will give concordant results.

HEATING VALUES HIGH AND LOW FOR THREE KINDS
OF GAS.

Unit Used.	Calories per Cubic Meter.		B.T.U. per Cubic Meter.		Calories per Kilogram.		B.T.U. per Pound.	
Water of Combustion as . .	Liquid	Gas	Liquid	Gas	Liquid	Gas	Liquid	Gas
Calorific Value .	High.	Low.	High.	Low.	High.	Low.	High.	Low.
Lighting Gas	5810	5154	650	577	11,350	10,070	20,430	18,126
Power Gas . .	1048	1048	117	117	838	838	1,508	1,508
Water Gas . .	3054	2813	332	315	4,558	4,199	8,204	7,558

16. Fuel Calorimeters. Mahler Bomb.—The calorific power of a fuel is a matter of experimental observation. The general method used in determinations is to cause a known weight of the fuel to burn in a closed vessel into which oxygen is introduced and the fuel ignited in the atmosphere of oxygen. The closed vessel is surrounded by an observed weight of water at an observed temperature, which is usually made to circulate so as to maintain a constant temperature in order that no variation in the value of the specific heat may occur. The number of heat-units absorbed by the rise of that weight of water through its observed range of temperature gives the calorific power of the fuel tested, so that the apparatus is correctly called a calorimeter or measurer of heat.

One of the best known of the calorimeters is that of Mahler, sometimes called the Mahler Bomb. In a very usual form of this apparatus, shown in Fig. 1, *B* is a thick steel chamber lined with porcelain in order to prevent any chemical action between the steel of the vessel and the fuel burned within it. A weighed amount of the fuel, whether as a solid pulverized, or as a liquid in the form of an oil, is introduced into a platinum pan, *C*, into the bomb, and then a large excess of oxygen gas at a pressure of 300 pounds per square inch, or thereabout, is introduced to surround the pan. An electric circuit is completed through a wire of small cross-section where it touches the fuel, so that it shall become red-hot when the current meets the resistance of that small section. This brings the combustible to the firing-point, so that in the dense atmosphere of oxygen it burns completely. The water, *D*, which surrounds the bomb in an outer vessel is agitated by a stirring apparatus, *S*, in order to keep its temperature uniform throughout. Carefully calibrated thermometers of high accuracy, reading to the hundredth of one degree, record the temperature rise in the enveloping water, *A*, and the outer jacket is heavily felted so as to prevent loss of heat by radiation to the surround-

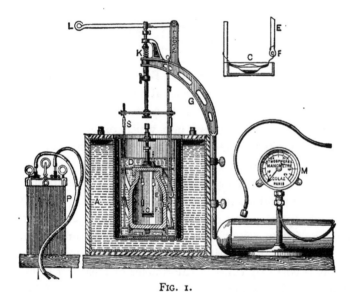

Fig. 1.

ing air. The rise in the temperature of the water is less than that due to the combustion of the fuel by the absorption of heat by the metal of the bomb itself in reaching the temperature of the combustible within it. This is determined, experimentally, by calibration and is usually called the constant of the calorimeter. It can conveniently be expressed as a quantity of water which would absorb the same quantity of heat for each single degree increase of temperature as the metallic parts of the calorimeter have been observed to absorb. The Mahler Bomb is more conveniently applied to measurements of the calorific power of solid and liquid fuels than of gases.

17. The Junker Gas Calorimeter.—The Junker gas calorim‑
eter has, as one of its advantages, the fact that readings can be

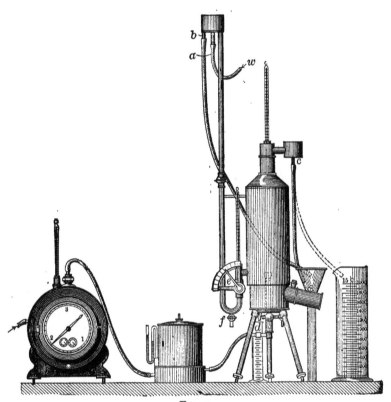

Fig. 2.

taken continuously with it over a considerable period of time, so
that the percentage of errors due to observation becomes less,
and also variations in the quality of the gas-supply can be de‑
tected, while any test is in progress. As ordinarily used, the
apparatus for the Junker calorimeter is represented in Fig. 2.
The gas to be measured is passed through the test meter at the
left hand of the cut, which should be finely graduated so as to
read down to thousandths of a cubic foot. Next to this is a

pressure-regulator so that all observations may be made without effect from pulsations in the gas-main or caused by the engine itself which is under test. This pressure-regulator is of the ordinary construction of a gas-holder. The tube on the outside of the regulator measures the pressure by which the inverted cylindrical vessel is raised in the water-seal which closes the open bottom of the inverted vessel. The reading in pressures is, of course, in inches of water-pressure. From the pressure-regulator the gas is led to the burner proper which is introduced into the bottom of the apparatus which forms the calorimeter itself. The necessary quantity of air for combustion to produce a Bunsen effect enters with the gas through regulated openings, and the additional air-supply comes in through the open bottom of the central tube of the calorimeter. The section of the calorimeter in Fig. 3 shows the gas-flame in position and the arrangement whereby the hot products of combustion ascend to the top into the space marked 29 and there descend around the central tube and pass out at the bottom through the tube 32 in small tubes which are completely surrounded by the enveloping cooling water. This cooling water is supplied to an overhead vessel through the pipe 1. The chamber 3 is open at the top so that any excess of supply beyond what passes through the calorimeter is discharged through the tube 5. A constant head for the flow is thus maintained in spite of variations in the supply. The cold water enters at the bottom and is discharged at the top by overflowing in a funnel 20. A thermometer, 12, in the inlet and a thermometer, 43, in the outlet measure the range of temperature caused by the combustion. The rapidity of the flow of the water is controlled by the plug-cock 9, with a view of having the rise in temperature kept within convenient limits. The outlet 35 from the bottom of the hot chamber is intended to remove any condensed water which may result from the combustion of hydrogen. The whole cylinder and top of the calorimeter are surrounded with an air-jacket to prevent radiation from the water used in absorbing

the heat of the gases. The use of the apparatus will be plainly
evident from the illustration. 'When the pointer of the meter

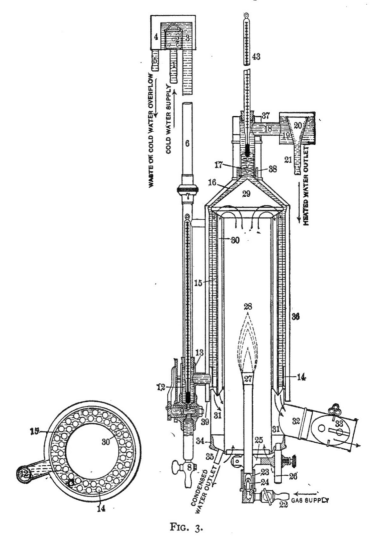

FIG. 3.

pásses the zero mark the discharge from the overflow *C* is transferred from the waste to the graduated vessel, and the tempera ture of the hot-water thermometer is observed at intervals while the glass is filling. The cold-water thermometer is not likely to change its reading. When the measuring-glass is filled to a designated point the meter reading is taken to determine the cubic feet of gas burned, and its heating value is computed by the following simple formula:

$$HG = WT,$$

in which H is the calorific value of one cubic foot of gas and G the quantity of cubic feet, by meter, burned during the experiment. Then if W is the weight of water passed through the apparatus while the volume of gas was burning, and T is the difference between the thermometer readings at the inlet and outlet ends of the apparatus, the equation can be solved for each directly. As the apparatus is continuous it does not need a correction for the calorimetric constant, since after the apparatus has once been heated this quantity is the same at the beginning and end of the experiment. Obviously, if the gas containing hydrogen deposits a certain amount of water in the annular space 31, the value given by the above formula will be a gross value, since the water formed by the combustion of the hydrogen will have given up its latent heat to the circulating water. Junker's calorimeter is usually constructed so that the measuring-glass reads in litres, making one litre weigh a kilogram. The result will therefore be given in this form of the apparatus in calories, which can be transformed to British thermal units by multiplying by the factor 3.9683, which is usually called 4. Since the combustion in the calorimeter chamber takes place at atmospheric pressure, the condensation of one pound of the watery vapor will set free 966 B.T.U. The weight of condensed water-vapor in pounds multiplied by 966 plus the range above 32° for the condensing water, usually bringing it to 1080, will give the absorption of B.T.U. for the weight of gas burned. If the gases were to be used hot, this condensation would not occur.

18. **The Lucke Gas Calorimeter.**—A form of calorimeter particularly adapted to measure the heat generated when explosive mixtures of gas and air are ignited and burn under constant pressure is the result of the investigations of Dr. C. E. Lucke concerning the conditions suitable and necessary for the combustion of such mixtures under constant pressure. The apparatus is illustrated in Fig. 4. A suitable vessel to contain

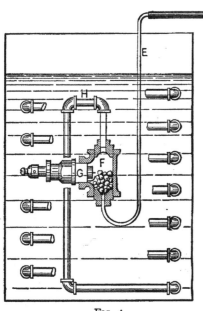

FIG. 4.

water surrounds an ordinary pipe fitting of the desired size *F* which is bushed at the bottom to receive a copper pipe *E* for the delivery of the explosive mixture. The mixture is ignited in the tee by a spark plug of the sort which is usual in gas-engine practice, inserted at *G* and connected with the necessary electric wires for the passage of a jump spark. The explosive mixture burns throughout the broken magnesite in the tee, and the hot products of combustion pass out through the connection *H* which leads them to a square coil of pipe whereby their temperature is withdrawn by means of the water circulating in the chamber. This water is measured as to weight and temperature as in the previous formula. The end of the square coil can be connected by rubber and glass tubes so that the products of combustion can be directed to any point in the water-chamber so as to act as a stirrer. The glass bottle B is to serve as a trap for convenience in the delivery of the entire quantity of explosive mixture as received from a measuring apparatus connected to B by the tube A.

19. Calorific Power of a Compound.—The calorific power of a compound will be the sum of the calorific powers of its components, provided that in the chemical reactions of the combustion there does not occur an absorption of heat at the expense of their surroundings. When heat is liberated as the outcome of the chemical change, the reaction is called *exothermic;* when heat is absorbed by the reaction it will be called an *endothermic* reaction. The discussion which follows is made on the assumption for simplicity that the compound has the exothermic property.

The proporsions of the components may either be the result of the known chemical combination for a true chemical compound, or may result from a chemical analysis which shall determine the percentage of the elements in the compound. For example: if the gas is a chemical compound such as olefiant gas, C_2H_4, which is made up of

$$C_2 + H_4 = 24 + 4 = 28 \text{ parts by weight,}$$

$\frac{4}{28} = \frac{1}{7}$ will be hydrogen, and $\frac{24}{28} = \frac{6}{7}$ will be carbon. If, then, $\frac{1}{7}$ of the calorific power of hydrogen be added to $\frac{6}{7}$ of the calorific power of carbon, their sum will be the calorific power of the compound. With analyzed hydrocarbons the percentage of each constituent will be used instead of the fraction above.

The accepted * formula for computing the calorific power from a fuel analysis is due to the physicist Dulong and is known by his name. It has the form

$$\text{Calorific power of } 1 \text{ lb. in B.T.U.} = 14,600C + 62,000\left(H - \frac{O}{8}\right).$$

In this, C, H, and O are the percentages respectively of carbon, hydrogen, and oxygen, divided by 100 to reduce them to actual fractions of one pound. This is often transformed by

---

* The Dulong formula is accepted as correct within limits of error of 5 per cent, the accuracy or error varying with the composition of the coal.

the expedient of factor'ng the constants denoting the respective calorific powers of carbon and hydrogen so as to read

$$\text{Calorific power} = 14{,}6\text{co}\left[C + 4.25\left(H - \frac{O}{8}\right)\right],$$

since $\frac{62000}{14600} = 4.246$.

If desired to take account of the sulphur present by analysis, or to express the formula in centigrade and metric units, the equation takes the two forms:

Heating value in B. T. U. $= \frac{1}{100}[14{,}600C + 62{,}000\left(H - \frac{O}{8}\right) + 4050\ S]$.

Heating value in calories $= \frac{1}{100}[8140C + 34{,}400\left(H - \frac{O}{8}\right) + 2250\ S]$.

Mahler's equation in parallel form is:

Heating value, calories $= \frac{1}{100}[8140C + 34{,}500H - 3000(O + N)]$.

In the above C = Carbon, H = Hydrogen, O = Oxygen, N = Nitrogen, S = Sulphur.

20. Computed Increase in Temperature Due to a Combustion.—It will appear in a later paragraph (par. 54) that each body requires a certain amount of heat to raise the temperature of one unit weight of that material by one degree on the thermometric scale. This quantity of heat, called its specific heat, seems to bear a constant ratio to its atomic weight (specific heat×atomic weight = 6.25 approx.) and is usually designated by the initial C. Hence, if one pound of air be raised from T_1 to T_2, the heat-units H to do this will have to be $H = C(T_2 - T_1)$.

If the calorific power of one pound of a fuel be denoted by Q in British thermal units, and y denote the weight in pounds, and x the weight of air to burn it (paragraphs 11–13), then $x + y$ will be the weight of gases present, and $(x + y)C$ will be the amount of heat required to raise this mixture one degree. But the total heat corresponding to H above will be yQ. Hence for $x + y$ pounds

$$yQ = (x + y)C(T_2 - T_1),$$

or

$$T_2 - T_1 = \frac{y}{x+y} \times \frac{Q}{C}.$$

From this if Q be known from experiments in calorimetry, and the combustion ratio $\frac{y}{x+y} = K$, be computed and C be given from the work of the physicist, the rise in temperature can be calculated. This is the theoretical temperature of combustion on the assumption that the actual specific heat is known and does not change during the process.

If the calorific power of carbon be called 14,000, and the ratio $\frac{y}{x+y} = \frac{1}{12.3} = .0813$, and C have an average value of .237, then

$$T_2 - T_1 = .0813 \times \frac{14,000}{0.237} = 4800°$$

when the heating is done under the condition of constant pressure, and the effective specific heat be called that of air under these conditions. If the figure .169 be used, as determined by Regnault for air at constant volume, then

$$T_2 - T_1 = .0813 \times \frac{14,000}{0.169} = 6735°.$$

It will appear in the later discussion that no such temperatures are realized in actual practice with engines and combustion in the cylinders, even when fuels rich in hydrogen are used, with higher calorific powers. Hence it becomes significant to ascertain what the value of the actual or effective specific heats of the gaseous mixtures are, and whether these are constant for all temperatures. A full discussion of this question will appear in par. 55. It becomes important to ascertain what effect the cooler metals of the piston and cylinder walls have in dissipating the heat due to combustion, and whether any other phenomena appear of chemical character which will account for this suppression of heat. The actual temperatures are deduced from maximum observed pressures by the formula

$$\frac{p_0}{T_0} = \frac{p_1}{T_1},$$

so that

$$T_1 = \frac{p_1 T_0}{p_0},$$

in which $p_0 T_0$ are the pressures and temperatures before the combustion occurred, and p_1 the observed pressure resulting from the combustion; but these computations give values much below the values computed by the foregoing method.

21. Dissociation.—Doubtless one cause for the lowering of the actual temperature of combustion below the computed theoretical value in an engine cylinder is a decomposition of chemical combinations by reason of the high temperature. Such a breaking up of gaseous compounds is called "the dissociation of gases," and absorbs as much heat as the formation of such combinations would liberate. Water-vapor, for instance, which is a product of the combustion of hydrogen, is broken up at between 1600° and 1800° Fahr. into component H and O; and while the hydrogen will recombine on a reduction of temperature during expansion, it may occur late enough to be incomplete before the exhaust opens. Or, the lowering of temperature by the cooled cylinder-walls may prevent complete recombination.

22. Sources of Gaseous Fuel for Gas-engines.—The hydrocarbon or carbon gases which are used in gas-engines are of three kinds. The first is natural gas, received from subterranean sources as the result either of distillation now in progress underground, or the accumulations of previous distillations which have ceased. The second would be designated as producer-gas, which is a manufactured article made by the distillation of solid fuel by heat. This gas is of two kinds. The first would be designated as fuel-gas and is a product rich in carbon but poor in hydrogen. The second group is the kind of gas which is made for illuminating purposes and is distributed through the mains of the cities. It is richer in illuminants than the true producer-gas, but producer-gas can be enriched so as to be made into illuminating-gas. Belonging to this producer group is the outflow of gas from the top of the blast-furnace used in the smelting of iron from its ores. This is a gas usually leaner than the other two in calorific power, containing little or no hydrogen. It carries with it a considerable quantity of finely divided dust from the limestone or other material in the furnace, whose removal must be provided for in the design of the engine which uses this gas. The third kind of gas is really a carburetted air, made by saturating atmospheric air with the volatile constituents of a liquid hydrocarbon, as discussed in Chapter X.

Where oil is the source of heat energy, it will be made into a gas and will be thus used in the motor. It may be made into a gas by a distilling process whereby the liquid oil is injected into a hot chamber or into a chamber so filled with heated air that the liquid becomes a vapor by the process of vaporization, which is analogous to a distillation by heat. Or a more volatile hydrocarbon liquid may be used and the air which is to serve as a medium and is to support the combustion of the hydrocarbon may be sent through thin layers of the hydrocarbon so that the air will pick up and carry with it a mist or vapor of the volatile hydrocarbon. This makes a species of air-gas having the properties of the foregoing with respect to ignition and other behavior, and is a form of gas much used in motors for automobile practice.

23. Natural Gas.—In certain parts of the United States of America, notably in Pennsylvania, Ohio, and Indiana, large accumulations of a natural fuel-gas are found in subterranean cavities or strata which can be reached by wells drilled from the surface of the ground. Such gas is usually under considerable pressure, so that it can be piped from its sources to industrial centres without too great loss of pressure, or artificial pressure may be secured by proper gas-pumps, which may themselves be conveniently operated as gas-engines.

The varying districts give varying constitutions of the gas and hence a calorific power which varies. In the neighborhood of Pittsburg, Pa., one pound of coal is considered to be equivalent to $7\frac{1}{2}$ to $12\frac{1}{4}$ cubic feet of gas. The following tables give some analyses:

VARIATION IN COMPOSITION OF NATURAL GAS.

Constituents.	1	2	3	4	5	6
Marsh-gas..................	57.58	75.16	72.18	65.25	60.70	49.58
Hydrogen..................	9.64	14.45	20.02	26.16	29.03	35.92
Ethylic hydride	5.20	4.80	3.60	5.50	7.92	12.50
Olefiant gas................	0.80	0.60	0.70	0.80	0.98	0.60
Oxygen	2.10	1.20	1.10	0.80	0.78	0.80
Carbonic oxide............	1.00	0.30	1.00	0.80	0.58	0.40
Carbonic acid..............	0.00	0.30	0.80	0.60	0.00	0.40
Nitrogen..................	23.41	2.89	0.00	0.00	0.00	0.00

Analyses from various wells in Indiana and Ohio are given in the table in parag. 29.

24. Producer-gas.—Gas made by distilling and volatilizing the separable elements in bituminous or anthracite coal in a closed furnace, using part of its own heat of combustion to effect the chemical reactions, is often called producer-gas, from the name given to the gas-generator. A thick bed of fuel rests upon properly constructed grates, and air or steam or both are forced from below the grates up through the bed of fuel. The first combustion is to carbonic acid (CO_2) with air alone, or to CO_2 and hydrogen if steam is used also. This carbonic acid gas, meeting the layers of carbon above where no free oxygen reaches, is decomposed by the carbon into two units of carbonic oxide (CO), which with the hydrogen passes up through the bed of fuel and outwards through a proper pipe to the place where it may meet the required oxygen and be burned at the point desired. Early producers of the Siemens type, operating with open ash-pits and no pressure below the grates, lost much

of their possible effectiveness in the cooling of the gases after
leaving the producer. This loss is estimated at 30 per cent.
To blow with air alone is to introduce inert nitrogen which dilutes
the gas and lowers its calorific power. On account of the loss
of heat in the producer itself in the distilling process, and some
loss in the dissociation of the water, which is not all recovered,
producer-gas usually carries only 87 per cent of the calorific
energy of the carbon. Some loss in unreduced CO_2 must be
allowed for, and the cost of making the steam used. 82 per cent
is a more usual figure when anthracite is used as fuel instead of
bituminous coal. Much inferior grades of fuel can be used in
the producer than could be used direct, however.

If an analysis of 85 per cent of solid carbon be assumed for
an anthracite stock, with 5 per cent of volatile hydrocarbons and
10 per cent of ash, and the further assumption be made of a combustion of 80 pounds to CO and 5 pounds to CO_2, the following
calculated statement of process, products, and resulting energy
may be agreed to:

Process.	Products.		
	Pounds.	Cubic Feet.	Anal. by Vol.
80 lbs. C burned to CO......................	186.66	2529.24	33.4
5 lbs. C burned to CO_2	18.33	157.64	2.0
5 lbs. vol. HC (distilled)...................	5.00	116.60	1.6
120 lbs. oxygen are required, of which 30 lbs. from H_2O liberate H.......................	3.75	712.50	9.4
90 lbs. from air are associated with N..........	301.05	4064.17	53.6
	514.79	7580.15	100.0

For quantitative values, analysis of the gas and other data,
reference should be made to the combination table under
parag. 29.

Energy in the above gas obtained from 100 pounds anthracite:

186.66 lbs. CO	807,304	heat-units
5.00 " CH$_4$...........................	117,500	"
3.75 " H.............................	232.500	"
	1,157,304	"
Total energy in gas per pound	2,248	"
" " " 100 lbs of coal..............	1,349.500	'
Efficiency of the conversion.	86 per cent	

If the gas-stock be a bituminous coal with 55 per cent of fixed carbon, 32 per cent of volatile matter, and 13 per cent of ash, and the calorific power of the hydrocarbons be taken at 20,000 heat-units to the pound, the table below results under the same assumptions.

	Products.		
Process.	Pounds.	Cubic Feet.	Anal. by Vol.
50 lbs. C burned to CO......................	116.66	1580.7	27.8
5 lbs. C burned to CO$_2$	18.33	157.6	2.7
32 lbs. vol. HC (distilled).	32.00	746.2	13.2
80 lbs. O are required, of which 20 lbs., derived from H$_2$O, liberate H..................	2.5	475.0	8.3
60 lbs. O, derived from air, are associated with N ..	200.70	2709.4	47.8
	370.19	5668.9	99.8

Energy in 116.66 lbs. CO	504,554	heat-units
" " 32.00 lbs. vol. HC..............	640,000	"
" " 2.50 lbs. H....................	155,000	"
	1,299,554	"
Energy in coal.	1,437,500	"
Per cent of energy delivered in gas......................	90.0	
Heat-units in 1 lb. of gas..................................	3484	
Heat-units in one cubic foot of gas......................	229.2	

Fig. 5 illustrates the old type of Siemens producer without artificial blast, and Figs. 6 and 7 the more modern Taylor producer with forced steam-blast and revolving grates.

These computations, however, make no allowances for the variable losses in ashes, soot, tar, and pitch. These losses as low as 1 per cent in anthracite practice may rise to 10 per cent

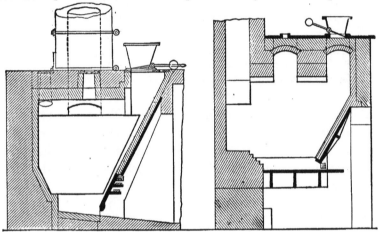

Fig 5

with bituminous fuels. When the gas is not cooled, the tarry vapors are burned with the gas; in gas-engine practice, the tar must be removed.

Ordinary producer-gas has a calorific value of 110 to 125 B.T.U. per cubic foot. 80 cubic feet of gas should be given from one pound of coal.

A bituminous producer-gas process has been perfected by Dr. Ludwig Mond, of England, intended to operate on slack and to recover the most important by-products in the form of ammonium sulphate. An excess of steam is blown into the producer with the air (see water-gas). This excess increases the hydrogen component of the gas, and the balance is recovered later. The output from a ton of slack should be from 140,000 to 160,000 cubic feet of gas having a calorific power of 140 to 145 B.T.U., with a heating value of from 80 to 86 per cent of the total energy resident in the fuel. Under favorable conditions ammonia will be recovered equivalent to 90 lbs. of ammonium sulphate. About 60 cubic feet of Mond gas seems to be required per H.P. in the gas-engine, so that a plant large enough to gasify one ton of slack per hour will supply from 2000 to 2500 H.P. of engines, making the cost for fuel per H.P. very low. Otherwise stated, this gas will furnish one H.P. per $14\frac{60}{0000}$ of $2000 = \frac{9}{10}$ of

FIG. 6.   FIG. 7.

a pound of coal. Fig. 245 shows a type section of such a producer and its connections. The following table shows an average analysis and calorific value of Mond compared with illuminating gas:

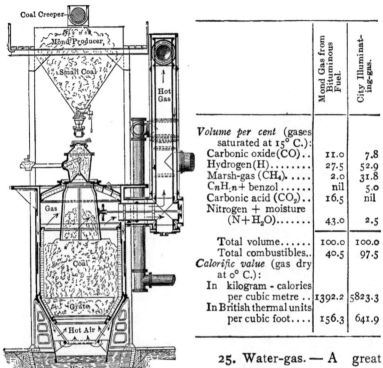

	Mond Gas from Bituminous Fuel.	City Illuminating-gas.
Volume per cent (gases saturated at 15° C.):		
Carbonic oxide (CO)..	11.0	7.8
Hydrogen (H)........	27.5	52.9
Marsh-gas (CH_4).	2.0	31.8
C_nH_2n + benzol	nil	5.0
Carbonic acid (CO_2)..	16.5	nil
Nitrogen + moisture (N + H_2O).......	43.0	2.5
Total volume......	100.0	100.0
Total combustibles..	40.5	97.5
Calorific value (gas dry at 0° C.):		
In kilogram - calories per cubic metre ..	1392.2	5823.3
In British thermal units per cubic foot....	156.3	641.9

Scale of Feet

FIG. 245.

25. Water-gas. — A great deal of gas for illuminating and power purposes is now made by the process of intermittent and alternate blowing of air and steam through a thick bed of fuel in a cylindrical producer of boiler-plate lined with refractory material. While the process may be conducted also continuously, the product of the continuous process is more often called producer-gas. The fuel is blown by air from below until it becomes highly incandescent; the producer may be open at the top, and waste the lean carbonic oxide which comes off from the top, or the latter can be caught and used. After blowing the air as long as necessary, in what is called the "intermittent" process, the air is shut off, and steam is similarly blown from below, with the producer closed except at its delivery to a gas-holder. The steam is dissociated by the incandescent carbon into hydrogen and oxygen, and the latter unites with the carbon as in the air-producer, to be reduced to carbonic oxide. The hydrogen passes out without further chemical reaction. The process may be made continuous by blowing air and steam together. Since the usual steam-jet blower will carry the necessary air with it, this

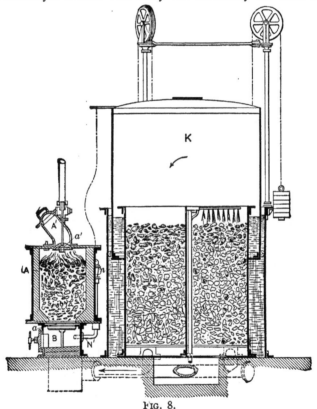

Fig. 8.

method is the one in more general use. This process was introduced in 1874 by Mr. T. S. C. Lowe, and is often known generally as the Lowe process. For illuminating purposes this fuel-gas is more highly carburetted by sprays of hydrocarbon vapors (such as naphtha or similar petroleum products) which are made a fixed gas by later heating in a superheater.

Fig. 8 illustrates what is called in England the Dowson gas-producer, which belongs to this class. Its product is sometimes

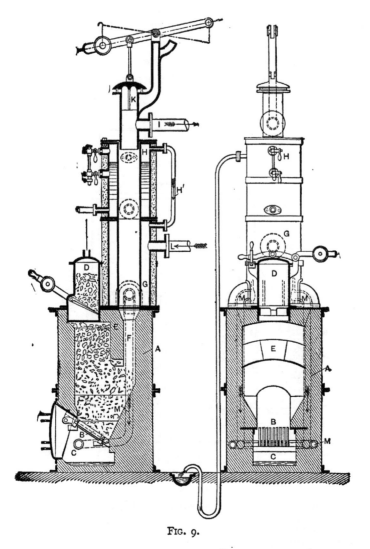

FIG. 9.

known in America as semi-water-gas. Its analysis is given **in** parag. 29

The ash-pit *B* is closed and air and steam are forced through N and up through the mass of anthracite or coke which fills the producer-chamber. The feeding is done through the hopper *A'* by means of its double lid and air-lock action. The gas passes up through the coke-scrubber into the holder *K*.

A French form of water-gas producer is known as Lencauchez'. Its object is to improve on the Dowson type by saving waste heat, and render it available for coals having some tendency to fuse together from the presence of tarry matters (Fig. 9). The hanging bridge *E* forces the gases above the middle of the fuel-bed to pass downwards before escaping to the flue *F*, and so out to the holder through the passage *I*. The annular chamber *H* is a steam-boiler, whose water cools the outflowing gases, and whose steam entering the chamber *G* meets with the air from a blower through the pipe *L*, and the combined air and steam are forced through the pipes *M* into the closed ash-pit and so up through the fuel. The descent of the distilled gas through the hot fuel before passing out is the feature which is expected to break up the tarry elements of the distillation. Lencauchez' gas analysis shows:

Hydrogen,	H	18.34
Olefiant gas,	C_2H_4	1.25
Hydrocarbons,	C_4H_4	1.55
Carbonic oxide,	CO	27.32
Carbonic acid,	CO_2	3.60
Sulphur dioxide,	SO_2	0.04
Hydrogen disulphide,	H_2S	0.06
Nitrogen,	N	47.84

Dowson gas has a calorific value averaging 150 B.T.U. per cubic foot, while the true water-gas should have 290.

Fig. 10 is a general elevation of a complete producer plant for power purposes, with economizer, scrubber, and gas-holder.

A comparison of water-gas and anthracite producer-gas might take the following form:

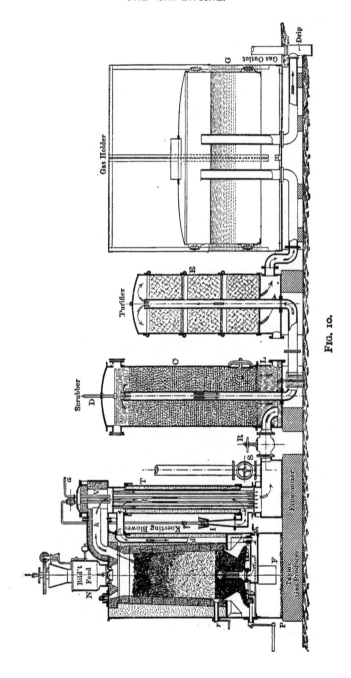

FIG. 10.

First-class carburetted water-gas, made with 4½ gallons of Lima oil per 1000 feet of gas, C.P. 26½, contains 730 H.U. per cubic foot.

One pound of anthracite coal (C 85 per cent, HC 5 per cent, ash 10 per cent) will make about 90 cubic feet of gas of following composition: CO 27 per cent, H 12 per cent, CH_4 1.2 per cent, CO_2 2.5 per cent, N 57 per cent. This gas contains about 137 H.U. per cubic foot. Therefore 17 cubic feet of carburetted water-gas are equal in heat-units to gas from one pound of anthracite.

1000 feet C.W. gas equals gas from 59+pounds anthracite.

The Lencauchez feature of drawing the products of distillation through the bed of fuel to gasify the tarry matters may be carried further, so that the gas passes downward through the entire fuel mass except where free oxygen is present for combustion. Such producers are called inverted-combustion producers and to it belong the Deschamps and the Fangé-Chavanon types illustrated in Fig. 250. In the sections of European designs given in Figs. 248–250 will be noted the differences in detail involved where the attention of the creator has been directed to secure advantages not realized in competing forms.

25a. Aspirating Producers.—The gas producers discussed in paragraphs 25 and 24 operate by pressure of air or steam or both in the closed ash-pit, and the tension within them is above the atmospheric pressure. This principle of action entails an auxiliary plant for making steam or producing air-pressure, and the heat or fuel to operate this auxiliary equipment should properly be charged to the heat or fuel account of the producer itself. Such auxiliary plant makes the apparatus cumbrous, and when the quantity of gas required is small, as for small gas-engines, it becomes disproportionately costly. The system also requires a gas-holder, as shown in Figs. 8 and 10, since the production of gas is controlled by the rapidity of combustion at the base of the producer, and this action in turn by the jet in the ash-pit and not by the consumption of gas by the motor or motors. The holder must therefore act as an accumulator between the producer and a variable demand for gas.

For isolated plants of relatively small capacity, or where but one gas-motor is to be supplied by such a generator, it appeared at once of great importance to simplify the plant as to bulk and cost on the one hand, and on the other to make the demand of the engine the governing factor in the generation of the gas from the solid fuel. This has given rise to the producer acting by aspiration, directly connected to the suction inlet of the gas-motor, and operating under a tension less than atmospheric pres-

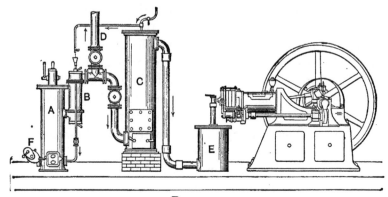

FIG. 74.

sure. These are called aspirating producers, and Fig. 74 will illustrate a typical form.

In this arrangement, as designed for an isolated plant, the producer *A* at the left operates by air and steam, which are drawn into the ash-pit below the fire by the diminished pressure above the bed of combustible. The fuel is charged through the top as required. To start the process before the motor begins to make aspirating strokes a small fan or other air-driving apparatus, as at *F*, must be run by hand or by stored or other power to start and maintain the combustion at the base of the producer. The poor and lean gas of the starting process will be wasted through an air-vent pipe D to the open air. The generated gas with the heat of the producer and its combustion process is led into a steam-generator B, which is in effect a multitubular boiler. Water is delivered into the top of this boiler at atmospheric pressure, and is made into steam at low pressure by the heat of the flowing gas. The steam passes by a pipe downward into the ashpit of A by the aspiration effect therein, and any excess of water flows to waste through an overflow. The gas passes from B to the scrubber C, which is filled with coke and through the interstices of which the gas rises, while the washing water descends. The water takes up and removes the dust from the producer, and catches the ammonia liquor and other impurities which water will absorb. The gas then passes to a receiver E of relatively large cross-section as compared to the gas suction-pipe of the motor, so that the strokes of the motor shall not cause pulsations between the receiver and the producer. The blower F has to be run from ten to fifteen minutes in a plant of small size, and after a further fifteen or thirty minutes of light or empty running of the engine itself the producer will be making gas enough regularly for its full load. Of course the vent D is closed as soon as the gas is rich enough in quality and sufficient in amount to start the engine by the usual procedure (par. 162). Such aspirating producers can be applied from the smallest capacities up to 300 horse-power of the engines. It is plain that in the smaller sizes, and with the gentle pressures which prevail, the pitchy and caking coals are at a disadvantage, and the producer works better when the fuel is of about a standard size, about that of a walnut. When the coal goes to powder also with heat, the passages for gas are clogged, and the back pressure on the motor-cylinder is increased for its aspirating stroke.

Many modifications of this typical form are possible, such as the use of an open water-jacket around the incandescent zone of the producer, from which the steam generated by the heat shall

be led under the fire, and the leading of the water from the
motor-water-jackets to both producer and scrubber.

An American arrangement of aspiration or suction producer
is illustrated in Fig. 75. In this type, the producer *A* is a cylindrical steel shell lined with fire-brick, and fitted with a revolving
grate. Between the shell and the brickwork is an annular space
through which the air for combustion and reaction with the fuel
can pass from the element *B*, together with the steam vaporized
therein, so as to enter the producer at the bottom. This jackets
the producer on the one hand, and warms the air and heats the
low-tension steam on the other. The charging of fuel is effected

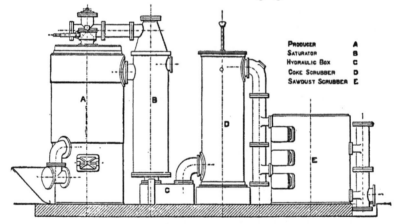

PRODUCER   A
SATURATOR   B
HYDRAULIC BOX   C
COKE SCRUBBER   D
SAWDUST SCRUBBER   E

FIG. 75.

by one or two charging-hoppers, through tubes and valves. A
central collecting-bell hanging from the top receives the gas, and
serves also to keep the fuel at a constant level. The bottom
of the producer is closed by a water-seal, so that the fire can be
cleaned and the ashes and clinker removed without interrupting
the continuous operation of the plant. The element *B* is called
the saturator, and is a water-jacketed pipe through which the
hot gas passes, and evaporates the water in the jacket. This
water is kept at a constant level, and the steam which is formed
is entrained by the entering air and carried with it to the base of
the producer. In modifications of this type, the evaporator is

put on the top of the producer. At *C* is a hydraulic box acting as a check-valve or seal to prevent gas from backing up into the producer. The elements *D* and *E* are scrubbers. *D* is the coke scrubber, necessary in any case, and *E* is a sawdust scrubber, filled with trays on which sawdust or similar material acts to remove the last vestiges of fine ash or other solid matter which may have passed the coke scrubber *D*. If the gas is clean enough

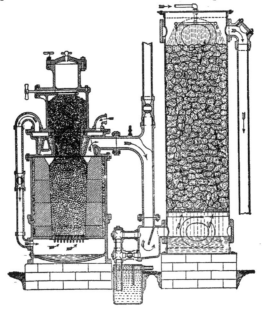

FIG. 246 —Tangye Suction Producer.

without the use of the second scrubber, its use can be omitted. From here the gas is piped to the engine.

If a typical producer of this sort be applied to supply gas to an engine of 40 horse-power with anthracite or coke as fuel, it can reasonably be expected to give a horse-power on 1 pound of fuel. In ten hours it will therefore use only 400 pounds of fuel or one-fifth of a ton, and at ordinary prices of coal this will be much less than a dollar a day for fuel expense of the plant.

Fig. 246 shows an English type of aspirating producer with attached scrubber, the motor drawing gas from the down-take at the right hand.

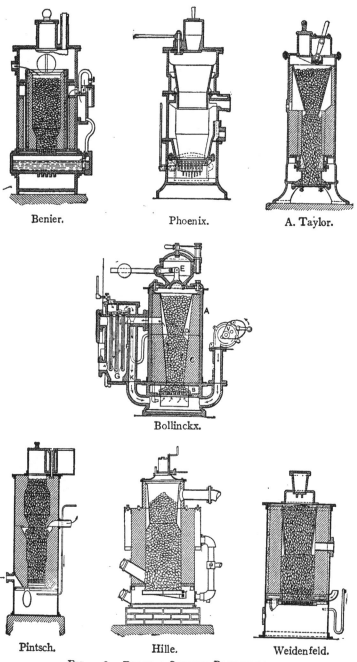

Benier. Phoenix. A. Taylor.

Bollinckx.

Pintsch. Hille. Weidenfeld.

FIG. 248.—FOREIGN SUCTION PRODUCERS.

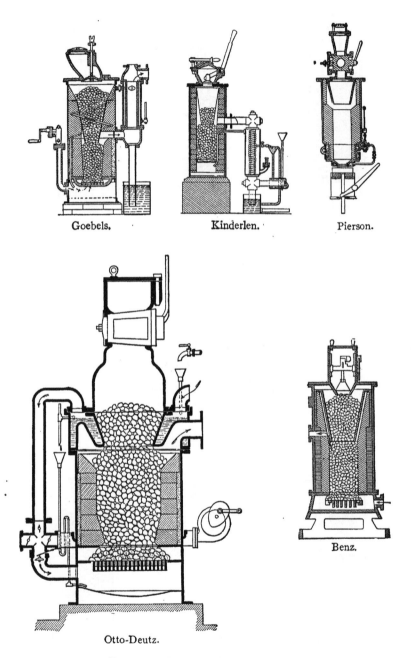

Goebels. Kinderlen. Pierson.

Otto-Deutz.

Benz.

Fig. 249 —Foreign Suction Producers.

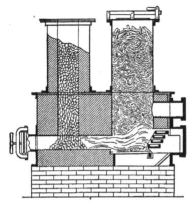
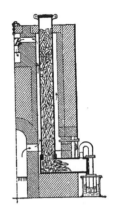

Riche Combustion.　　　　　Riche Wood Distillation

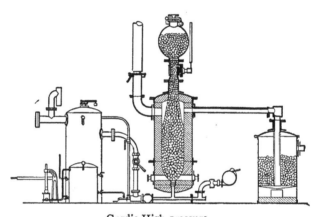

Gardie High-pressure.

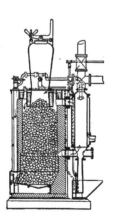
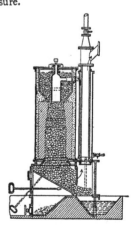

Fangé Chavanon Inverted Combustion.　　Deschamps Inverted Combustion.

FIG. 250.—FOREIGN PRODUCERS.

26. Coal-gas or Illuminating-gas.—The ordinary gas used in cities and large towns, and which was universal previous to the introduction of water-gas, is made by distilling bituminous coal in retorts. These retorts are long semi-cylindrical tubes holding each from 160 to 300 pounds of caking bituminous coal —often enriched by some cannel coal—under and around which the heat from a coke fire is maintained. The vapors distilled off become a fixed gas by being passed through that part of the distilling apparatus which is kept at a white heat. Other features of the process involve the methods for condensing tarry and offensive vapors and for cleansing, which are aside from the present purpose. The products of distillation of 100 pounds of ordinary gas-coal are usually·

Coke......................	64	to 65	pounds
Purified gas...............	15	" 12	"
Ammonia liquor............	10	" 12	"
Tar.....................	6.5 "	7.5	"
Loss and impurities........	4.5 "	3.5	"
	100.0	100.0	"

27. Acetylene Gas.—The gas C_2H_2 released from calcium carbide by addition of water is as yet of no significance for large-scale heating, but has been much examined for use in motor carriages and elsewhere where gas-power in small bulk is the prerequisite. One pound of calcium carbide with a little over half its weight of water will liberate $5\frac{3}{4}$ cubic feet of gas. It has a heat capacity of 18,260 to 21,492 B.T.U. per pound or 1500 to 1624 per cubic foot, and weighs 0.0725 pound per cubic foot or occupies $14\frac{1}{2}$ cubic feet to the pound. It requires $12\frac{1}{2}$ volumes of air to burn it, which is usually raised to 14 or 15 in practice. It has been compressed to a liquid at 68° F. by a pressure of 600 pounds per square inch.

Acetylene ignites at 510° F. in proper mixtures with air, and has a specific heat about 0.245 for its products of combustion at constant volume. Some very complete experiments by Mr.

Frederick Grover of England between 1898 and 1901, to be referred to later (Chapter XIX), showed that whereas a mixture of 9 parts coal-gas to 1 of air was required to give an initial pressure of 180 pounds in the cylinder, after compression to 30 pounds, the same pressures resulted from acetylene with 30 to 1. With 11.7 to 1 and the same compression he observed 352 pounds pressure to result from ignition. If the thermal efficiency for acetylene be taken at 30 per cent, 6.1 to 6.3 cubic feet of gas would be required per H.P. per hour. The ignition temperature is low; the transmission of flame in the mixture is rapid; the combustion temperature and consequent mechanical energy are high. For quantitative results the reader is referred to Chapter XIX.

28. Blast-furnace Gas.—Since the first experiments in 1895 an increasing use is continually found for the gas discharged from the top of the blast-furnace as a source of motive power. Less than one-third of the carbon introduced into the blast-furnace can be allowed to reach the state of CO_2 in order to maintain the reducing action demanded for the chemical reactions on the iron ore. Hence the discharged gas consists largely of carbonic oxide (CO), although it is probably the leanest form of fuel-gas which is used, running about 100 B.T.U. per cubic foot. There is a small proportion of CH_4 from the dry distillation of the coal before it gets far down the shaft, and hydrogen from moisture in the charge. The CO_2 is due to imperfections in the reduction process upon the ore, from calcination of the limestone used as flux and from union of CO with oxygen from the ore at temperatures too low to be decomposed again. German analyses give 25 to 30 per cent of CO, 55 to 60 per cent of N, 12 per cent of CO_2, 3 per cent of hydrocarbons. The CO burns with a transparent blue flame of low calorific power, taking therefore a large volume of gas when used to make steam for power. When the gas is used in gas engines on the other hand the ineffectiveness disappears, and a given gas from a ton of pig-iron made will supply five times the power obtained when used indirectly under boilers. German computations state that the gas per ton of pig-iron when used to make steam, will give an average of 400 H.P. The usual computation calls for 28 per cent of the gas for preheating the blast for the furnace and 10 per cent in loss and waste, leaving the balance or 62 per cent for power. The power gas may be all used in gas-engines, or part in making steam and part in gas-engines, or all may be used in making steam. Taking as an accepted value for the cubic feet of gas per ton of pig-iron the total of 163,590 with a calorific power of 101 B.T.U. per cubic foot, the proportions and distribution would be as follows:

I. ALL GAS ENGINES.

	Per Cent.	Cubic Feet.	Cubic Meters.
Heating the blast................	28.06	45,903	1,300
Waste in pipes, etc...............	10	16,348	463
Driving blast-engines	10.87	17,796	504
Balance available for other purposes	51.07	83,543	2,366
	100.00	163,590	4,633

II. PARTLY STEAM, PARTLY GAS ENGINES.

Heating the blast................	28.06	45,903	1,300
Wastes	10	16,348	463
Driving blast engines	39.28	64,264	1,820
Balance available	22.66	37,075	1,050

III. ALL STEAM ENGINES.

Heating the blast................	28.06	45,903	1,300
Wastes	10	16,348	463
Driving blast engines	39.28	64,264	1,820
Balance available	22.66	37,075	1,050

The financial possibilities of the first table will be governed by the possibilities of sale in the neighborhood of the surplus power in the form of electrical or other energy.

The objections which have been urged against using the blast-furnace gases as producer gases for power have been:

1. The dirt which they contain.

2. Their high temperature.

3. The danger to the men from their asphyxiating effect at leaks and joints.

4. Their low heating value.

5. The irregular supply of heat energy due to fluctuations of composition and quantity.

The dust from the blast furnace seems to be of two sorts. One is heavy, consisting of particles of ore, coke, and lime carried over in the current and which will be dropped by the expedient of lowering the velocity of flow at some point or by the passage of the gas over water at such point. These dust-catchers have been used and have been successful in the plants where the gas is burnt under boilers. The other dust is fine and impalpable, resembling flour in consistence, and is deposited by the metallic vapors, containing also clay and lime. In amount it varies from 3 to 5 grams in the cubic meter, in ordinary gases, the difficulty increasing greatly when zinc dust or similar oxides are also present. Two general systems are in use, which may be distinguished by the two names of static and dynamic. The static systems use stationary devices such as scrubbers and purifiers, in which coke, sawdust, or moss are used as with producer gas, in connection with sieves or trays, acting as filters, with water to wash the gas in transit and to remove the inconvenient excess of temperature. In the dynamic system, centrifugal fans are used with jets of water, either singly or in series with a long pipe for settling purposes between, or a coke filter may be used after the fan. In such a plant, in England, the dust is 1.8 grams per cubic meter before reaching the fan, and 0.4 gram after leaving it. Five hundred gallons of water per hour are injected, but the water can be used over and over again. The fan has to be cleaned once in two weeks. The trouble from the dust at the engines is both from clogging of passages and the combination with the lubricating material, but even more than this the presence of the grit causes annoying abrasion of valve faces and cylinders and pistons.

Vertical engines offer advantage over horizontal ones from this cause because the grit does not lie on the surface over which the piston moves.

The solution of the cleaning problem which seems to be satisfactory in the dynamic type, is also the solution of the inconvenient heat problem. The cooling injection water can be used over and over again with adequate reservoirs or tanks.

The physiological effects of blast-furnace or producer gas rich in carbon monoxide (CO), are of special difficulty to meet, because on coming into contact with the blood in the lungs, the necessary process of oxygenation not only does not take place, but a curiously stable compound seems to be found in the blood itself which is reacted upon by the oxygen of the air with reluctance. As this reaction of the carbon monoxide takes place even in the presence of oxygen, a poisoning can occur even in the open air. Hence the physiological effects are of two classes: the slow progressive type and the rapid though gradual asphyxiation. The slow poisoning evidences itself in pains in the head; pallor, due to anæmia; lassitude; muscular weakness, and ultimate paralysis, general and local. The end is an asphyxiation, if the man is not removed from the influence of the gas. The rapid asphyxiation process is revealed by headache and nausea, vertigo, congestion at the temporal blood-vessels, dimness of vision, and mental depression. A strong desire to sleep succeeds, and if the victim can get fresh air at this stage he usually comes out all right. A little later under the influence of the poison, paralysis of the lower limbs follows, and then coma and death. If the asphyxiation is discovered and antagonized in time, only a severe headache remains after it in most cases. A little later than this, however, the oxygen of the air is too feeble to recover the victim, and a current of pure oxygen from a pressure tank delivered to a respirating orifice in a mask over the face is required. If necessary the same methods of producing respiration artificially or mechanically must be used, as would be applied to those apparently drowned: the chest cavity is alternately expanded and contracted by drawing the arms upward over the head and they are then forced downward and pressed inward upon the chest as the patient lies upon his back. The movements are made slowly and decisively fifteen or twenty times per minute, and kept up if necessary for several hours, and heart action stimulated by external heat, friction, and pungent vapors in the nostrils. These difficulties with blast-furnace gas should be no worse than with producer gas except from the scale on which they are carried on, and from the danger that in furnace plants the pressure of production and the financial loss due to stoppage will postpone the overhauling necessary to keep the plant up to conditions as respects the cleanness of the engines and the prime quality of joints and packings. Operatives object to respirators and will discard them; incessant vigilance and frequent inspections from the superintendence class above them is the only safeguard. The gas is colorless and without effect on the nerves of smell or taste when pure.

The fourth alleged difficulty from low heating value is without significance, or does not exist. The cylinder volume of the engine is greater, but the poor gas is as efficient per heat unit as the richer, and a greater compression ratio is possible without inconvenient pre-ignitions. Gases as lean as 87 B.T.U. per cubic foot have been successfully used.

The irregular quality and quantity of the fuel supply disappears as the number of furnaces in the plant is increased. It has not been found a significant factor of trouble. If thought advisable a holder of sufficient capacity may be introduced between blast furnaces and engine plant, acting as an accumulator and equalizing both quality and quantity. An independent producer may also be made to serve this same accumulator if thought desirable. Lighting and power gases from any source are not perfectly uniform at all times.

If the average heat capacity or calorific power per pound of blast-furnace gas be called 1283 B.T.U. and the percentage of heat energy effective in the gas be from 20 per cent to 30 · per cent, as is usual, then there will be required

$$\frac{33,000 \times 60}{778} \times \frac{1}{1283 \times .25} = 7.93$$

pounds of gas per H.P. per hour if an efficiency of 25 be assumed. Reducing this to cubic feet per minute, with a piston speed of 800 linear feet per minute in a two-stroke cycle engine, with a stroke one and a half diameters, it gives an accepted figure of four cubic feet per minute per H.P. Fig. 11 shows a design of engine for utilizing blast-furnace gas, originated by Mr. Chas. H. Morgan of Worcester, Mass. By using vertical cylinders he diminished the abrasive action of grit or dust in the gas, and by the beam mechanism he diminished the tendency of the piston to "cock" or produce oblique pressure in the cylinder.

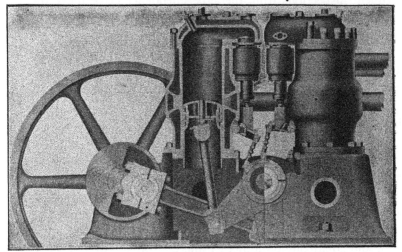

FIG. 11.

Analyses of blast-furnace gas will be found among the following tables.

Gases from coke-ovens are also available sources of fuel-gas for engines. After the first eight hours of the coking process gas appears from the top of the oven, and continues to be evolved until it reaches a maximum at the end of the second day, and then gradually diminishes until the oven is discharged at the end the third or on the fourth day. With the maximum output of gas at 40 per cent from one oven, the average will be 20 per cent, with a heating value per cubic foot averaging 60 B.T.U., with the maximum of 120 and the minimum of zero at the end when the constituents are CO_2 and N. Both blast-furnace and coke-oven gases permit and demand much higher compressions to secure certain ignition of the lean mixture. (Pars. 152, 202.)

Retort coke-oven gas should also be mentioned here, which has a calorific value from 550 to 750 B.T.U. Plants of considerable size are in operation both in America and Europe, particularly in Westphalia and in Austria.

29. Tables of Compositions and Properties of Gases.— Of the various kinds of gas referred to in the foregoing paragraphs, water-gas has the highest theoretical temperature of combustion—4850° F. Producer-gas gives 3441°. The natural gas and coal-gas give nearly the same theoretical combustion temperature, but vary greatly in calorific value with varying conditions.

COMPOSITION OF GASES BY VOLUME.*

	Findlay, O. Natural Gas.	Coal-gas.	Water-gas.	Penna. Steel-works Producer-gas.
Hydrogen.	2.18	46.00	45.00	6.00
Marsh-gas.............	92.60	40.00	2.00	3.00
Carbonic oxide.	0.50	6.00	45.00	23.50
Olefiant gas...........	0.31	4.00	0.00	0.00
Carbonic acid	0.26	0.50	4.00	1.50
Nitrogen..............	3.61	1.50	2.00	65.00
Oxygen...............	0.34	0.50	0.50	0.00
Water-vapor	0.00	1.50	1.50	1.00
Sulphydric acid........	0.20			
	100.00	100.00	100.00	100.00

* The values for water-gas in this table are for the uncarbureted product, usually known as "blue" gas. The analysis for producer-gas is for an unusually lean product In modern practice it is made richer.

COMPOSITION OF GASES BY WEIGHT.

Hydrogen.	0.268	8.21	5.431	0.458
Marsh-gas.	90.383	57.20	1.931	1.831
Carbonic oxide.	0.857	15.02	76.041	25.095
Olefiant gas.	0.531	10.01	0.000	0.000
Carbonic acid.	0.700	1.97	10.622	2.517
Nitrogen.	6.178	3.75	3.380	69.413
Oxygen.	0.666	1.43	0.965	0.000
Water-vapor.	0.000	2.41	1.630	0.686
Sulphydric acid.	0.417			
	100.000	100.00	100.000	100.000

TABLE OF RELATIVE COSTS OF GASES PER MILLION B.T.U. WHICH THEY ARE THEORETICALLY ABLE TO PRODUCE.

Cents per
Million B.T.U.

Coal-gas .... 734,976 units, costing 20.00 cents......27.21
Water-gas............... 322,346 " " 10.88 " 33.75
Producer-gas .... 117,000 " " 2.58 " 22.05

Approximately 30,000 cubic feet of gas have the heating power of one **ton** of coal.

COMPARATIVE COMPOSITION OF GAS.

	Natural Gas Findlay, O.	Coal-gas.	Water-gas.	Producer-gas.	
				Anthrac.	Bitum.
CO	0.50	6.0	45.0	27.0	27.0
H	2.18	46.0	45.0	12.0	12.0
CH_4	92.60	40.0	2.0	1.2	2.5
C_2H_4	0.31	4.0	....	....	0.4
CO_2	0.26	0.5	4.0	2.5	2.5
N	3.61	1.5	2.0	57.0	56.2
O	0.34	0.5	0.5	0.3	0.3
Vapor.		1.5	1.5		
Pounds in 1000 cu. ft.	45.60	32.0	45.6	65.6	65.9
Heat-units in 1000 cubic feet	1,100,000	735,000	322,000	137,455	156,917

PRODUCER-GAS FROM ONE TON OF COAL.

Analysis by Vol.	Per Cent.	Cubic Feet.	Lbs.	Equal to
CO..........	25.3	33,213.84	2,451.20	1,050.50 lbs. C + 1400.7 lbs. O
H.............	9.2	12,077.76	63.56	63.56 " H
CH$_4$..........	3.1	4,069.68	174.66	174.66 " CH$_4$
C$_2$H$_4$..........	0.8	1,050.24	77.78	77.78 " C$_2$H$_4$
CO...........	3.4	4,463.52	519.02	141.54 " C + 377.44 lbs. O
N (by difference)	58.2	76,404.96	5,659.63	7,350.17 " air
	100.0	131,280.00	8,945.85	

RELATIVE CALORIFIC VALUES.

	By Weight.	By Volume.	Sp. Gr.*
Natural gas........	1,000	1,000	
Coal-gas..........	949	666	0.400
Water-gas.........	292	292	0.570
Producer-gas.......	76.5	130	0.970

* With air as 1.00.

The heating value of New York City illuminating-gas, as given by Mr. E. G. Love, per cubic foot at 60° F. and barometer at 30 inches will range 715, 692, 725, 732, 691, 738, 735, 703, 734, 730, 731, 727, which will average at 721. Probably 710 would be more nearly representative of average good quality. The coal-gas of London, with 16 to 17 candle-power, has a calorific power of 668 units per foot and costs from 60 to 70 cents per thousand cubic feet. It ignites at temperatures of 750° to 800° F. with proper mixtures of air.

In the accompanying large table is gathered a summary of data on the gaseous fuels which have been treated in the foregoing paragraphs. These will be useful in computing quantities hereinafter required, and for convenience of reference it can be made to include also columns deducible from the data recorded. This table should form the basis of intelligent and thorough design, and is believed to be of the importance which is given to it by its comprehensive character. In comment thereon and to demonstrate its utility it should be observed:

1. It has been arranged with some reference to the economic significance of the various sources of gas. The retort or city gas made primarily for illuminating purposes has been put quite low in the list, since producer or fuel gas proper will be made for power purposes and distributed by mains in the cities of the future, when the development of the internal-combustion motor shall justify the necessary investment of capital.

2. The atomized carburetor gas is the carbureted-air gas used in car-lighting, in the motor vehicle for propulsion and elsewhere, where the hydro-carbons come from alcohol, gasoline or kerosene. It might be called "oil-gas" with equal propriety. The only limitations to be observed are that the carburetors should not require vacuum to make them effective with gasoline or kerosene, and the kerosene should not be injected in liquid state to be vaporized by heat or compression.

3. If the fuel actually used in any case has only a partial analysis made, as by the Orsat or other analytical method (par. 38), no considerable error will be made by using the data for the gas which comes nearest to these known figures in the table. Where CO, CO_2 and O have been found, and the gas process of manufacture is known, the other values will lie within the limits given. Where only the process of manufacture is known but no analysis is available, then the high and low limits should be taken for the gases of that sort, and these computations used as the two limits within which correct results will fall.

4. In using the table for computations on any gas or compound not listed in it, the method to be followed is identically that used in making the table itself. From some authentic and reliable source of reference and record the constants for each element are to be taken. Then by the method given in paragraph 14, the quantity by weight of oxygen required for complete chemical union or combustion is to be computed, and the volume of air to furnish that weight of oxygen will be 4.75 times the volume of oxygen required at 0° C. At 62° the more usual and convenient figure is 4.8. Having then the analysis given in

cubic feet of each element in one cubic foot or one hundred or one thousand cubic feet, this figure is multiplied by the B.T.U. of each such element from the standard data, both high values and low. These elementary values added together give the total B.T.U. high and low per unit of the compound gas under computation. Then the cubic feet of air per element is found by multiplying the volume of each elementary gas by the cubic feet of air required for its combustion. Adding these together the total is the cubic feet of air for the combustion of the unit of gas. It must be noted that any oxygen present in the gas acts negatively on the cubic feet of air required, since by its presence it renders unnecessary the admission of external air to satisfy the hydrogen and carbon. The volume of oxygen multiplied by 4.8 is to be added therefore with a minus sign, or is to be subtracted.

5. The ratio of the mixture of air and gas to the calorific power is found by dividing the figure in columns 11 and 12, respectively, by the figure in column 13 increased by one: that is the total cubic feet of air is increased by one cubic foot of gas to form the mixture, and the ratio this bears to the B.T.U. per cubic foot high and low are the quantities in columns 14 and 15. The lower value of these in column 15 has a very important significance in design of cylinders for a given power, since the mixture in the cylinder must have the quantity of oxygen computed and supplied in order that the combustion of fuel may be complete. No more fuel can be gotten in per cubic foot of cylinder volume without wasteful (because incomplete) combustion from lack of air: the fuel supply can be diminished without trouble, but the power of the motor will be diminished, since it receives less energy to be converted into work in each cubic foot of mixture.

6. If now the quantity in column 15 be multiplied by 778, there results the figure in column 16 which is the theoretical foot-pound or work capacity of such cubic foot of gas mixture received into the cylinder of the motor if it could be completely

Number.	Name of Gas or Source.	Authority Quoted.	Analysis of Constitution, Volumes in 100 Volumes.							Calorific Values. B.T.U. per Cu. Ft.		Cubic Feet of Air per Cu. Ft. Gas.	Combustion Mixture B.T.U. per Cubic Foot.		Max. theoret. ft. lbs. per cu. ft. of cylinder vol. col. 15×778.
			H	CO.	CH_4	C_2H_4	O	CO_2	N	High.	Low.		col.11/(a+1) High.	col.12/(a+1) Low.	
1	Atomizing Carburetor Gas	......	5.6	8.9	54.9	28.9	0.21	0.9	...	1085	994	9.78	100.7	92.2	71,732
2	" "	......	24.3	...	58.3	17.4	...	...	...	967	875	8.68	100.0	90.4	70,531
3	Blast Furnace Gas	Donkin	3.0	27.5	0.75	...	...	10.0	59.4	100	99	0.81	55.3	54.8	42,634
4	" " Scotch	"	2.33	24.75	...	...	...	5.75	66.42	97.1	95	1.12	45.7	44.8	34,854
5	" " English	"	1.5	27.3	...	...	...	6.0	65.2	95	94	0.69	56.1	55.7	43,334
6	" " Belgic..	"	1.02	27.90	7.0	...	...	13.95	50.12	168	160	1.36	71.1	67.8	52,748
7	" " German	"	2.5	32.0	...	...	...	8.5	57.0	114	112	0.83	62.1	61.4	47,769
8	Coke Oven Gas	......	53.0	8.2	34.5	...	...	2.0	4.9	560	485	4.76	97.4	84.2	65,508
9	"	......	53.0	6.0	35.0	2.0	...	2.0	2.0	620	524	5.06	102.3	86.4	67,219
10	Natural Gas Mie, Ind.	......	2.5	0.4	92.67	0.25	0.35	...	3.53	975	876	9.00	97.5	87.6	68,153*
11	" Findlay, O.	......	2.18	0.5	92.6	0.31	0.34	3.6	3.61	974	876	9.46	93.1	83.9	65,274
12	" St. Mary's, O.	......	2.14	0.44	93.85	0.20	0.35	...	2.98	986	885	9.03	98.3	88.2	68,620
13	" Sun, Ird Findlay	......	2.01	0.73	93.07	0.47	0.42	0.26	3.02	982	883	9.06	97.6	87.7	68,308
14	" Findlay, O.	......	1.84	0.41	93.35	0.35	0.39	...	3.41	982	882	9.06	97.6	87.7	68,231
15	" Kokomo, Ind.	......	1.7	0.55	94.16	0.30	0.30	0.29	2.80	989	888	9.13	97.6	87.7	68,231
16	" Marion, Ind.	......	1.4	0.60	93.57	0.15	0.55	0.30	3.42	980	880	9.04	97.3	87.6	68,153
17	" Pittsburgh, Pa.	......	22.0	0.60	67.0	1.0	0.8	0.6	3.0	786	703	7.51	92.5	82.7	64,341
18	Arithmet. Average of above series	......	1.97	0.53	90.46	0.38	0.44	0.81	3.22	980	881				
19	Producer Gas, dry, bl u.	......	5.3	20.0	3.0	0.20	0.40	3.6	67.5	118	113	0.91	61.2	59.2	46,058
20	"	......	6.8	22.1	3.74	0.34	0.40	4.84	61.68	139	131	1.08	66.9	63.0	49,014
21	"	......	6.9	20.0	2.2	0.20	0.40	4.6	64.9	118	111	0.90	62.1	58.3	45,357
22	"	......	8.37	22.74	2.56	0.36	0.54	5.3	60.13	135	127	1.02	66.8	62.8	48,858
23	"	......	8.5	24.8	5.2	0.40	0.40	5.6	55.1	172	160	1.35	73.2	68.1	52,982
24	"	......	9.2	25.3	3.1	0.8	...	3.4	58.2	160	150	1.24	75.0	67.0	52,126
25	"	......	9.8	24.0	3.4	0.4	0.8	6.0	55.6	153	145	1.18	70.2	66.5	51,737

No.	Description	Authority													
26	" Siemens	Donkin	8.6	24.40	2.40			5.20	59.40	133	126	1.00	66.4	62.8	48,858
27	" from anthracite	Mathot	9.0	25.0	2.0			5.0	59	133	126	1.00	66.5	63.0	49,014
28	" " from anthracite		12.0	27.0	1.2			2.5	57.3	141	134	1.05	68.8	65.3	50,803
29	" " from bituminous		12.0	27.0	2.5			2.0	56.5	154.5	146	1.17	70.7	60.9	47,380
30	Fichet-Heurtey (mixed)	Witz	20.6	23.0	2.1		0.1	6.6	50.4	156	143	1.17	72.8	66.1	51,426
31	Gas-Motoren Fabrik-Deutz	Dawson	17.0	25.0	2.0			6.0	52.0	153	142	1.15	71.0	65.5	50,959
32	Producer Gas, Dowson		18.0	17.55	3.0		0.23	7.0	47.0	119	115	1.32	51.4	49.6	38,589†
33	" "	Donkin	24.36	23.8	1.31			6.07	50.48	152	138	1.17	72.6	64.2	49,948
34	" "	J. E. Dowson	19.8	18.2	1.3		0.47	6.3	48.8	157	146	1.12	73.7	67.1	52,204
35	" " semi-water gas	"	26.55	25.07	1.11		0.03	11.3	42.28	159	144	1.36	75.6	66.7	51,893
36	" "	"	28.73	21.00	0.31		0.5	6.57	48.98	181	166	1.34	73.3	70.1	54,538
37	" " Lencauchez	Donkin	20.0	27.32	4.00			5.00	49.50	177	162	1.21	74.1	69.3	53,915
38	" "	Richards	18.34	26.0	1.25			3.6	47.84	164	161	1.24	71.9	72.4	56,327
39	" "	Dawson	18.0	12.0	2.0			7.0	47.0	166	154	1.17	71.3	68.7	53,449
40	" " Korting		29.0	11.5	2.1			14.5	42.5	156	139	1.16	70.0	64.0	49,792
41	" " Mond		28.5	13.2	2.3			15.0	42.9	154	137	1.12	65.6	63.4	49,655
42	" "	Taylor	24.5	12.0	2.0		3.0	12.9	46.8	148	133	0.98	68.9	62.8	48,858
43	" " H. A. Humphrey	"	21.0	27.0				6.0	57.0	130	116			59.0	45,902
44	" " Taylor	Donkin	12.0		1.20			2.50	57.30	141	134	1.05		65.6	51,037
45	Producer Gas, Loomis-Pettibone, coal		14.0	20.0	2.0	0.20	0.10	8.2	55.5	135	125	1.02	66.8	62.1	48,314
46	Producer Gas, Loomis Pettibone, wood	Mathot	14.0	20.0	2.0	0.20	0.10	16.0	47.7	135	125	1.02	66.8	62.1	48,314
47	Producer Gas, Riché, wood		44.0	29.0	15.0			11.0	1.0	397	358	3.19	94.7	85.4	66,441
48	" "	Donkin	44.2	22.0	12.40			21.3		347	311	2.78	91.8	80.3	47,073
49	Rich retort gas, from English cannel-coal		27.7	6.8	50.0	13.0		0.1	2.4	843	762	6.50	98.2	85.7	66,675
50	Common retort or illuminating gas		39.78	7.04	45.16	6.38	0.06	1.08	0.50	727	651	6.38	98.5	83.1	64,652
51	" "		44.4	5.2	37.1	2.3	1.1	1.3	8.6	586	521	5.05	96.7	86.1	66,986
52	" "		46.0	7.5	39.5	3.8	0.1	0.6	2.5	650	579	5.62	98.2	85.2	66,675
53	" "		47.73	6.15	35.6	4.88	0.31	1.41	3.9	628	569	5.73	93.3	84.7	65,897
54	" "		47.9	6.0	33.3	12.3	0.5			723	650	6.29	99.1	89.1	69,320

Number.	Name of Gas or Source.	Authority Quoted.	Analysis or Constitution, Volumes in 100 Volumes.							Calorific Values. B.T.U. per Cu. Ft.		Cubic Feet of Air per Cu. Ft. of Gas.	Combustion Mixture B.T.U. per Cubic Foot.		Max. theoretical ft. lbs. per cu. ft. of cylinder vol. col. 15×778.
1	2	3	4	5	6	7	8	9	10	11	12	13	14	15	16
			H	CO.	CH$_4$	C$_2$H$_4$	C	CO$_2$	N	High.	Low.		$\frac{\text{col.11}}{a+1}$ High.	$\frac{\text{col.12}}{a+1}$ Low.	
55	Common retort or illuminating gas		50.1	6.0	38.0	4.0	...	...	2.0	644	514	5.51	98.05	87.08	68,534
56	" " " "		52.5	5.0	34.0	4.0	0.05	...	4.0	608	540	5.11	99.5	88.4	68,775
57	" " Paris	A. Witz.	52.8	5.60	32.30	5.5	...	...	3.80	617	550	5.21	98.1	87.1	68,534
58	" " "	"	52.9	7.18	31.8	5.0	...	...	2.5	610	544	5.21	98.2	87.7	68,231
59	" " (carbureted)	G. E. Moore	46.20	8.88	34.02	5.09	0.65	3.01	2.15	650	577	5.29	103.3	90.3	70,253
60	" " London, G.L. & Coke Co.	Dawson.	47.9	3.75	37.64	4.41	0.26	...	5.95	644	464	5.48	99.3	72.0	56,016
61	" " average of above series		47.54	6.27	36.61	5.22	0.38	1.48	3.34	642	573	5.54	98.2	87.7	68,231
62	" " range of average values	W. H. Fowler	34–53	2.7–6.0	36–43	3–13	...	0.3–1.0	3–5	710–	735	...	...	...	...
63	Carbureted Water Gas, Worcester.	G. E. Moore	37.20	28.26	18.88	12.88	0.06	0.14	2.64	650.1	597	5.33	102.7	94.3	73,365†
64	Carbureted Water Gas, Lake.	"	35.88	23.58	20.95	15.43	0.01	0.30	3.85	688.7	646	5.66	103.4	97.0	75,466§
65	Carbureted Water Gas.	"	34.0	33.0	15.0	12.5	...	...	6.0	579	532	4.85	98.9	90.9	70,720
66	" " New York	Dawson.	30.30	26.50	24.30	15.00	0.50	1.00	2.40	682	625	5.86	99.6	91.3	71,031
67	" " Lowe process	Donkin.	30.0	28.0				34.0	8.0	192	177	1.39	80.5	73.9	57,494
68	" " "	"	29.6	28.33	24.42	16.12	0.21	...	4.0	608	540	5.11	99.5	88.4	68,775
69	" " "	"	29.2	29.2	31.9	3.4	0.6	...	2.3	714	659	6.22	96.1	91.2	70,954
70	" " "	"	21.8	28.1	30.7	12.9	0.5	3.8	2.2	702	635	6.00	100.3	90.7	70,565
71	" " "	"	18.4	14.8	30.7	21.2	...	4.6	9.3	782	704	6.80	100.2	90.2	70,176
72	Arithmetical Average of above series		28.5	26.46	24.65	13.50	0.27	2.21	4.32	675	616	...	...	...	...

73	Water Gas uncarbureted, (blue)	51.8	43.4	...	...	3.5	1.30	314	287	2.28	95.8	87.5	68,075
74	" " "	49.55	45.89	...	...	3.87	0.71	315	289	2.33	94.6	86.8	67,530
75	" " "	49.50	35.93	1.05	...	4.25	8.75	295	265	2.10	92.2	82.8	64,418
76	" " "	49.17	43.75	0.31	...	2.71	4.06	310	284	2.26	96.3	87.1	68,534
77	Arithmetical Average of above series	50.0	42.2	0.68	...	3.77	3.70	308	280	...	...	...	...
78	Retort Gas from English lignite......Donkin	20.0	17.50	1.50	...	15.00	20.00	239	211	1.76	83.9	76.5	59,517

* Indiana gas averages .045 pounds per cubic foot.
† Computed from averages.
‡ Av. CP = 22.06.
§ Av. CP = 26.3.

utilized. This is the same, therefore, as the work capacity of each cubic foot of cylinder volume which is filled with the mixture in question at atmospheric pressure, to undergo the transformations of the cycle of operations. It can be reduced to horsepower per minute by dividing by 33,000, or per second by dividing by 550. This, however, is a purely theoretical maximum, and must be carefully so understood. To make it practical the figure in column 16 must be multiplied by a factor which shall take account of the temperature limitations necessarily imposed, because the mechanical work done is never equal to the potential total of heat supplied. Calling the ratio of the output of work to the input of heat by the accepted term "efficiency" and denoting it by E (par. 47), then

$$\text{Efficiency} = E = \frac{\text{Work utilized}}{\text{Heat supplied}}.$$

Whence,

$$\text{Work utilized} = (\text{Heat supplied}) \times (E).$$

There are also practical losses in friction, and in the other wastes referred to in the analysis of paragraph 7a. When a factor is known or found to cover these losses, the value of column 16 multiplied by it will give the power capacity for such gas per cubic feet of cylinder volume supplied with the proper mixture per unit of time. Since the efficiency value will be shown to be greater as the compression possible in the cylinder is greater, and compression is greater with the lean or poorer mixtures, it will be found that the dynamic value of the various fuels differs little per unit of weight. The leaner mixtures require larger cylinder volume to secure the same power, as compared with the richer fuels. The significance of column 16 will appear more fully in the sequel (par. 40).

8. The values in columns 15 and 16 are also too high when each cubic foot of the cylinder volume must contain also a volume of neutral gases, not supporters of combustion, remaining there as burnt gases from the previous combustion stroke of the motor.

These appear as adding a volume in cubic feet to the denominator of the fraction $\dfrac{H}{a+1}$ so that it becomes $\dfrac{H}{n+a+1}$ and therefore diminishes the volume available to receive fuel energy. The clearance volume may be usually considered to be full of burnt gas without error, and the value of n computed therefrom. Any mechanical defects of leakage, mal-adjustment, excessive friction and the like, will still further lower the final efficiency.

9. Since the value in column 16, when multiplied by the factor E, gives the mechanical work of a cylinder having a cubic foot of volume, it must also express the work done by the mean efi.ctive pressure over a piston area of one square foot, when this pressure moves through one foot of length. Hence the tabular value multiplied by E can be considered as the mean effective pressure for one stroke of a 12-inch stroke motor, having one square foot of piston-area, which will become the M.E.P. for pounds per square inch, by dividing by 144. This is the same then as multiplying the value in column 15 by $\dfrac{778}{144}E = 5.4$, E, to give the theoretical M.E.P. in pounds per square inch for a 12-inch stroke one-foot-area engine using this fuel, and with no neutrals in the mixture.

10. If the cylinder stroke is not one foot but a number of feet, L, then the M.E.P. will be the value for one foot of length divided by L or

$$\text{M.E.P.} = \frac{5.4\,E}{L}$$

for a cylinder L feet long in which L is greater or less than one foot. This will be used in detail in paragraph 40.

11. Columns for calories per cubic meter and of B.T.U. per pound might have been added and left blank for the convenience of any to whom these computations may at any time prove desirable to make for their own use. It has not seemed worth while to burden the working table with them. To convert B.T.U. per cubic foot into calories per cubic meter, multiply the former by 8.9. The temperature used in most cases has been

DATA AND COMPUTATIONS ON ELEMENTARY GASES.

1	2	3	4	5	6	7	8	9	10	11	12	13	14
Name of Gas.	Chemical Symbol.	Molecular Weight $H=2$.	Atomic Weight $H=1$.	Weight per Cu. Ft. at 0° C. Col. 4 × .005591	Weight per Cu. Ft. at 62° F. Col. 5 × 0.9425.	Calorific Values per Cu. Ft. at 0° C. in B.T.U.	Calorific B.T.U. per pound. High.	Calorific Values per pound. Low.	Calorific in B.T.U. High Col.6 × Col. 8.	Values at 62° F. per Cu. Ft. Col. 6 × Col. 9 Low.	Cu. Ft. of Air per Cu. Ft. Gas.	Vols. of CO_2 produced.	Vols. of H_2O produced.
Hydrogen	H	2	1	0.005591	0.00527	293.5	61,524	51,804	331.0	278.7	2.4	0	1
Me. (Marsh-gas)	CH_4	16	8	0.04728	0.04215	1666	24,021	21,592.8	1037.2	932.4	9.6	1	2
Ethylene (Olefiant gas)	C_2H_4	28	14	0.0783	0.07363	1678	21,222	19,834.2	1619	1513.5	14.4	2	2
Benzene (Benzole vapor)	C_6H_6	78	39	0.2181	0.2055	4023	17,654 / 18,954	16,934 / 17,847	3627.8	3480	35.7	6	3
Butylene (Tetrylene)	C_4H_8	56	28	0.1566	0.1476	3275	20,857	19,345	3078.5	2855	28.6	4	4
Carbonic oxide (Carbon monoxide)	CO	28	14	0.0783	0.07363	342.3	4395.6	4395.6	329.6	329.6	2.4	1	...
Carbonic acid (Carbon dioxide)	CO_2	44	22	0.1230	0.1159	...	...	...	...	...	...	...	...
Nitrogen	N	28	14	0.0783	0.07363	...	...	...	...	...	...	...	...

17° C., or 64° F. In any case where a special temperature standard may be used or may have been used, a column should be left for the record of such temperature basis.

12. To exemplify the use of the table and its method of compilation from the fundamental data on the elementary gases, as given above, the basal table herewith entitled " Data and Computations on Elementary Gases " may be used as the starting point. The data are from J. Thompson, or other reliable sources.

The atomized gas given as No. 1 in the large table is taken for the illustration below. Column 2, below, repeats the analysis from the large table. Columns 3 and 4 of that table result when the values in column 2 are multiplied by the values in columns 10 and 11 of the element table preceding. Column 5 results when the analysis figures in the second column are multiplied by the cubic foot requirement of that element taken from column 12 of the preceding table. Totaling these constituent computations the total B.T.U. and the total cubic feet of air are given; and dividing these total B.T.U.'s by the cubic feet of air increased by one cubic foot of gas-fuel, the results for the mixture given in columns 14 and 15 of the large table are obtained, as in columns 6 and 7.

TYPICAL COMPUTATION TABLE.

1	2	3	4	5	6	7
Constituents of Carbureter Gas Line No. 1.	Analysis or Volume in Cubic Feet in 100.	B.T.U. in 100 Cubic Feet.		Cubic Feet of Air per 100 Cubic Feet Gas.	Col. 3 ÷ Col. 5 + 100.	Col. 4 ÷ Col. 5 + 100.
		High.	Low.			
Hydrogen ...	5.6	1853.6	1560.7	13.44	...	...
Marsh gas......	54.9	56942.3	51188.8	527.0	...	...
Olefiant gas ...	28.9	46789.1	43725.7	416.2	...	...
Carbonic oxide..	8.9	2933.4	2933.4	21.4	...	...
Carbonic acid...	0.9	...	...	...	...	...
Oxygen........	0.21	...	...	...	...	...
Nitrogen.......	...	...	...	...	...	...
Totals	100.	108518.4	99408.6	978.04	100.7	92.2

30. Liquid Fuel. Petroleum. — Another great source of hydrogen and carbon as fuels for industrial purposes comes from the oils which are pumped up from the earth or which flow under pressure from subterranean reservoirs and which are designated by the general name of petroleum. There are oils of animal origin, but they are now supplied to such a limited extent for fuel purposes as scarcely to deserve consideration, and the cost of extracting vegetable oils from the seeds or other products which carry them precludes the use of such oils for fuel. Hence the mineral oil, or petroleum, is the principal source of heat from liquids either in its crude form as it comes native from the oil-well, or after a part of the constituents of the natural oil have been eliminated by the refining process. In the present state of the art of using liquid fuel in motor-engines, the use of crude oil is so difficult as to be practically prohibited. The difficulties arise from the fact that the mineral oil is not a homogeneous chemical substance, but is a mechanical mixture of several constituents having varying temperatures at which they change their state from a liquid to a gas. The consequence of this mechanical mixture of constituents is that the more volatile elements form a gas first and are eliminated from the mixture, leaving behind the thicker and more viscous components, which presently form a gum or a solid mass in the generating chamber which is difficult to handle. It is much easier to use the refined products of the refining process rather than the crude oil in its entirety. The average composition of crude petroleum is usually given as:

	From	To	Average.
Carbon....	82	87.1	85
Hydrogen	11.2	14.8	13
Oxygen and impurities.	0.5	5.7	2
			100

Its specific gravity is from 0.79 to 0.82. Lima oil from the Ohio wells is of a dark green color, is quite fluid and volatile, and has a disagreeable odor. Its volatility makes it flame easily and give off an explosive vapor in a confined space. These two properties have resulted in restrictions upon its use in many cities; the health boards object to the odor, and the fire departments to the danger of fire from explosions. Hence the refining companies have introduced what is called fuel-oil. This is the residue after a part of the fractional distillation process has been completed. A tabular summary of this process is as follows:

No.	Tempera-ture Fahr. degrees.	Distillate.	Prob-able Per Cent.	Specific Gravity	Density Baumé.	Boiling Point Fahr.	Flashing Point.
					degrees	*degrees*	
1	113	Rhigolene ⎱ Petroleum	traces	.590 to .625	85–80	104–158	
2	113 to 140	Chymogene ⎰ ether		.660 to .670	80–78	158–176	
3	140 to 158	Gasolene	1.5	.680 to .700	78–68	176–214	14
4	158 to 248	Benzine, naphtha C	10.0	.714 to .718	68–64	212–248	
		" " B	2.5	.725 to .737	64–60	248–302	32
5	248 to 347	" " A	2				
		Polishing-oils					
6	338 +	Kerosene	50	.753 to .864	56–32	302–572	100–122
7	482	Lubricating-oil	15	.864 to .960	32–15	572 up.	230
8	....	Paraffine wax	2				
9	....	Residuum and loss	16				

The distillation for fuel-oil is stopped after the kerosene has been obtained. In many refineries only the three products of crude naphtha, burning oil, or kerosene, and the distillate are recognized, the latter being the fuel-oil. Its average specific gravity is about .818 or 40 Baumé at 60° F., so that a gallon weighs 7.3 pounds, as against 6.81 pounds for the crude oil. It flashes at 218° F., or just above the boiling-point of water. It is thick in consistency. The calorific power of crude oil is from 20,000 to 21,000 British thermal units, and that of the fuel-oil is from 17,000 to 19,000 heat-units. Fuel-oil is called "astatki" by the Russians. Thos. Urquhart of Russia, in considering the use of petroleum for locomotives, gives the following table of the theoretical evaporative power of petroleum in comparison with that of coal, as determined by Messre. Favre and Silbermann:

Fuel.	Specific Gravity at 32° F., Water= 1.000.	Chemical Composition.			Heating Power, British Thermal Units.	Theoret. Evap., Lbs. Water per Lb. Fuel from and at 212° F.
		C.	H.	O.		
Penna. heavy crude oil.....	0.886	84.9	13.7	1.4	20,736	21.48
Caucasian light crude oil ..	0.884	86.3	13.6	0.1	22,027	22.79
" heavy crude oil .	0.928	86.6	12.3	1.1	20,183	20.85
Petroleum refuse	0.928	87.1	11.7	1.2	19,832	20.53
Good English coal, mean of 98 samples.............	1.380	80.0	5.0	8.0	14,112	14.61

The further details of refining for elimination of coloring matter, and the steps of acid and alkaline agitation, are aside from the present purpose.

31. Pintsch Oil-gas.—The Pintsch oil-gas (Julius Pintsch, Berlin, 1871) is a true gas made in retorts by the vaporization of crude petroleum. From 70 to 85 cubic feet of a 50- to 60-candle-power gas result from the distillation of one gallon of oil. The gas is rich in illuminating properties and does not lose so much of its illuminating power by compression as a coal-gas would. This system is much used for the lighting of railway cars, but air-gas and acetylene systems compete with it. With the air-gas system the gas is heated before being delivered to the burner by passing through a spiral coil of fine copper pipe above the lamp itself.

32. Kerosene.—Kerosene has already been referred to as "burning oil" or No 6 in the process of fractional distillation of crude petroleum (par. 30). Usually 3½ parts of crude oil render one part of kerosene. The heat of combustion will depend on the composition, but will range between 22,000 and 24,000 B.T.U. per pound. The quicker the distillation the poorer the product, albeit more abundant; but the more abundant the lighter elements the less safe is the kerosene.

The flashing-points at which an ignitible vapor is given off by heating will range from 115° to 125° F.; the oil will itself ignite and burn when heated to between 130° and 140° F. This is called its burning-point. It boils anywhere between the limits of 300° F. and 500° F., giving a vapor density five times that of air, and requiring for its combustion nearly 190 cubic feet of air per pound. The principal component of kerosene in the hydro-carbon series is the element decane, whose composition by the formula $C_nH_{2(n+1)}$ (see p. 64) will be $C_{10}H_{22}$. If the kerosene be regarded as composed entirely of this compound, its combustion will be given by the equation

$$\underbrace{C_{10}H_{22}}_{120+22} + \underbrace{O_{20} + O_{11}}_{320+176} = \underbrace{10CO_2 + 11H_2O,}_{440+198}$$

so that

$$\tfrac{496}{142} = 3.5 \text{ pounds of oxygen,} \quad \text{or} \quad \tfrac{100}{23} \text{ of } 3.5 = 15.21$$

pounds or $15.21 \times 12.387 = 188.4$ cubic feet of air will be required when in its combustion it is treated as a gas. Since the vapor of kerosene is five times as heavy as that of air, a pound of vapor will occupy $\dfrac{12.387}{5} = 2.47$ cubic feet. The ratio of volumes will therefore be $\dfrac{188.4}{2.47} = 76.2$ volumes of air to one volume of kerosene vapor. This computation can be applied in a later paragraph to compute the temperature increase due to combustion. When kerosene is used as a source of heat for internal-combustion engines it is usually atomized or broken into a mist, and is then vaporized by heat so as to form a gas. (See Chapter VI, on Kerosene-engines, and Chapter X, on Carbureters.) It is cheaper than the more usual gasoline, but by reason of the inconveniences in starting up and some easily superable difficulties in regulation du to its variable composition it has not received the attention which has been given to gasoline. It does not waste or change its quality in storage; the supply is practically unlimited; it is everywhere obtainable; the fire-risk or insurance rate is not increased by its presence. The wide range of the boiling-point of the commercial article has been the source of one difficulty in using it. Being a mixture in any case, an isolated mass decomposes or "cracks" by heat into various components, and when heated too hot it decomposes, depositing its carbon in the form of a hard cake like coke. For the complete combustion of liquid kerosene the conditions of a lamp-wick are ideal by reason of its presenting four essential conditions:

1. The heat for gasifying the oil is graduated from that of the containing vessel to the highest temperature of combustion of the oil.

2. The combustion is slow enough to allow complete union with necessary oxygen for combustion.

3. With a proper exposure of distilling or gasifying surface to the oxygen, the whole of the combustible is consumed.

4. The gasification at the flame is at just the proper temperature, and the delivery of fuel is at just the required rate.

If the wick is turned too high, more fuel is supplied above the burner top than can be gasified by the heat of the flame, and the excess of carbon unconsumed appears as a smoke with a suffocating odor. If the wick is turned down, however, the heat is again too low to consume the distilled gas and some carbonic oxide is formed. Of course, if the oxygen supply is cut off while other conditions are retained normal, both of the above evil consequences present themselves. In the chapter on kerosene-engines (Chap. VI) some further points will be discussed concerning the conditions for its use in motors.

33. Gasoline.—Gasoline is the next higher or more volatile distillate from crude petroleum, having a specific gravity ranging from the highest grade of about 88° Baumé little used for power purposes down to 68° B., or with a specific-gravity range in the ordinary hydrometer scale between 0.680 and 0.710. The commercial names differ in different places, but in general the qualities and names will be:

> (1) 88° to 86° B. or .640 . . . . light volatile oil
> (2) 76°   B. or .682 . . . . stove-gasoline
> (3) 68° to 73° B. or .692 to .709 . . . . . . benzoline
> (4) 62°   B. or .730. . . . . benzine

No. 2 in the series is the usual internal-combustion motor fuel, but frequently this changes on storage for any length of time to No. 3, which is sometimes called prime city naphtha. Stove-gasoline is also locally known as boulevard gas-fluid. It will be observed from the table in paragraph 30 that the crude petroleums usually yield but 8 to 10 per cent of gasoline, so that a definite limit is set upon the amount of it available at any one time. This will have a notable effect upon its price as the use of gasoline for fuel becomes more extended and the number of motors using it increases.

The boiling-point of gasoline ranges from 120° to 250° Fahr., with an average range between 149° and 194° Fahr. Its principal hydrocarbon constituents are the elements hexane and heptane of the series, which on the formula $C_nH_{2(n+1)}$ gives a composition of $C_6H_{14}+C_7H_{16}$. The range is between C_6H_{12} and C_7H_{16}. Taking C_6H_{14} as the average of all, the combustion computation will be

$$C_6H_{14} \; + \; O_{12}+O_7 \; = \; 6CO_2+7H_2O.$$
$$\underbrace{72+14} \quad \underbrace{192+112} \quad \underbrace{264+126}$$

This will require $\dfrac{304}{86} = 3.53$ pounds of oxygen or $3.53 \times \dfrac{100}{23} =$ 15.3 pounds of air or 189.52 cubic feet of air.

The vapor of gasoline is 3.05 times as heavy as that of air. Hence a pound of vapor will occupy $\dfrac{12.387}{3.05} = 4.06$ cubic feet at 32° and one atmosphere of pressure. Hence a pound of gasoline vapor occupying 4.06 cubic feet will require 189 cubic feet of air, or the ratio of volumes of vapor to air will be

$$\frac{189.52}{4.06} = 46.6.$$

This computation can be later used to compute the temperature increase. The calorific power of gasoline is between 18,000 and 20,000 B.T.U. per pound or 690 per cubic foot. Redwood found that gasoline vapor with air in proportions ranging between 5 in 100 up to 12.5 in 100 were explosive. The mixture of 11 per cent of vapor gave the strongest effect. Incomplete combustion of gasoline in mixtures results in the formation of a smoky mixture in the products, with an offensive odor. As in the case of kerosene, this phenomenon results from either too much fuel in the mixture, or from too little. The former is the more usual and the more objectionable.

The volatile elements in gasoline tend to escape from it even at ordinary temperatures, so that the liquid alters and deteriorates in storage unless in very tight metallic vessels. The volatile elements mixing with the air in a vessel partly emptied form an explosive mixture which will ignite readily from an open flame. The heavy vapor seeks the bottom levels in confined places before becoming diffused. Neither kerosene nor gasolene acts like water in swelling the staves of wooden barrels and keeping such receptacles tight against leakage and loss. When gasoline is entrained by a current of air, as in a carburetor (Chapter X), it forms an air-gas, or an atmosphere saturated with hydrocarbon mist. Gasoline is sufficiently volatile to evaporate completely without additional heat when thus finely divided, and the carbureted air can be burned for power or lighting purposes. Much car-lighting by gas is done on this system, using compressed air from the train-brake supply. There is no considerable storage of gas, since such a carbureted atmosphere is liable to deposit its liquid hydrocarbon by a sort of liquation in any storage tank where it may be at rest, particularly in cold weather. On the other hand, to pass this mixture through a hot chamber will cause a deposit of fixed carbon or coke as in the case of kerosene.

The arguments for gasoline as a fuel for motor vehicles are the ease of its vaporization in starting and running, in spite of difficulties of maintaining proper proportions of the mixture of fuel and air for varying conditions of load and speed.

34. Alcohols.—There are two kinds of alcohol used in the arts and as sources of heat: methylic alcohol or wood-alcohol, which has the chemical symbol $C_2H_4O_2$, and ethyl alcohol, the ordinary form, which is represented by $C_4H_6O_2$.

Wood-alcohol is formed by dry distillation of wood in iron retorts (usually horizontal) at a heat not above 900° F. It has a strong characteristic odor and boils at 150° F. In the United States a considerable internal-revenue taxation is levied upon alcohol, which operates with some hardship upon pro-ducers of corn at considerable distances from their market. The transportation charges on the grain may preclude an attractive profit upon the raw material, whereas in manufactured and concentrated form as alcohol the profit from an acreage would be a handsome one. The French and German ministries of agriculture have been encouraging the development of alcohol-motors with a view of stimulating production of alcohol-grain among the farming districts. Their interest has developed a process for deriving alcohol from the electrically manufactured carbides.

Ethyl alcohol is obtained by distillation from the fermented infusions of the cereal grains, which contain either sugar or starch. It has a specific gravity of 0.792 and boils at 173° F., but will freeze only at 200° below zero when pure. It expands $3\frac{1}{2}$ times as much as water between 32° and 173° F.

Hydrated alcohols contain water ranging from 50 per cent by volume (proof spirits) to 93 per cent (cologne spirits). The affinity for water is very strong. The table on page 91 is of convenient reference.

For the range of percentage contained in that table the correction for temperatures different from 60° F. should be made as follows:

If the density is measured at a temperature above 60°, 0.0005 should be added to the measured density for each degree which the temperature at the time of the measurement differs from 60°. When the temperature at the time of measurement is below 60°, the same correction should be subtracted from the measured density. The corrected density should then be used in the table for finding the true percentage of alcohol.

SMITHSONIAN TABLE OF SPECIFIC GRAVITIES OF ETHYL ALCOHOL.

Specific Gravity at 60° F. Compared with Water at 60° F.	Percentage of Alcohol.		Specific Gravity at 60° F. Compared with Water at 60° F.	Percentage of Alcohol.	
	By Weight.	By Volume.		By Weight.	By Volume.
0.834	85.8	90.0	0.822	90.4	93.4
.833	86.2	90.3	.821	90.8	93.7
.832	86.6	90.6	.820	91.1	94.0
.831	87.0	90.9	.819	91.5	94.2
.830	87.4	91.2	.818	91.9	94.5
.829	87.7	91.5	.817	92.2	94.8
.828	88.1	91.8	.816	92.6	95.0
.827	88.5	92.1	.815	93.0	95.3
.826	88.9	92.3	.814	93.3	95.5
.825	89.3	92.6	.813	93.7	95.8
.824	89.6	92.9	.812	94.0	96.0
.823	90.0	93.2	...	...	...

The percentage of alcohol found in a sample is always likely to be greater when determined chemically than when determined by the hydrometer, because the presence of impurities in the way of solids dissolved in the alcohol or as any of the series of higher alcohols tends to make the specific gravity of the sample greater, and hence make it indicate too low a percentage of alcohol.

Pure alcohol is very inflammable and burns with a pale-blue smokeless flame. Its calorific power is about 28,500 B.T.U., which runs down to 12,000 with greater hydration.

For motor purposes the custom has prevailed very widely to mix the alcohol with some other hydrocarbon, usually from the petroleum group. Such mixtures become undrinkable and are known as "denatured" alcohol. For example, a prevalent French mixture is

$$
\begin{array}{ll}
\text{Ethyl alcohol .....} & 100 \quad \text{volumes.} \\
\text{Methyl " .....} & 10 \quad \text{"} \\
\text{Hydrocarbon .....} & \underline{0.5} \quad \text{"} \\
& 110.5
\end{array}
$$

The hydrocarbon is defined only by its boiling-boint, which should be between 350° F. and 440° F. It will have a specific gravity of .832 to .835 referred to water, or about 38° B., and a calorific power of 9300 B.T.U. per pound. An alcohol mixture known as electrine has a composition of equal parts of the above mixture with a benzol, resulting in a specific gravity of .835 and a calorific power of 13,150.

A motor which is to operate with alcohol in internal combustion should work with a higher compression of the charge before igniting than is satisfactory for gasoline or kerosene. The carbureting apparatus has also to be kept hotter, particularly if the alcohol is considerably hydrated. Some tests by Delahaye with the same motor gave the following results in fuel consumption:

Fuel.	H. P.	Pints per Hour	Pints per H. P.
Gasolene	6.23	5.4	0.86
Carbureted alcohol, 50%	6.29	6.22	0.99
" " 75%	6.23	7.6	1.22
Alcohol, 100%	6.32	10.5	1.70

An American test with a Westinghouse engine gave the economy of 1.2 pints per H.P. per hour; Japy of Beaucourt in France reports 1.03 pints per H.P. In some results of trials in Paris (1902) the thermal efficiency of the four fuels in motors were given as:

For gasoline.............. 14 to 18 per cent
" kerosene .... 13 " "
" gas................. 18 to 31 " "
" alcohol.............. 24 to 28 " "

These results might have been foreseen to some degree, since the alcohol contains by weight a certain proportion of oxygen, which is not a fuel, so that weight for weight the gasolines have more heat and power value than alcohols. The Delahaye experiments give the relation of gasoline to alcohol to be $\dfrac{10.5}{5.4} = 1.9$, or the same engine requires 1.9 times as much alcohol as gasoline. This has been confirmed by recent tests made for the United States Department of Agriculture,* where the relation of 1.8 was found, which corresponds so closely with the relative heating value as to indicate that practically the thermal efficiency of the two fuels was the same when vaporization is complete. The advantage of alcohol is the fact of having part of its oxygen necessary for combustion in its own composition, so that intimate mixture of fuel and oxygen is secured in part no matter what unfavorable arrangements may exist for a proper mechanical admixture. Alcohol on the other hand is a very variable aggregate, hydrating itself when exposed to damp air, so that comparisons of two alcohol tests need to be carefully made, lest the two fuels of the same name be really quite different in heating value. This is true also of the volatile gasolines. Alcohol is particularly liable to partial vaporization in carburetors, an excess going through the motor as a liquid, greatly increasing the apparent consumption.

In comparison with gasoline, alcohol offers the following advantages:

1. Greater safety in storage, and in its handling by careless or ignorant persons, both ashore and afloat. For launches and yachts alcohol should supplant gasoline for this reason.

2. The exhausted products of combustion and the exhaust pipe are cooler, diminishing danger from fire, discomfort in the engine room, and the danger of burning the lubricating oil.

3. The exhausted products are less offensive as respects odor, and the presence of smoke. The engine passages do not clog with soot or deposits of carbon so soon, or at all. Both these difficulties are made less or greater by the degree of skill in operating, particularly if combustion is allowed to be incomplete, or excesses of fuel and lubricating oil are permitted.

On the other hand, to secure equal thermal efficiency, motors using alcohol should work with a higher degree of compression of the mixture before ignition. If the maximum compression pressure with gasoline be taken at 65 pounds, with an average nearer 40 pounds, the pressure with alcohol should be over 80

* "The Use of Alcohol and Gasoline in Farm Engines," by C. E. Lucke and S. M. Woodward, Government Printing Office, 1907.

pounds per square inch; and in engines to be started by hand, as in usual motor-vehicle practice, such compression values form an obstacle to change, or in any use where the cost of the fuel is made a secondary consideration. This condition will be changed again should the cost of petroleum derivatives, such as gasoline and kerosene, become higher by the exhaustion of the supplies or the increasing demands relative to the supply. The sources for alcohol are practically inexhaustible and are self-renewing under powerful natural law. The following table taken from the Bulletin referred to above, gives the estimate of its authors as to the relative quantitative values of these considerations:

COST OF ENERGY IN FUELS.

Kind of fuel.	Cost of fuel.	British thermal units (B.T.U.)	Number of B.T.U. bought for $1.
Small anthracite	$2.50 per ton	12,500 per pound	10,000,000
Large anthracite	6.25 per ton	14,000 per pound	4,500,000
Illuminating gas	1.00 per 1,000 cubic feet	550 per cubic foot.	550,000
Natural gas	.10 per 1,000 cubic feet	1,000 per cubic foot.	10,000,000
Crude oil	.04 per gallon	20,000 per pound	3,650,000
Kerosene	.10 per gallon	20,000 per pound	1,200,000
Do	.30 per gallon	20,000 per pound	400,000
Gasoline	.10 per gallon	20,000 per pound	1,200,000
Do	.30 per gallon	20,000 per pound	400,000
Grain alcohol	.30 per gallon	12,000 per pound	270,000
Do	.40 per gallon	12,000 per pound	200,000

FUEL COST OF POWER.

Fuel and type of Plant.	Fuel required per horse-power per hour.	British thermal units required per horse-power hour.	Thermal efficiency.	Cost of fuel.	Cost of fuel per horse-power per hour.
			Per cent		*Cents.*
Anthracite coal :					
Large steam plant	2 pounds	25,000	10	$2.50 per ton	0.25
Do	2 pounds	25,000	10	6.25 per ton	.57
Small steam plant	7 pounds	100,000	2½	2.50 per ton	1.00
Do	7 pounds	100,000	2½	6.25 per ton	2.20
Producer gas plant	1½ pounds	14,000	18	2.50 per ton	.14
Do	1½ pounds	14,000	18	6.25 per ton	.31
Do	2 pounds	25,000	10	2.50 per ton	.25
Do	2 pounds	25,000	10	6.25 per ton	.57
Illuminating gas	24 cubic feet.	12,000	20	1.00 per 1,000 cubic feet	2.20
Crude oil	1.4 pints	25,000	10	.04 per gallon	.68
Gasoline	1.1 pints	13,400	19	.15 per gallon	1.70
Do	1.1 pints	13,400	19	.30 per gallon	3.40
Alcohol			*a*19	.30 per gallon	5.00
Do			*a*19	.40 per gallon	6.70

a Efficiency of alcohol is assumed to be the same as that of gasoline for identical conditions of use.

The United States government by enactments of its Federal Congress in 1905–1906 removed part of the restrictive revenue exactions from alcohol to be used in industry, keeping, however, the procedure for its manufacture, storage and distribution under close official surveillance and control, and compelling its denaturization to be perfect and satisfactory to designated inspectors, and to be done by them. This step will doubtless lead to a wider use of denatured alcohol. Their requirement by law is that to 100 volumes of ethyl or grain alcohol of a strength not less than 90 per cent there must be added either 10 volumes of methyl or wood alcohol and one-half of one volume of benzine; or 2 volumes of methyl alcohol and one half of one volume of pyridin bases. A proof gallon in the United States contains 50 per cent of alcohol by volume and the balance is water. In Germany the strength is stated in percentage by weight. A technical commission appointed under the above act made very thorough experiments, and reported in the Bulletin above referred to as follows:

(*a*) Any engine on the American market in 1907 operating with kerosene or gasoline can operate with alcohol fuel without any structural change whatever, with proper manipulation.

A fair average or approximate relation for the three fuels to each other would give:

Fuel.	Specific Gravity.	Pounds per Gallon.
Gasoline	0.71	5.9
Kerosene	0.80	6.7
95 per cent grain alcohol	0.82	6.8
90 per cent grain alcohol	0.83	6.9

(*b*) It requires no more skill to operate an alcohol engine than one intended for kerosene or gasoline.

(*c*) There is no reason to suppose that the cost of repairs and lubrication will be any greater for an alcohol engine than for one built for kerosene or gasoline.

(*d*) The different designs of gasoline or kerosene engines are not equally well adapted to the burning of alcohol, though all may burn it with a fair degree of success.

(*e*) The thermal efficiency of such engines can be improved when they are to be operated with alcohol by altering the construction and functioning of the carburetor so as to give complete vaporization, and particularly in cold weather; and secondly by increasing the compression very materially.

(*f*) An engine designed for gasoline or kerosene, without any material alterations to adapt it to alcohol, gives about 10 per cent more power than when operated with gasoline or kerosene, but at the expense of greater consumption of fuel. This increase can be raised to 20 per cent by alterations adapting the engine to the new fuel.

(*g*) Because of this increased output of power without corresponding increase in size, alcohol engines should sell for less per horse-power capacity than gasoline or kerosene engines of the same class. Until the interest of builders is drawn to this phase of the industry by action of buyers this will not take place.

(*h*) Alcohol may immediately compete with gasoline as a fuel for engines in localities where the supply of cheap raw material for manufacture of denatured alcohol exists, which are at the same time remote from the sources of supply of gasoline.

(*i*) In most localities it is unlikely that alcohol power will be cheaper than gasoline for some time to come.

An objection which has revealed itself with motors using alcohol is the excess of water-vapor formed in the cylinders, and that this water-vapor, absorbing the acetic acid which forms with the occasional incomplete combustion of the alcohol, attacks and corrodes metal surfaces and scores valves and seats. Proper manipulation reduces this to an inconsiderable difficulty.

35. Products of Combustion of a Gas.—In the discussion of par. 11 it was made apparent that the combustion of carbon and hydrogen gave the weight of the products resulting from such combustion. It will be apparent that when the volume is to be considered this will vary with the temperature of such products of combustion. Anticipating, for the moment, a later discussion which will define the term "absolute temperature" it may be said here that if V_0 denote the volume at the temperature of melting ice, and T_0 the corresponding absolute temperature, while V and T are the volume and absolute temperature corresponding to the state of the hot and expanded gases, the volumes will be proportional to these absolute temperatures, whence

$$V = \frac{V_0 T}{T_0}.$$

Similarly if the initial volumes be observed or taken at 62° F., the final or expanded volumes can be calculated. For example,

CO_2 at 62° occupies 8.594 cu. ft. to the pound;
H " " " 190 " " " " "
SO_2 " " " 5.848 " " " " "
N " " " 13.501 " " " " "

whence .0366C×8.594 = .315C = cu. ft. of CO_2 at 62°;
.090H ×190 = 1.9H = " " " H " "
.2S ×5.85 = .117S = " " " SO_2 " "
$\left.\begin{matrix} .0859C \\ .2584H \end{matrix}\right\}$ N×13.501= $\left\{\begin{matrix} 1.16C \\ 3.49H \end{matrix}\right\}$ = " " " N " "

Adding and neglecting the smaller weight of SO_2, the volume at 62° F. becomes

$$V_1 = 1.475C + 5.39H,$$

and the volume V_2 at any greater temperature will be found by multiplying the above expression by the fraction $\frac{T_2}{T_1}$, in which T_1 is the absolute temperature corresponding to 62° F., and T_2 the absolute temperature at which the volume is sought.

36. The Dilution of the Mixture.—The discussion in paragraphs 11 and 12 made it apparent that for complete combustion of hydrogen and carbon in air a certain minimum weight or volume was required in order to supply the necessary quantity of oxygen for that combustion. Experience has shown that in the combustion of solid fuels under constant pressure in air, the distillation process preceding the true combustion, a certain excess of air is desirable above that required for chemical combination. This will be often one and a half times or even as much as twice the theoretical or computed amount from the combining weights. In gas combustions, however, either in fires or in the cylinders of engines, this dilution process is not required to the same extent, and is prejudicial to the production of high temperatures. In the slow process of distillation-combustion of solids the products of combustion are not supporters of combustion, but the latter must be deported in a current of excess of air in order to bring necessary additional oxygen to reach the burning surfaces. With gaseous combustions, and particularly when so mixed as to be self-propagating, the flame will carry through the mass of gas without action from the burnt gases present. Hence an excess of air merely means an increase in the denominator of the second member of the equation.

$$T_2 - T_1 = \frac{yQ}{(x+y)C},$$

in par. 20, whereby the rise in temperature due to yQ will be diminished. The unnecessary weight of air and its inert nitrogen has to be heated by the expenditure of the fuel energy, and for a given amount of such energy the less the mass to be heated the hotter it will be.

On the other hand, upon the economic side, for a given volume of engine-cylinder, it will be cheaper to fill the latter with as lean a mixture of gas and air as is consistent with positive ignition of the charge, since the gas is the element which costs and the air is free. When an excess of air is used above what the gas needs for combustion, there will be oxygen found in the products of combustion other than that in combination with the carbon as CO_2. Hence it becomes a matter of great importance to mix the proportions of fuel and air for the gas-engine, with a given compression pressure, so as to avoid incomplete combustion on the

one hand, with its deposit of soot or lampblack in the cylinder
and passages, and the offensive odor of the exhaust; and on the
other the use of excess of fuel entailing waste and unnecessary
cost, and diminishing the effective power of the motor. If the
compression pressure can be increased with safety, then a more
dilute mixture can be made to give the same card area and the
same mean effective pressure. Some interesting and useful
experiments with a gas whose analysis showed a requirement
of 5 or 5.5 volumes of air to 1 of gas in an apparatus to be described
in Chapter XIX gave results which are shown graphically in Fig.
12. The points on the horizontal line indicate the mixtures of
gas and air, and the vertical ordinates the pressures caused by
igniting such mixtures. The highest pressures belong to the
correct proportions, but the engine would operate all right with
nearly twice as much fuel as was necessary, and such extravagant
working might go undetected for some time. At the other end
of the series, the danger appears of a mixture so poor
in fuel elements that it will ignite with difficulty or not
at all. It will be shown in Chapter XIX that excess of
products of combustion may so dilute a mixture as to produce
the same effect, but this excessive dilution will not be normal in
the internal-combustion motor. The tests in this same chapter will
show why the relations of gas to air of 1:12 or 1:13 for rich gas

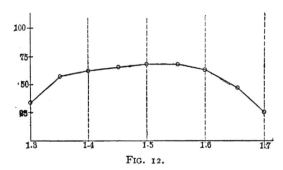

Fig. 12.

have been chosen. The student is referred to further consideration of these questions under the heads of ignition, carburation,
governing, and manipulation, in their respective chapters. The

experienced operator of an internal-combustion motor will have observed the effect of stronger propelling pressures as he increases the ratio of air to gas, but will also have noted the approach to the limit where the charge of mixture does not always ignite under these conditions.

37. Gas Analysis. Elliot's Gas Apparatus.—It will be aside from the present purpose to go at all exhaustively into the problem of gas analysis, or that of the products of combustion. As some general knowledge may be useful, however, the Elliot apparatus and the Orsat are illustrated for their respective uses. Fig. 13 shows the Elliot apparatus. The constituents for whose determination it is adapted are CO_2, CO, O, H, N, CH_4, and illuminants. The candle-power of a water-gas is about twice the percentage of illuminants in it; with a coal-gas the multiplier is about 3.5.

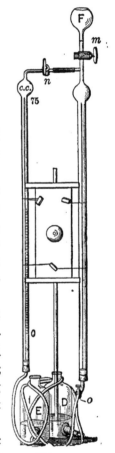

FIG. 13.

The three glass tubes in the apparatus may be called the laboratory- or reagent-tube, on the right, surmounted by the funnel for the introduction of the absorbent liquid; the measuring-tube, on the left, for measuring after each reaction the volume of gas not acted on by the test; and the explosion-tube.

The usual quantity in a test is 100 cubic centimetres. The CO_2 is determined by introducing the measured volume from the measuring-tube into the laboratory-tube, and introducing through the funnel 5 cubic centimetres of strong potassic hydrate (KOH). The gas transferred back to the measuring-tube shows by a diminished volume the amount of CO_2 which it has lost. For the illuminants a few drops of bromine in water are allowed to enter the laboratory-tube after the gas has been transferred back to it from the measuring-tube. The tube fills with red fumes which are absorbed by a second introduction of potassic hydrate. The gas when back into the measuring-tube will show a second reduction in volume. For oxygen a strong KOH solution is mixed with pyrogallic acid. For CO a saturated solution of cuprous chloride in strong HCl is used. For hydrogen and marsh-gas the explosion-tube is used. The volume of the sample is mixed with twice its volume of oxygen and an equal volume of atmospheric air, and the mixture fired by the electric spark. The CO_2 formed by the explosion is absorbed by KOH as in the first step, and the hydrogen which formed water goes out with the displacing liquids. When the gas is cool, measure what remains in the measuring-tube, and it may be called nitrogen to make up the full one hundred per cent. The above treatment has not referred to detail nor precautions, and has not elaborated the computations by formula to determine the proportions of CH_4 and H resulting from noting the proportionate contraction of volume after explosion.

The interest attaching to such analysis as the foregoing to the student of the power problem is the relation which the composition of the gas bears to its computed or theoretical calorific power, or to the actual calorific power as observed in calorimeters.

38. Analysis of Products of Combustion. Orsat's Apparatus. —It is often convenient, also, to make a similar analysis of the products of combustion, in order to make sure that the combustion has been complete and the mixture wisely selected. An excess of oxygen would not only make the mixture difficult to ignite, but it would lower the temperature on ignition, and any considerable quantity of carbonic oxide in the products of combustion would not only mean a waste of carbon, but would indicate the danger of explosions in the exhaust pipe or passages, which are inconvenient and possibly dangerous. The apparatus which is most used in analyzing the products of combustion is known as the Orsat apparatus and is illustrated in Fig. 14. P''', P'', and P' are pipettes containing, respectively, solution of caustic potash to absorb carbon dioxide, pyrogallic acid and caustic potash to absorb oxygen and cuprous chloride in hydrochloric acid to absorb carbon monoxide.

At d is a cock to control the admission of gas to the apparatus; at B is a graduated burette for measuring the volumes of gas; and at A is a pressure-bottle connected with B by a rubber tube

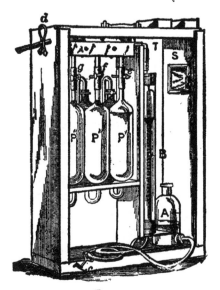

FIG. 14.

to control the gases to be analyzed. The pressure-bottle is commonly filled with water, but glycerine or some other fluid may be used when, in addition to the gases named, a determination of the moisture or steam in the flue-gases is made.

The several pipettes P', P'', and P''' are filled to the marks g, f, and e with the proper reagents, by aid of the pressure-bottle A. With a three-way cock to open to the atmosphere, the pressure-bottle A is raised till the burette B is filled with water to the mark *m*; communication is then made with the flue, and by lowering the pressure-bottle the burette is filled with the gas to be analyzed, and two minutes are allowed for the burette to drain. The pressure-bottle is now raised till the water in the burette reaches the zero mark and the clamp *c* is closed. The valve in the pipe to the flue is now opened momentarily to the atmosphere to relieve the pressure in the burette. Now open the clamp *c* and bring the level of the water in the pressure-bottle to the level of the water in the burette, and take a reading of the volume of the gas to be analyzed; all readings of volume are to be taken in a similar way. Open the cock *g* and force the gas into the pipette *P'''* by raising the pressure-bottle, so that the water in the burette comes to the mark *m*. Allow three minutes for absorption of carbon dioxide by the caustic potash in *P'''*, and finally bring the reagent to the mark *a* again. In this last operation, brought about by lowering the pressure-bottle, care should be taken not to suck the caustic reagent into the stop-cock. The gas is again measured in the burette, and the diminution of volume is recorded as the volume of carbon dioxide in the given volume of gas. In like manner the gas is passed into the pipette *P''*, where the oxygen is absorbed by the pyrogallic acid and caustic potash; but as the absorption is less rapid than was the case with the carbon dioxide, more time must be allowed, and it is advisable to pass the gas back and forth, in and out of the pipette, several times. The loss of volume is recorded as the volume of oxygen. Finally, the gas is passed into the pipette *P'*, where the carbon monoxide is absorbed by cuprous chloride in hydrochloric acid.

The solutions used in the Orsat apparatus are:

P'''. Caustic potash, 1 part; water, 2 parts.

P''. Pyrogallic acid, 1 gram to 25 cc. of caustic potash.

P'. Saturated solution of cuprous chloride in hydrochloric acid having a specific gravity of 1.10.

These reagents will absorb per cubic centimetre:

P'''. Caustic potash absorbs   40 c.c. of CO_2;

P''. Pyrogallic of potash absorbs 22 " " oxygen;

P'. Cuprous chloride absorbs   6 " " CO.

Improvements in the Orsat apparatus and its manipulation have been made by Hempel, Carpenter, Hale, and others, and the student is referred to Hempel's treatise for further detail.

CHAPTER III.

THE MECHANICAL ENERGY FROM EXPANSION OF GAS AND AIR.

39. Introductory. Mechanical Equivalent of Heat.—The unit of work for industrial purposes is the foot-pound. It means the amount of energy required or developed when one pound moves through a space of one foot in one unit of time, which is usually the second. It was found by James Watt, as the result of experiment, that the average high-powered draught-horse could do an amount of work in foot-pounds which was represented by the product of 330 pounds into 100 feet per minute so that 33,000 foot-pounds became established as the horse-power per minute

It was ascertained by the physicist Joule, as corrected by later determinations, that the amount of heat necessary to raise one pound of water 1° F. was equivalent to the mechanical energy represented by 778 foot-pounds, and that the quantity of heat to raise a unit weight of water 1° was always convertible quantitatively into foot-pounds, and vice versa.

The quantity of heat was designated as a heat-unit, and the equivalent in foot-pounds has been designated as the mechanical equivalent of heat.

In countries using the metric system of weights and measures the unit of force is the kilogram, and the unit of path is the metre. The product of effort multiplied by its path is called a kilogram-metre in these units. The following table shows the relations of the three most usual values for the horse-power:

104

Horse-power.	English Foot-pounds per Minute.	French Kilogrammetres per Minute.	Austrian Foot-pounds per Minute.
English and American	33,000	4,572.9	25,774
French	32,470.4	4,500	25,561
Austrian	33,034.2	4,549.5	25,800

Convenient transformations of the British and metric heat-unit values are given in the following:

A French Calorie = 1 kilogram of H_2O heated 1° C. at or near 4° C.

A British Thermal Unit (B.T.U.) = 1 lb. of H_2O heated 1° F. at or near 39° F.

A Pound-calorie Unit = 1 lb. of H_2O heated 1° C. at or near 4° C.

1 French Calorie = 3.968 B.T.U. = 2.2046 pound-calories.

1 British Thermal Unit = .252 French calories = .555 pound-calories.

1 calorie per cubic meter = 0.113 B.T.U. per cubic foot.

1 B.T.U. per cubic foot = 8.91 calories per cubic metre.

1 Pound-calorie = 1.8 B.T.U. = .45 French calories.

1 B.T.U. = 778 ft.-lbs. = Joule's mechanical equivalent of heat.

1 H.P. = 33,000 ft.-lbs. per minute.

$$= \frac{33000}{778} = 42.42 \text{ B.T.U. per minute.}$$

$$= 42.42 \times 60 = 2545 \text{ B.T.U. per hour.}$$

40. The Piston-motor. Mean Effective Pressure. — The only continuous motion in industry is the rotary motion. The most convenient form for utilizing pressure to produce continuous rotary motion is to have a piston travel back and forth in a cylinder and transform its reciprocating motion by means of a crank and connecting-rod into such continuous rotary motion. In motors of this class, which will be called piston-motors, the work done in the cylinder per minute will be the product of the pressure upon the piston in pounds multiplied by its travel in feet. That is, if the pressure, constant or variable, upon **the**

head of the piston be denoted by P in pounds per square inch and the area in square inches be denoted by A, the product PA will be the total effort in pounds pushing or pulling that piston. If the stroke of the piston in the cylinder be designated in feet by L, and the number of times that the piston makes this traverse by N, it will be apparent that the

$$\text{Work per minute} = PA \times LN.$$

If both members of this equation be divided by 33,000, the first member becomes horse-power and the expression reads:

$$\text{H.P.} = \frac{PA \times LN.}{33,000}$$

In the gas-engine, N will not be the number of traverses which the piston makes per minute, but will be the number made under the effort of the working medium. If the engine is single-acting and operates through the Otto cycle (par. 62), N will be the number of explosions or ignitions per minute.

If this pressure denoted by P is constant and uniform throughout the stroke, the expression needs no correction or revision. If that pressure, however, is a variable, then it is apparent that P must be the mean of the varying pressures throughout the length of the stroke, and the value of that mean pressure must be found either by observation with proper instruments or by calculation.

If the area A in the foregoing equation be expressed in square feet, and L in linear feet, the product AL becomes the volume of the cylinder in cubic feet and can be designated by V. If N be one traverse and the pressure P be expressed in pounds per square foot, we have the expression

$$\text{Work per stroke} = PV.$$

This equation is the general form used by physicists in their formulæ and discussions. V is the final volume filled by the medium as the piston makes its traverse from initial to final position in the stroke; if it be the volume filled by one pound of the medium used, it denotes the work of such unit weight of medium acting under the mean effective pressure throughout that stroke.

In the gas-engine, however, by reason of the directness of the transfer of heat energy into work in the cycle of the cylinder working, an expression of the same truth in a more detailed form leads to some very practical results. If the volume of gas mixture filling the cylinder volume at atmospheric pressure at the end of the stroke which has drawn it in be called the initial volume, and be denoted by V_i, and the volume after the compression is complete be called its final volume and be denoted by V_f, then the piston-displacement per stroke will be

$$\text{Piston-displacement} = V_i - V_f,$$

and the work per stroke as above

$$W = [\text{M.E.P.}] \times (V_i - V_f).$$

But it is also true, and will be further discussed for quantitative results in paragraph 47, that the efficiency of a conversion of heat into work is measured by the relation of output to input, or the ratio of work done in foot-pounds to heat energy supplied in heat units; or, $W = 778\,QE$ since,

$$\text{Efficiency} = E = \frac{\text{Work done in foot-pounds}}{\text{Heat supplied in units}}.$$

But the computations for the data in columns 15 and 16 of the large table in paragraph 29 made it clear what heat energy Q could be drawn into the cylinder per cubic foot in the mixture of gas and air so proportioned as to burn completely; hence the quantity $(V_i - V_f)$ in cubic feet, when multiplied by 778 should give the total work units capable of being supplied in one stroke; whence the work done should equal both

$$\text{Work done} = Q \times E, \times 778;$$

and $$\text{Work done} = (\text{M.E.P.})\,(V_i - V_f).$$

Whence, equating, and changing the latter into pounds per square inch for convenient comparison,

$$\text{M.E.P.} = \frac{778\,QE}{144\,(V_i - V_f)} = \frac{5.4\,Q}{(V_i - V_f)} \times E.$$

This enables columns 15 and 16 of paragraph 29 to be used to find the M.E.P. with any fuel whose analysis is known or assumed, when E also is computed as in paragraph 47, where it will be shown to be a function of the compression pressures admissible with that fuel. The reader is referred to that section for quantitative values for this factor. If $(V_i - V_f)$ be one cubic foot then $\dfrac{Q}{V_i}$ = B.T.U. per cubic foot of the mixture of gas and air, and the quantity in the table in paragraph 29 may be used directly in the form of $\dfrac{H}{a + 1}$, giving in pounds per square inch,

$$\text{M.E.P.} = 5.4 \frac{H}{(a + 1)} \times E,$$

as therein discussed. (See also par. 200.)

An effort has been much in evidence recently to simplify the horse-power for the gas-engine, by reducing the variable factors which are nearly uniform for most motors of the same class into one factor which is assumed constant for the time and to apply to all engines of its class and type. The errors of this are plainly due to

1. Neglecting the variation in the value of E with different compressions.

2. Assuming that Q or its derivative H are the same for all fuels.

3. Assuming all motors to have practically the same ratio between their net or brake horse-power at the shaft, and the effort in foot-pounds upon the piston head, or that all engines have the same mechanical efficiency.

4. Assuming that all motors attain the same possible maximum efficiency due to their design because there are no leakages or poor adjustments or other defects tending to lower the output of power.

An example of such a formula has the form:

$$\text{Brake horse-power} = \frac{nd^2 L N}{K},$$

in which d is the cylinder diameter, in inches, L the length in inches (or feet), n the number of cylinders, and K carries in it the constants 33,000 and π and an assumption of the mean pressure. For engines of a special class, as in motor cars, the apparent simplification may be carried further, as in a recent motor-car standard where

$$\text{H.P.} = \frac{d^2 N}{2.5},$$

in which d is the diameter, N the number of such cylinders. Such motors usually have the stroke L about $1.2d$ and turn at 1200 revolutions per minute, but the formula makes no allowances for variation of mean effective pressure with fuel variations.

Frederic Grover proposed to recognize the relation between the compression pressure and the mean effective pressure by calling

$$\text{M.E.P.} = 2\,p_b - 0.01\,p^2{}_b$$

in which p_b is the compression pressure in pounds per square inch. (Fig. 205, see also paragraph 201.) It will be true when different motors are compared, which are working on the same fuel, but the variation of the values in columns 15 and 16 of the table in paragraph 29 will show the inaccuracy of its indiscriminate use. It will be correct for a compression pressure of 100 pounds per square inch, but its results will be too high below this, and too low above it. In the intermediate class, and applicable only to motors using the same fuel, is a formula given by several authorities for the indicated horsepower:

$$\text{I.H.P.} = \frac{d^2\,(l + c) \times \left(\left(\frac{l + c}{c} \right)^{\frac{1}{3}} - 1 \right) \times N}{9160}.$$

In this, d is the cylinder diameter in inches; l the stroke in inches; c the length of the clearance space in per cent of stroke, or $\dfrac{\text{clearance volume}}{\text{piston area in inches}}$, so that $\dfrac{l + c}{c} = \dfrac{\text{total cylinder volume}}{\text{clearance volume}}$.
N = Rev. per min.

This attempts a closer approximation by recognizing the exponent of the expansion and compression curve.

40a. Computed Cylinder Volume, Diameter and Stroke.* —
To use the foregoing discussion for computing the necessary
area of cylinder and length of stroke, it should be noted that
the gas-engine cylinder requires a clearance volume between the
piston head at its inner dead centre, and the cover or head of the
cylinder. This is to act as a combustion chamber to receive
the compressed charge and to provide for igniting it. This
clearance volume has only an effect upon the mechanical capacity
of the motor so far as it affects the compression pressure of the
mixture of fuel and air before ignition, and thereby the mean
pressure. It must be computed by itself, and with special
reference to the pressures permissible with this fuel without
pre-ignition. It is treated fully in paragraph 152. The diameter
and stroke which together form the volume displacement of
the piston are directly computable from the foregoing data
already found.

The design is practically always made with an output of work
in foot-pounds or horse-power · demanded. The fuel also is
given, so that a value for the mean effective pressure with that
fuel is derivable from the table in paragraph 29 and the preceding
discussion. The piston speed usual in motors of its class is a
conventional outcome of experience and may be safely assumed
within the following limits:

Horse-Power Capacity.	Usual Piston Speeds, in Feet per Minute.	Average Piston Speed in Feet per Minute.
Motor-car cylinders 2–10 H.P. per cylinder	600–1000	750
Small stationary engines less than 30 H.P....	450–700	550
50–100...............................	500–700	600
150–250...............................	600–800	650
500–600...............................	650–850	700
700–850...............................	700–900	750
1000 up...............................	700–1000	800

* See also paragraph 202.

These standards will fix the value of the quantity LN in the horse-power formula, where L is the length of the cylinder and N the number of traverses of the cylinder which the piston makes per minute; this is twice its number of revolutions per minute. In the four-cycle or Otto type of single-cylinder, single-acting engine (par. 62), only one stroke in four traverses is made under working or effective pressure: the formula for indicated horse-power may therefore be written

$$\text{I.H.P.} = \frac{PLAS}{4 \times 33000},$$

in which S is the piston-speed in feet per minute as above, and the factor 4 appears in the denominator. The factor A is the effective area, or the gross area of the piston diminished by the area of any piston-rod which may reduce the area receiving the pressure. It is equal to

$$A = \frac{\pi d^2}{4} = .7854\, d^2,$$

in which d is the diameter in inches if the mean effective pressure P is in pounds per square inch. Hence, if there are n cylinders, single-acting, or n cylinder ends, in operation of a double-acting type

$$d = \sqrt{\frac{\text{I.H.P.}}{\text{M.E.P.}} \times \frac{4 \times 33000}{.0.7854 \times S} \times \frac{1}{n}}.$$

In the two-cycle type (par. 73), an effort of the pressure is exerted once in every two traverses of the piston, and the factor 4 becomes 2 giving a smaller diameter in the relation,

$$d = \sqrt{\frac{\text{I.H.P.}}{\text{M.E.P.}} \times \frac{2 \times 33000}{0.7854 \times S} \times \frac{1}{n}}.$$

In most cases, the designer is called on to provide that his engine shall be capable of overcoming an overload demand for power. This means that the cylinder diameter shall be increased, so that under such demand for an increased horse-power which is a percentage f of the normal load, the engine at the same piston-speed should be able to respond. In other words the normal output with no overload is to be increased from

$$\text{I.H.P. to} \left(1 + \frac{f}{100}\right)\text{I.H.P.}$$

for an overload of f per cent as an overload factor. Then the formula for diameter becomes

$$d = \sqrt{\frac{\left(1 + \frac{f}{100}\right)\text{I.H.P.}}{\text{M.E.P.}} \times \frac{4 \times 33000}{0.7854 \times S} \times \frac{1}{n}}$$

for the four-phase type, and the corresponding change made for the two-phase cycle.

The **stroke** is twice the crank-arm length. In giving the value to S the piston-speed in feet per minute, the length of the stroke is at once fixed if the number of revolutions has been specified. Or, the conventions of a relation of the stroke to the cylinder diameter may be observed, and with the length thus determined the factor N may be computed and the revolutions per minute will be twice N. In practice the latter method governs, since it has been found desirable to give the crank-arm a considerable leverage, to give a large turning moment, and to keep the diameter less than the stroke, so that the shock of the ignition increase of pressure shall not be excessive and introduce inconvenient vibration. On the other hand, excessive crank length makes a long and bulky engine, and the power is applied obliquely to turn the crank during a large fraction of the stroke. Hence the ratio $\frac{l}{d}$ is never made less than one, nor more than two; in motor-car practice it is from 1.0 to 1.3; in small stationary engines it is nearer the larger limit, while in large stationary practice the ratio is settled by convenience anywhere within the extremes. A study of the engines tested by various experimenters and reported under paragraph 177, will show the trend of practice. For example:

Type of Engine.	Builder.	Cylinder.		Ratio $\frac{l}{d}$	H.P	Rpm.	No. Cyls.
		Diam. Inches.	Stroke Inches.				
Motor-car	......	4½	5½	1.22	10	900	4
Stationary	Otto.............	8.5	14	1.64	10	158	1
"	"	6.0	12.5	2.08	8	281	1
"	Westinghouse.....	13 .	14	1.08	86	270	3
"	Crossley	17	24	1.41	118	156	1
"	Schleicher Schumm	14.	25	1.72	100	160	1
"	Cockerill	51.18	55.12	1.08	550	80	...
	......	25	25	1.0	...	180	...
	......	29	40	1.4	...	110	...
	......	20.08	25.56	1.3	...	150	...
	......	40.55	51.24	1.3	...	90	...

41. Graphical Representation of the Work of a Piston-motor. The PV Diagram.—Since the work of a piston-motor is the product of the two factors, pressure in pounds multiplied by feet of traverse, it is obvious that a closed figure can be drawn enclosing an area which, upon an assumed scale of units, shall be the same as the given product in foot-pounds. Furthermore, whatever the shape of that figure, a rectangle of equivalent area can be drawn, the product of whose base into its altitude shall

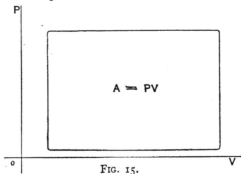

Fig. 15.

represent that same number of foot-pounds of work. If, then, a horizontal line be drawn from an assumed origin on which may be measured distances in feet on any scale, and from that same origin a vertical line on which may be measured pressures on an appropriate scale, and these horizontal units be designated by V and vertical units by P, the area of an enclosed figure upon these lines as coordinate axes will reproduce a work diagram of a piston-motor. The simplest case where the pressure P was constant would give a simple rectangle (Fig. 15). If, however, as is the general case, the pressure is not constant, the curve which forms the upper line of such a diagram (Fig. 16) will be a curve of varying ordinates, and it will be necessary by means of convenient methods to get an ordinate which shall be the mean of all ordinates to be multiplied by the length, in order to give

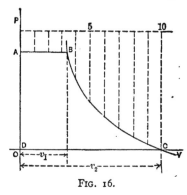

FIG. 16.

the actual area. If the appliance known as a planimeter is at hand, the area of the diagram can be ascertained and that area divided by the measured length will give the height by which the length is to be multiplied in order to give an area. If a planimeter is not at hand, the length of the diagram can be divided into a convenient number of equal parts (say ten) (Fig. 16) and the height of each partial area measured by the scale. The total of these partial heights divided by their number gives the mean height by which the length is to be multiplied. Or, what is known as Simpson's rule may be used.

To apply Simpson's rule for determining an area the diagram is divided vertically by ordinates. The first one is called p_0 and the last one p_n. Then the area A is given by the formula

$$A = \frac{1}{n}l\left(\frac{p_0+p_n}{2}+p_1+p_2 \cdots p_{n-1}\right),$$

where l is the measured length. Dividing this area by the length l, the mean pressure p_m results, or

$$\frac{A}{l}=p_m=\frac{1}{n}\left(\frac{p_0+p_n}{2}+p_1+p_2 \cdots p_{n-1}\right).$$

Such a diagram (Fig. 16) will be called a *PV* diagram, inasmuch as its coordinate factors are the pressure and the volume corresponding to one stroke.

42. Gay-Lussac's Law for Air.—It was found by the physicist Gay-Lussac that atmospheric air increased in volume by $\frac{1}{273}$ of itself for each degree centigrade. In the Fahrenheit scale this fraction becomes $\frac{1}{493}$. If this be expressed in symbols and v is the volume at any temperature, it will bear to its volume v_0 at the temperature of melting ice the relation

$$v = v_0(1 + at),$$

in which a is the above fraction, and t is the range above the temperature of melting ice. The equation expressed decimally is

.00365 on the centigrade scale

and

.002035 on the Fahrenheit scale.

By the expedient of heating the air behind a piston in a piston-motor by injecting a volume of gas and igniting that volume in the air, it will be apparent that a great increase of volume tends to occur, and that this increase in the confined space will be accompanied by the increase of pressure above that which existed before the gas was ignited. It is this principle which is used in the ordinary types of piston gas-motor to produce the pressure which gives the desired work. It is obvious, therefore, that the gas-engine derives its capacity for doing work by the expansion of air caused by heat. The fundamental conception, therefore, of the *PV* diagram above must be extended to take account of the influence on the air which results from changes of temperature.

43. Law of Mariotte.—It was announced by Mariotte in France, in 1640, and by Robert Boyle in England, independently, at about the same date, that: The temperature of the gas remaining constant, the volumes of the same weight of gas at different pressures will be inversely as the pressures. Expressing this law by symbols, if p_0 be an initial pressure expressed in any unit of pressure on a unit of area, and v_0 the corresponding initial volume of the gas, then for any other pressures and volumes p and v which come together it will be true that

$$p_0 : p :: v : v_0;$$

or, more conveniently,

$$p_0 v_0 = pv = \text{a constant,}$$

provided no change of temperature or heat energy occurs by reason of processes connected with such change of volume. It follows further, that since for a given weight of gas the density will vary inversely as the volume, the pressures must vary directly as the densities, and will be directly proportional to them at the same temperatures. Or, in symbols,

$$p_0 : p :: D_0 : D; \quad \text{or,} \quad \frac{p}{D} = \frac{p_0}{D_0} = \text{a constant.}$$

44. Mariotte and Gay-Lussac Law Combined.—If a given weight or volume of gas be enclosed in a cylinder behind a piston and the pressure and volume be made to vary by moving the piston, it will follow from Mariotte's law alone, using the same symbols as above, that

$$p_0 v_0 = p_1 v_1 = p_2 v_2.$$

But by the Gay-Lussac law the volumes varying by change of temperature of the gas in that cylinder would give

$$v_1 = v_0(1 + at_1);$$
$$v_2 = v_0(1 + at_2).$$

Hence if the second member be multiplied by the appropriate value for the pressure, the first member will become

$$p_1' v_1 = p_0 v_0(1 + at_1)$$

and

$$p_2' v_2 = p_0 v_0(1 + at_2).$$

It will be observed that the first members of these last equations are not the same as held for the Mariotte relation alone.

Dividing one by the other, and transposing the factors p, and dropping the primes, since they both correspond to the changed condition caused by the addition of heat, we get

$$\frac{v_1}{v_2} = \frac{p_2(1+at)}{p_1(1+at_2)}.$$

For a its value in either thermometric scale may be substituted, so that the equations take the form

$$\frac{v_1}{v_2} = \frac{p_2}{p_1} \times \frac{(273+t_1)}{(273+t_2)} \text{ for centigrade readings,}$$

or the form

$$\frac{v_1}{v_2} = \frac{p_2}{p_1} \times \frac{(461+t_1)}{(461+t_2)}$$

when the computation is made from the zero point of the Fahrenheit scale.

45. Absolute Temperature.—It is an immediate deduction from the law of Gay-Lussac that air increases by $\frac{1}{273}$ of its volume at zero centigrade for each degree increase of temperature to infer that with each degree of temperature below zero the volume of the gas should be diminished by that same fraction of its volume at zero. It follows, therefore, that when the temperature has been lowered by $273°$ the equation for the volume will read

$$v = v_0(273 - 273).$$

This is equivalent to saying that at this temperature the energy resident in the gas to cause it to increase its volume has disappeared, or has become zero. Such a temperature, therefore, is an ideal point from which all temperature can be counted as a zero, and for this reason is called the absolute zero. Temperatures on the ordinary centigrade scale become absolute temperatures by adding 273 to the reading of the thermometer. Similarly, for. Fahrenheit degrees they become absolute readings when 461 is added to the thermometer reading. It is usual to designate the reading in absolute degrees by the capital letter T. If this substitution be made in the equations of paragraph 44, they become

$$\frac{v_1}{v_2} = \frac{p_2}{p_1} \times \frac{T_1}{T_2},$$

which may be transformed so as to read

$$\frac{p_1 v_1}{T_1} = \frac{p_2 v_2}{T_2}.$$

Each of the members of this equation must be equal to the expression

$$\frac{p_0 v_0}{T_0},$$

which may be translated to say that at constant pressures the volume varies directly as the absolute temperatures, or at constant volumes the pressures will vary directly as the absolute temperatures. The law of Mariotte says that when the absolute temperatures are constant, the product of pressure multiplied into volume is a constant for any given condition of the gas at starting with respect to pressure and volume.

The advantage of the use of the absolute temperature in computations and formulæ is that it makes every temperature reading a positive reading throughout the entire range of experience and practice, and eliminates the negative reading which is the result of the location of the zero of the ordinary scale at the point where water freezes.

It is also of advantage in enabling the energy due to heat in a gas to be compared directly with the energy of that gas under other conditions.

46. Total or Intrinsic Energy. Available Energy.—It will be apparent from the discussion in the previous paragraph that the capacity of a given weight of gas for doing work against a mechanical resistance will be measured, first, by the weight of the gas or the amount of matter present in it, which, by definition, is its mass. It is, secondly, conditioned by the amount of heat which it contains when that amount of energy is measured by counting from the zero on the absolute scale. It is, thirdly, measured by the capacity of the gas for the absorbing of heat which is measured by the quantity of heat required to raise the temperature of a unit weight of the gas by one degree on the thermometric scale. This heat capacity of the gas in units is called its *specific heat*, and is usually designated by the symbol C, which is the initial of the French word *chaleur*. If, then, the weight of the gas be multiplied by its specific heat and by the temperature which it has above the absolute zero, a product results which is the expression for the intrinsic energy which that gas has under those conditions and without having that energy artificially increased. In symbols, this total energy is the product of

$$W \times C \times T = \text{intrinsic energy.}$$

It will be observed that this expression does not contain the pressure under which the gas is maintained which is the practical shape in which the energy is made manifest as discussed in paragraph 40. The reason for this is that there are only two ways in which the pressure can be increased. The first is by the addition of heat energy to the given weight or mass of gas which will, of course, increase its intrinsic energy by increasing the value of the factor T. The other way is by a mechanical compression due to a force exerted to compress the gas. By

the principles of the conservation of energy and the mechanical
equivalent of heat (par. 39) this mechanical pressure is a manifestation of heat energy in another form or can be replaced by
such heat energy; and the mechanical compression, if no loss
were experienced, would reappear in the compressed gas in the   ·
form of an increase of its temperature. For this reason when
comparing two states of the gas, it is their difference in temperature which is significant as respects their difference in energy
and not their difference in pressure.

It may easily happen that an amount of intrinsic energy is
not available for the doing of mechanical work. It is necessary
in the continuous operation of a piston-motor that on one side of
it shall be a forward pressure driving the piston and overcoming
resistance, while on the other side, which may be called the negative side of the piston, is a pressure less than the impelling pressure,
due to the fact that that side of the piston is in communication
with a vessel in which is maintained a pressure less than the
impelling pressure. In other words, if the pressure on both sides
of the piston were the same, there would be no impelling energy
to overcome the external resistance. The lowest pressure which
can be produced in nature is that which results when the atmospheric pressure or the tension of the atmospheric air is removed
from a vessel by the creation therein of a Torricellian vacuum.
Under ordinary circumstances the negative side of the piston
(and in gas-engine practice universally) the pressure on the
negative side is that of the atmospheric air and the absolute
temperature that of the atmospheric air, as counted from the
absolute zero. The temperature of the impelling medium must,
therefore, be much higher than the temperature of the air, in
order that it may have an energy sufficient to do the required
work by the difference in temperature. The available energy
of a given mass of gas will be expressed by the equation

$$W \times C \times (T_1 - T_2),$$

In which W is the weight in pounds of the fuel, T_1 is the temperature of the heated air, and T_2 is the temperature of the atmosphere, or that to which the heated air can be conveniently cooled on leaving the motor.

The more applicable form of this for gas-engine practice is to take the weight of the mixture of fuel and air, the latter being the quantity computed as necessary for the complete combustion of the former by the methods of paragraphs 13 and 29. The calorific power of the fuel is supposed to go into raising the temperature of its own mass and that of the air to the high temperature T_1 without loss or transfer to other or surrounding objects, and to do this without doing external work during the process, so that the specific heat factor is that for a constant volume change. Hence the mixture will be $1 + a$, and if its heat-energy be called H, it will be true that

$$H = (1 + a)\, c_v\, (T_1 - T_o).$$

The relation therefore

$$\frac{H}{1 + a} = c_v\, (T_1 - T_o)$$

may conveniently be compared to the value and its computation method in columns 15 and 16 of the table in paragraph 29, where the unit is the cubic foot of volume, and not the weight.

47. Efficiency. Thermal Efficiency. — The term efficiency applied to a machine or to an engine is the ratio between the available energy put into the apparatus and the energy actually utilized by it. This ratio is expressed by a fraction whose numerator is the energy utilized and whose denominator is the total available energy. When this is expressed in symbols for a given weight, W, and a given value for the specific heat, C, it takes the form

$$\text{Efficiency} = \frac{WCT_1 - WCT_2}{WCT_1},$$

because the engine rejects at the exhaust an energy WCT_2 and can therefore only have utilized the difference between the energy at the beginning and at the end of the stroke. Dividing out the common factors for the weight and specific heat, the equation appears,

$$\text{Thermal efficiency} = E_t = \frac{T_1 - T_2}{T_1}.$$

This is an expression for the efficiency of a heat-engine first deduced by Carnot, and is the expression for the best result theoretically obtainable from an engine operating between the two temperatures T_1 and T_2. In its application to the internal combustion engine, however, it must be noted that, as discussed in the previous paragraph, the specific heat of the combustion process is c_v or the value at constant volume, while the cooling is at varying volume, or at the constant pressure conditions; so that in reality the formula should read:

$$E_t = \frac{c_v(T_1 - T_0) - c_p(T_2 - T_0)}{c_v(T_1 - T_0)}.$$

If the ratio of these two specific heats (par. 54) be designated by n, this becomes on simplification,

$$E_t = 1 - n\left(\frac{T_2 - T_0}{T_1 - T_0}\right).$$

But this is applicable only to a type of combustion motor in which no energy is imparted by a previous compression. This is the cycle discussed in detail under paragraph 183 in the sequel, and is little in modern use by reason of the better economy of those in which compression occurs.

For the latter, another temperature is of greater significance in determining the efficiency, which may be called T_c, and is due to the work of compressing the gas from the initial volume V_i, to the final or compression volume V_f. The computations in paragraph 52 will show that this temperature volume-pressure relation will be given by the equation:

$$\left(\frac{V_f}{V_i}\right)^{n-1} = \left(\frac{p_2}{p_1}\right)^{\frac{n-1}{n}} = \frac{T_c}{T_1},$$

whence

$$T_c = T \left(\frac{V_f}{V_i}\right)^{1.41-1} = T_0 \left(\frac{p}{p_f}\right)^{.29}$$

and

$$E_t \; \frac{c_n (T_1 - T_c) - c_p (T_2 - T_0)}{c_v T_1 - T_c)} = 1 - \gamma \left(\frac{T_2 - T_0}{T_1 - T_c}\right).$$

There is a third type in which the addition of heat is at constant pressure instead of at constant volume, and compression is used. Its efficiency equation will be

$$E_t = \frac{c_p (T_1 - T_c) - c_p (T_2 - T_0)}{c_p (T_1 - T_c)} = 1 - \frac{T_2 - T_0}{T_1 - T_c}$$

These will be more fully and exhaustively discussed in the sequel (pars. 183 to 200), and therefore for the present, attention is directed only to the second type in general industrial use. If the hypothesis be made that the final volume after expansion is complete is the same as the mixture had before compression began, then the cooling may be considered as taking place also at constant volume, simplifying the equation so that it becomes

$$E_t = \frac{c_v (T_1 - T_c) - c_v (T_2 - T_0)}{c_v (T_1 - T_c)} = 1 - \frac{T_2 - T_0}{T_1 - T_c}.$$

But since the two curves of temperature range may both be practically considered as adiabatic (par. 50) and with the volume-range assumed the same

$$\frac{T_2}{T_1} = \frac{T_0}{T_c}.$$

Multiplying both terms of the fractions in the **efficiency** equation by T_c and substituting for $T_2 T_c$, its equal $T_1 T_0$, factoring **and** dividing both terms by $T_1 - T_c$, there results

$$E_t = 1 - \frac{T_0}{T}.$$

For this can be substituted its equal from the foregoing discussion in either of the forms:

$$E_t = 1 - \left(\frac{V_f}{V_i}\right)^{.4} \quad \text{for volumes}$$

or

$$E_t = 1 - \left(\frac{P_i}{P_f}\right)^{.29} \quad \text{for pressures.}$$

Substituting this value for E in the equations established in paragraph 40, the expression for the mean effective pressure becomes

$$\text{M.E.P.} = 5.4\, \frac{H}{a+1}\left(1 - \left(\frac{P_i}{P_f}\right)^{.29}\right)$$

or

$$\text{M.E.P.} = 5.4\, \frac{H}{a+1}\left(1 - \left(\frac{V_f}{V_i}\right)^{.4}\right)$$

From this the mean effective pressure for design may be computed when the compression ratios of volumes or pressures are fixed or assumed, and the tabular value computed in paragraph 29 inserted for the fuel to be used.

47a. Mechanical Efficiency. — It will be apparent also from an inspection and study of the energy analysis of paragraph 7a, that a certain amount of the energy of the motor is absorbed in operating its own functions, and in mechanical imperfections causing loss. These absorptions of energy will always make the energy computed or realized in the cylinder of the motor more than that actually delivered for useful work at the revolving shaft. There will therefore be a ratio which is another efficiency, existing between the indicated or gross or cylinder horse-power and the net or actual or brake horse-power, realized when the engine is tested by the method described in paragraph 170. If this efficiency or ratio of input to output be called the mechanical efficiency, and denoted by E_m, then

$$E_m = \frac{\text{Brake Horse-Power at Shaft}}{\text{Indicated Horse-Power in Cylinder}},$$

whence

$$\text{B.H.P.} = E_m \times (\text{I.H.P.}).$$

The losses which cause E_m to have a considerable value are, first, those due to fluid losses in friction in valve-passages, through carburating apparatus, exhaust pipes and the like; in inertia of the fluid in pipes, passages and valves; and from leakage past piston rings and valves. Second, those due to mechanical friction of moving parts. These latter are lessened by effective lubrication, and in smaller engines such as used in motor cars by the use of ball or roller bearings. The fluid friction should be inconsiderable, say from 5 to 6 per cent in engines which are skilfully designed, with small leakage and low velocity of flow through valves, passages and parts. By throttling, however, and the action of governors it may rise to 30 or 35 per cent, and at the limit become equal to the entire capacity of the engine reduced intentionally by this throttling process.

The pre-compression of the charge which is practised in the two-cycle type of engine (par. 73) compels a mechanical loss to overcome the resistance to compression, which is not recovered in the working stroke. This loss is less when the pre-compression is done by the motor piston, than when a second cylinder is used with its crank-pin and wrist-pin losses. In small engines with simple methods, it is probably 10 per cent; in larger ones with separate pump it may rise to 30 per cent, and forms part of the price paid for the apparent simplification of the working of the system.

Values for E_m will probably range in engines of different sizes within the values of the following table.

Capacity of Engine in I. H. P.	Four Cycle.		Two Cycle.	
	Percentage Loss.	Value of E_m.	Percentage Loss.	Value of E_m.
4–25	20–26	74–80	27–30	63–70
25–500	19–21	79–81	24–26	64–66
500 and over	14–19	81–86	27–30	63–70

The loss from fluid friction effect increases necessarily with light loads; the mechanical friction on the other hand usually increases with the loads and as pressures increase. If the mechanical fluid and unrestored compression losses be indicated by L_m, L_f, and L_c, then

$$E_m = 1 - (L_m + L_f + L_c)$$

and $$\text{B.H.P.} = (1 - L_m - L_f - L_c)\ \text{I.H.P.}$$

47b. Combined Mechanical and Thermal Efficiency. The Guarantee. — A desirable tendency of recent contracts between the maker of internal combustion engines and their buyers has been to base the agreement on the output of horse-power per heat unit of the fuel to be supplied. The unit of fuel multiplied by its price is the investment which the user is to make, the interest on which is the return in horse-power which he is to receive. If both parties use the units in the same sense this understanding is in every way defensible, eliminating all questions of opinion, assertion or controversy, and leaving only a possible discussion on facts which are easily proven to the satisfaction of both parties. But it will be apparent that a guarantee of economy based on the relation of output to input as discussed in the foregoing sections must be clearly understood by both buyer and seller to prevent regrettable confusion. For if the formula on which the agreement is based be merely stated:

$$\text{Efficiency} = \frac{\text{Horse-power or foot-pounds per minute}}{\text{B.T.U. per minute of fuel} \times 778},$$

it is apparent that there may be two understandings as to the value of the numerator, and two interpretations of the value of the denominator. If, for example, the indicated horse-power be used in the numerator by one party and the brake horse-power by the other; or the low fuel value be used in the denominator by one and the high fuel value be used by the other, there are four different guarantees all correct in fact and theory. That is, it is true to place

$$E_l = \frac{\text{B.H.P.} \times 33000}{\text{B.T.U. in fuel per minute (high)} \times 778} \qquad \textbf{(1)}$$

$$E_{m_1} = \frac{\text{B.H.P.} \times 33000}{\text{B.T.U. in fuel per minute (low)} \times 778}$$

$$E_{m_2} = \frac{\text{I.H.P.} \times 33000}{\text{B.T.U. in fuel per minute (high)} \times 778}$$

$$E_h = \frac{\text{I.H.P.} \times 33000}{\text{B.T.U. in fuel per minute (low)} \times 778}.$$

From this, the last or E_h is the highest guarantee and will therefore be that preferred by the seller in evaluating his product; the first one is the lowest, and will be that naturally used by a dissatisfied buyer. The seller put on the defensive will claim his product satisfactory if it meets the value of E_l; while the buyer really supposed he was to get the value of E_h. It would appear that justice to both parties was best secured by the use of one of the intermediate ones, and there is less ground for controversy where tests can be made by absorbing or transmitting the power actually delivered, by the use of E_{m_1}. Where the engine is too large to test, or is to be applied to uses where power-tests would be inconvenient, then the value in E_{m_2} can be agreed to. The lower fuel value is that actually available in the hot-engine cylinder, and is the fairest for both parties.

If, for example, an engine be assumed, working on a fuel of 600 B.T.U. per cubic foot (high) and of 550 B.T.U. (low) (pars. 15 and 29), and that it uses for facility in computation 1000 cubic feet per hour, giving a brake-horse-power of 40, and an indicated horse-power of 50, then the ideal computations for the four efficiencies would be:

$$E_l = \frac{40 \times 33000 \times 60}{1000 \times 600 \times 778} = \frac{132}{778} = 17 \text{ per cent nearly.}$$

$$E_{m_1} = \frac{40 \times 33000 \times 60}{1000 \times 550 \times 778} = \frac{144}{778} = 18 \text{ per cent.}$$

$$E_{m_2} = \frac{50 \times 33000 \times 60}{1000 \times 600 \times 778} = \frac{165}{778} = 21 \text{ per cent.}$$

$$E_h = \frac{50 \times 33000 \times 60}{1000 \times 550 \times 778} = \frac{180}{778} = 23 \text{ per cent.}$$

It will have been noted in paragraph 39 that the ratio of the B.T.U. to the foot-pound was 778, so that one horse-power per hour or $60 \times 33000 = 1980000$ foot-pounds per hour should be the output of $\frac{1980000}{778} = 2545$ B.T.U. if there were no losses. But from paragraph 47 the inevitable loss from defective temperature range was:

$$E_t = 1 - \left(\frac{P_i}{P_f}\right)^{.29}$$

Hence the

B.T.U.(low) per ⊥.H.P. per hour $= \dfrac{2545}{1 - \left(\dfrac{P_i}{P_f}\right)^{.29}}$

if the theoretical mean pressure is realized in the cylinder. But the actual engine never succeeds in reaching this, but reaches instead a pressure $P' = fP$. That is, the actual work diagram falls inside of the theoretical one by a percentage f. If this includes the difference also between the indicated and brake-horse-power results, the formula for the guarantee becomes:

B.T.U. (low) per B.H.P. per hour $= \dfrac{2545}{1 - \left(\dfrac{P_i}{P_f}\right)^{.29}} \times f.$

The computation of the value to be given to the factor f in design will be taken up further in paragraphs 201, 202. The quantity $f = \dfrac{P'}{P}$ applied to a work or indicator diagram is called the diagram factor, and indicates how much larger its area should be than that imposed by pure computation from theoretical data.

48. Expansive Working of Media Compared with Non-expansive Working.—It will be apparent from the foregoing equation that the efficiency increases with the difference between

the initial state of the gas and its terminal state as to temperature. When these temperatures are accompanied, as is usually
the case with a corresponding pressure, it becomes apparent

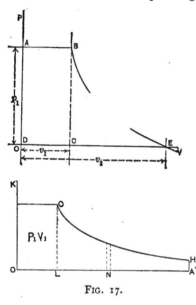

FIG. 17.

that with a diminution of temperature comes a diminution
of pressure. The diagram
(Fig. 17), similar to that in
paragraph 40, shows that at
the end of the stroke of the
piston in the cylinder the pressure has materially fallen, so
that when the exhaust-valve
opens and empties the contents
of the cylinder into the open
air there is less energy rejected
than if that terminal pressure
were more nearly that which
prevails at the beginning of the
stroke. This indicates, therefore, that it is of manifest
advantage to cause the medium to expand in the cylinder while
driving the piston so that it shall change from an amount of
intrinsic energy at the beginning of the stroke to one which is as
far reduced at the end of the stroke as is consistent with a margin
of impelling force to overcome any resistance caused by back
pressure on the negative side. Such an operation of the medium
secures a more complete utilization of the heat energy by the
considerable change in the amount of such energy from the
beginning to the end of the stroke, which energy should, of course,
appear in overcoming the resistance at the crank-pin of the
engine. Not only is the energy more completely utilized and a
less amount of it rejected with the exhausted air, but the noise
incident to the discharge of the exhausted gases is diminished
and there is a tendency to diminish the back or negative pressure
for the succeeding stroke. It will be found that expansive working is a feature of all important heat-engines.

49. Isothermal Expansion.—The most natural condition for expansion is that in which the fall of pressure occurs with increase of volume in the PV diagram, accompanied with a drop in temperature incident to the external work which the gas is doing as it acts upon the piston. If, however, by surrounding the cylinder with a provision to maintain its temperature the gas expands without drop in temperature, due to the external work, but has the same amount of intrinsic energy at the end of the stroke as it had at the beginning, it will be exhausted from the cylinder at the same temperature at which it came in. An expansion which takes place without change of the temperature is called an "isothermal" expansion, since the heat is equal at all points of the stroke. The heat necessary to do the mechanical work of that stroke has been supplied from the appliance outside the cylinder which maintained its heat, and not by the heat of the expanding gas which is within the cylinder.

The mass working in the cylinder carried out with its exhaust as much heat as it took in, and so far as heat is concerned that heat is wasted. Such expansion follows the Mariotte law by definition, so that (Fig. 18)

$$p_1 v_1 = p_2 v_2.$$

This law may be expressed graphically by the curve of an equilateral hyperbola referred to the coordinate axes of pressure and volume as asymptotes. The work

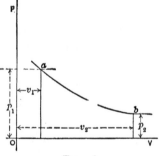

FIG. 18.

done by the PV diagram under the curve of the hyperbola will be expressed by the differential equation

$$W = \int_{v_2}^{v_1} p\,dv = p_1 v_1 \int_{v_2}^{v_1} \frac{dv}{v_1}.$$

If this expression be integrated by the methods of the calculus between the limits v_1, corresponding to the initial volume, and v_2, corresponding to the final volume at the end of the stroke,

$$W = p_1 v_1 \text{ hyp. log. } \frac{v_2}{v_1}.$$

It will be obvious that isothermal expansion will be of limited significance in gas-engine practice.

50. Adiabatic Expansion.—The more natural and usual form of expansion takes place when there is no means of keeping up the temperature of the gas in the cylinder as it expands, but in which the gas cools by an amount equivalent in heat-units to the external work overcome by the piston when driven by such expanding gas. This kind of expansion is called adiabatic since there is no transfer or passage of heat through the cylinder-walls to the gas, but it operates by the expenditure of its intrinsic energy in overcoming the resistance. It is obvious that such expansion will be accompanied by a change in the ratio of the pressure to the volume, so that at the end of the stroke the pressure will be less than it would have been with isothermal expansion by the withdrawal of the heat represented by the overcoming of the mechanical resistance. This is expressed in symbols by giving to the factor v an exponent greater than unity. If that exponent be designated by n, then the expression pv attaching to isothermal expansion becomes pv^n when applied to adiabatic expansion. A corresponding integration of the expression

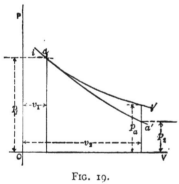

FIG. 19.

$$W = \int_{v_1}^{v_2} p\,dv$$

becomes

$$W = p_1 v_1 \int_{v_1}^{v_2} \frac{dv}{v^n} = \frac{p_1 v_1 - p_2 v_2}{n-1},$$

which is more conveniently written

$$W = \frac{p_1 v_1}{n-1}\left[1 - \left(\frac{v_1}{v_2}\right)^{n-1}\right].$$

If the ratio between the initial volume v_1 and the final volume v_2 be denoted by r, then this becomes

$$W = \frac{p_1 v_1}{n-1}\left[1 - \frac{1}{r^{n-1}}\right]$$

The condition in the internal combustion engine is always that in which an initial volume filling the clearance or combustion chamber partakes also of the expansion. The real ratio of expansion is therefore that of $\dfrac{v_1}{v_2 - v_1}$ whose reciprocal is $\dfrac{1}{r-1}$.

The area of the diagram being the work done in one stroke, and this area being equal to one whose length is that of the stroke of the piston and whose height is the mean pressure exerted over that area, the mean pressure will be found by dividing the work area by the length which is v_1; hence, when there is no back pressure

$$\text{M.E.P.} = \frac{W}{v_2} = \frac{p_1}{n-1} \times \frac{1}{r-1}\left[1 - \frac{1}{r^n - 1}\right]$$

51. Adiabatic Work in Terms of Pressures.—It is sometimes convenient, instead of expressing the work in adiabatic expansion in terms of volumes, to express this work in terms of the range of pressures between the beginning and the end of the stroke. The computation for this is as follows:

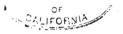

Since

$$p_1 v_{1n} = p_2 v_2{}^n, \quad \text{then} \quad \frac{p_2}{p_1} = \frac{v_1{}^n}{v_2{}^n},$$

whence by extracting the nth root

$$\frac{v_1}{v_2} = \frac{(p_2)^{\frac{1}{n}}}{p_1},$$

and by raising both members to the $n-1$ power,

$$\left(\frac{v_1}{v_2}\right)^{n-1} = \left(\frac{p_2}{p_1}\right)^{\frac{n-1}{n}}.$$

Hence the equation for work of expansion of the preceding paragraph becomes

$$W_1 = \frac{p_1 v_1}{n-1}\left[1 - \left(\frac{p_2}{p_1}\right)^{\frac{n-1}{n}}\right].$$

52. Temperature Change in Adiabatic Expansion.—Since in adiabatic expansion

$$p_1 v_1{}^n = p_2 v_2{}^n,$$

$$\left(\frac{v_1}{v_2}\right)^n = \left(\frac{p_2}{p_1}\right).$$

Multiplying both sides by $\dfrac{v_2}{v_1}$, we have

$$\left(\frac{v_1}{v_2}\right)^{n-1} = \frac{p_2 v_2}{p_1 v_1}.$$

But (par. 54)

$$\frac{p_2 v_2}{p_1 v_1} = \frac{RT_2}{RT_1};$$

hence

$$\left(\frac{v_1}{v_2}\right)^{n-1} = \frac{T_2}{T_1} = \left(\frac{1}{r}\right)^{n-1}.$$

But the previous paragraph has shown

$$\left(\frac{v_1}{v_2}\right)^{n-1} = \left(\frac{p_2}{p_1}\right)^{\frac{n-1}{n}};$$

hence

$$\left(\frac{p_2}{p_1}\right)^{\frac{n-1}{n}} = \frac{T_2}{T_1}.$$

Which can be substituted in either of the previous expressions, giving

$$W = \frac{p_1 v_1}{n-1}\left[1 - \frac{T_2}{T_1}\right].$$

53. Other Thermal Lines. Isometric. Isopiestic. Isobars.
—Since the pressure, volume, and temperature are the three attributes of a gas which can be caused to vary by variation in the heat-energy, the isothermal and adiabatic curves are not the only lines which may be used upon the PV plane to represent changes in the gas. It is apparent that a given volume or weight of the gas may be enclosed in a vessel or chamber, and without increase in its volume its pressure may be increased by the addition of heat. Such increase in pressure, without increase of volume, would be represented on the PV plane by a vertical line at right angles to the axes of volume and parallel to the axes of pressure. It would be designated as an isometric line, and is that which is traced when the gas in a gas-engine cylinder is ignited while the piston stands at the dead-centre. It is called

an isometric line (Fig. 20). When, on the other hand, the change
of condition in the gas is a change in its volume, without change
in its pressure, then a horizontal line parallel to the axes of volumes
at a height proportional to the constant pressure above that

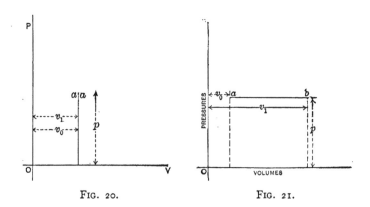

FIG. 20.   FIG. 21.

axis will represent the variations in volume as it increases or
decreases. Such a line is called an isopiestic line or an isobar
(Fig. 21). Such a line represents the condition when the piston
is either drawing in its mixture of gas and air into the cylinder,
or is expelling the products of combustion after the working
stroke when the areas of the valves are sufficient so that no variation of pressure occurs during such change of volume occupied
by the gas.

It is further apparent that if neither pressure, volume, nor
temperature remain constant, but all are caused to vary, a curve
may be determined by experiment or observation which shall
represent on the PV plane the variations of pressure and volume,
even when these do not follow any law which is capable of graphical delineation in advance. The curves, in any case, are curves
of the relation between the values of two quantities and are,
therefore, capable of being expressed analytically by an equation
which will be either that of a straight line or of a curve. Applications of these thermal lines will appear in Chapter XVII.

54. Specific Heat at Constant Pressure and at Constant Volume.—In the equation resulting from the combination of the two laws of Mariotte and Gay-Lussac, the expression appears,

$$\frac{p_1 v_1}{T_1} = \frac{p_0 v_0}{T_0}.$$

It will be usual that the pressure, volume, and temperature corresponding to the condition of the subscript zero will be the volume occupied by a unit weight of the substance (probably one pound) at a pressure p_0, which will denote the pressure on a square foot when the barometer reads 30 inches of mercury at the sea-level and the temperature T_0 is that corresponding to the absolute temperature at which ice melts. These values for the second member of the equation are not variables, but are matters of definite observation and are constants of nature. If the second member of that equation, therefore, be designated by the symbol R, the equation can be written

$$p_1 v_1 = R T_1.$$

The value for R for atmospheric air is easily calculated when it is recognized that it represents the increase in the value of the product $p_0 v_0$ when T is raised from zero degree Centigrade to one degree. Let a cylinder be imagined having a square foot of area in which fits a weightless piston, loaded with a weight of 14.7 pounds per square inch, which is the pressure at the atmosphere at sea-level, when the barometer reads 30 inches of mercury. The total pressure will then be $14.7 \times 144 = 2116.5$ pounds. If the piston enclose below it a cubic foot of air and this cubic foot be expanded by heat until it occupies a space of two cubic feet, the work in foot-pounds done by the cubic foot of air will be 2116.5 foot-pounds. Since the cubic foot of air at these conditions weighs .080728 pound, the work done by one pound will be

$$\frac{2116.5}{.080728} = 26217.66 \text{ foot-pounds}$$

by one pound of air. The Gay-Lussac law says that to double the volume of the gas requires an addition of 273° Centigrade, or 493° F.; hence the outer work which will be expended when the temperature is raised one degree will be $\frac{1}{493}$ of that expended in raising it 493°, so that the outer work entailed by the rise of one degree temperature Fahrenheit will be

$$\frac{26217.66}{493} = \frac{p_0 v_0}{T_0} = 53.354 = R.$$

Values of R can be similarly calculated for any other medium when the weight per cubic foot and the coefficient of expansion by heat are known. It is customary to describe a gas for which the value of R remains constant throughout all usual ranges of temperature by the term permanent gas. Where the gas can be made liquid by pressure or lowering of its temperature, or both, the value of R becomes uncertain near the point of such liquefaction.

In the foregoing deduction the volume was supposed to be variable and the pressure constant when the heat was applied. It took a certain amount of heat to increase that volume overcoming that constant pressure and a certain amount of heat received by the gas was expended in doing that external work. If, on the other hand, the gas had been enclosed in an inelastic vessel so that the gas could not expand by action of the heat, it is obvious that its temperature for a given application of heat would have been higher, inasmuch as no expenditure of heat energy took place and disappeared in overcoming the external resistance. Since the specific heat of a substance is the amount of heat required to raise a given weight one degree, it becomes apparent that there are two specific heats for gases: the specific heat at constant pressure, which was concerned in the process described for obtaining a value for R and which is designated by the initial C_p, and the specific heat at constant volume with variable pressure which is represented by the initial C_v. It will be apparent that the specific heat at constant pressure will always be the larger value, since the external work in foot-pounds denoted by R divided by the foot-pounds corresponding to one heat-unit should be equal to the difference in the specific heats; or, in symbols,

$$C_p - C_v = \frac{R}{778}.$$

Regnault's experiments gave for C_v the value 0.1691; for C_p his value is 0.2375; hence

$$\frac{C_p}{C_v} = \frac{2375}{1691} = 1.408 = n.$$

In gas-engine practice in which the working medium is a mixture of air with other gases, the value of this ratio will be different and should be a matter of experiment. This ratio will be the exponent which should be used in computations involving the expansion or compression of the medium which exponent has been designated in the preceding paragraph by the symbol n.

The condition of increasing volume with the pressure constant is the more desirable condition in heat-engines since the value for the mean pressure in the formula for work in terms of horse-power is greater and the weight of gas in the cylinder carries more heat and more energy. The specific heat at constant volume has been called the real specific heat and the specific heat at constant pressure the apparent specific heat, since there is no means conveniently at hand of exactly evaluating the equivalent for the outside work done and expended in overcoming mechanical resistance.

When a gas is heated from a temperature absolute T_1 to another higher absolute temperature T_2, under a constant pressure, the external work done will be that of overcoming the pressure through a space represented by the difference between the volume v_1 at the temperature T_1 and the volume v_2 which corresponds to the temperature T_2. The mechanical work done will be, there-fore,

$$p(v_2 - v_1),$$

which must be equal to the expression

$$R(T_2 - T_1).$$

The heat taken in under this condition per pound of the gas will be

$$C'_p(T_2 - T_1) \text{ in heat-units,}$$

which can be transformed into work-units by multiplying by 778. The difference in intrinsic energy will be the difference between these two quantities, and may be written

$$(778C_p - R) \quad (T_2 - T_1).$$

When, on the other hand, the same weight of gas (one pound) was heated at constant volume from T_1 to T_2 the heat taken in is expressed by

$$C_v(T_2 - T_1),$$

since no external work is done, and the whole applied heat-energy goes to store up internal energy. But if it be assumed that the same amount of heat-energy was applied to the gas in the two cases, so that

$$C_v(T_2 - T_1) \text{ should equal } (778C_p - R) \quad (T_2 - T_1),$$

the expression simplifies into

$$778C_v = 778C_p - R$$

as was just shown above.

It may, therefore, be stated that the expression $C_v(T_2 - T_1)$ expresses or measures the change of internal energy in a unit weight of gas in changing its temperature from T_1 to T_2 in any manner, no matter how the volume or pressure may vary during the process.

It may be of interest to compute the temperature and pressure in a gas-engine cylinder due to the ignition of the weight of com-bustible mixture which has been drawn in at atmospheric pressure p_i; compressed to its final volume v_f and pressure p_f and is then ignited at constant volume.

From paragraphs 29 and 46, the calorific value of the cubic foot of such mixture at atmospheric pressure was $\dfrac{H}{1+a}$ in which a is the volume of non-fuel elements, which are added to secure combustion of the unit volume of fuel. The volume is changed by compression to v_f and the temperature at the end of the compression is called T_c (par. 47), and is (par. 52)

$$T_c = T_o \left(\frac{v_f}{v_i}\right)^{.41} = T_o \left(\frac{p_i}{p_f}\right)^{.29}$$

Therefore the value of T_c being known by computation, the heat of combustion when H and the quantity $(1 + a)$ are reduced from their volume values per cubic foot to the corresponding quantities in pounds, will be given by the equation:

$$\frac{H}{1+a} = C_v \left(T_1 - T_3\right)$$

whence
$$\left(T_1 - T_3\right) = \frac{H}{C_v\,(1+a)}.$$

The quantity $\dfrac{H}{1+a}$ is given in cubic foot values for a wide range of fuels in column 15 of the table in paragraph 29. The value of C_v is given above. Hence the range above the compression temperature is at once given when the data are reduced to pounds. The corresponding pressure is (par. 44):

$$p_1 = p_3 \frac{T_1}{T_3}$$

since the volume and weight do not change by the condition imposed.

55. Effective Specific Heat.—In the gas-engine problem, however, the gas is not a simple element or a stable chemical combination, but is a mixture of varying proportions and of varying constituents. Serious error will result from a disregard of the difficulties introduced by these phenomena. For example, if the combustion is direct, so that a pound of carbon burns to carbon dioxide, the CO_2 gas has a volume at 32° according to the following table of 8.102 cubic feet:

1 Line No.	2 Gas.	3 Specific Gravity.	4 Pounds per Cubic Foot.	5 Cubic Feet per Pound.	6 Specific Heat at Constant Pressure.	7 Specific Heat at Constant Volume.	8 Ratio $\frac{C_p}{C_v} = \gamma$.
1	Air...............	1.0000	.080728	12.387	0.2375	0.1689	1.406
2	Oxygen...........	1.1051	.08921	11.209	0.2175	0.155	1.403
3	Hydrogen..........	0.0695	.00561	178.23	3.409	2.406	1.417
4	Nitrogen..........	0.9714	.07842	12.752	0.244	0.173	1.409
5	Carbon monoxide, CO	0.9674	.07810	12.804	0.245	0.173	1.416
6	Carbon dioxide, CO_2 .	1.5290	.12343	8.102	0.216	0.171	1.165
7	Marsh-gas, CH_4	0.5560	.04488	22.301	0.593	0.467	1.27
8	Ethylene, C_2H_4	0.9847	.07949	12.580	0.404	0.332	1.144
9	Steam.	....	8.03794	26.42	0.480	0.369	1.302

Nos. 1–8 at atmospheric pressure and at temperature of melting ice. Steam at 212° F.

But if the gas was first made into CO, and an extra volume of oxygen supplied to burn the one pound of CO into CO_2, the proportions of mixture would be

$$CO \quad + O = CO_2 \text{ of which } CO = \frac{28}{44} = \frac{7}{11}$$

$$12 + 16 + 16 = \quad 44 \quad \text{and} \quad O = \frac{16}{44} = \frac{4}{11}$$

the volume of the mixture would be

$$\frac{7}{11} \text{ of } 12.804 = 8.148$$

and

$$\frac{4}{11} \text{ of } 11.209 = \underline{4.076}$$
$$\text{Total} = 12.224$$

making a volume quite different from that resulting from the single-step process. Hence the proper value to be inserted in the formula for the increase in temperature for the quantity $\dfrac{y}{x+y}$ in that expression does not always readily appear in advance from theoretical considerations. But as a matter of fact, the value to be inserted for the factor representing the specific heat in that formula (par. 14) is still more uncertain. The air supporting combustion is gradually passing through a series of changes throughout the working stroke, and these changes doubtless involve molecular rearrangement in transit. Furthermore, the experiments of the physicist have shown that the specific heat is not a constant for all temperatures of a gas or a mixture, but increases with the temperature according to some law whose form takes the shape for specific heat at constant volume:

$$C_{T_2} = C_{T_1} + \beta(T_2 - T_1).$$

So that at a higher temperature, T_2, the specific heat is greater than at the lower temperature, T_1, by a small but as yet undetermined amount proportional to the difference between the temperatures.

Hence the present practice is to approximate to the actual values of the factor which may be called the effective specific heat by one of five methods.

The first method will be by the use of what is designated as Grashof's formula, which assumes all gases to have the same chemical composition, and that the letter R denotes the ratio of gas to air in the mixture. For the specific heat at constant volume,

$$C_v = \frac{[0.169 \times R] + 0.286}{R + 0.48},$$

for specific heat at constant pressure,

$$C_p = \frac{[0.2375 \times R] + 0.343}{R + 0.48},$$

In which the factors 0.169 and 0.2375 are the specific heats for air. Applying these to various mixtures, the following table results. Col. 5 is computed from the formula

$$T = 32 + \frac{Q}{[(Rx) + y]C_v}.$$

COMPUTED SPECIFIC HEATS OF PRODUCTS OF COMBUSTION
(GRASHOF S FORMULAS).

1	2	3	4	5
Proportion of Air to Gas by Volume.	Specific Heat of Mixture at Constant Vol. C_v.	Specific Heat of Mixture at Constant Pressure C_p	Ratio $\frac{C_p}{C_v}$.	Calculated Temperature Fahr. of Combustion.
4 to 1	0.214	0.289	1.35	6010
5 " 1	0.206	0.280	1.36	6380
6 " 1	0.200	0.273	1.36	6300
7 " 1	0.196	0.268	1.37	5570
8 " 1	0.193	0.265	1.37	5000
9 " 1	0.190	0.262	1.37	4580
10 " 1	0.188	0 259	1.38	4140
11 " 1	0.186	0.257	1.38	3820
12 " 1	0.185	0.256	1.38	3530
13 " 1	0.184	0.255	1.38	3280
14 " 1	0.183	0.254	1.39	3090
15 " 1	0.182	0.253	1.39	2900

The second method is to analyze the products of combustion and to assume that the effective specific heat of such a mixture is the sum of the specific heats of the constituents. Each percentage by weight of the constituent gases is multiplied by the specific heat of that gas. and the sum of these products is called the effective or rather the mean specific heat. This assumption may or may not be correct.

The third method is to invert the use of the theoretical formula to calculate increase of temperature. and, by observing the values of the variables which enter into it, calculate the actual value. That is, in the equation

$$T_2 - T_1 = \frac{y}{x+y} \times \frac{Q}{C_{v_e}}$$

let both T_2 and T_1 be temperatures observed in an actual experiment, and Q the heating value of the gas in B.T.U., while y is the weight of fuel in the mixture and $x+y$ the total weight of such mixture. Then Cv_e is the only unknown factor. If this method be applied to CO, for example, burning to CO_2,

$$\begin{bmatrix} C+O \\ 12+16 \end{bmatrix} + \frac{O}{16} = \begin{bmatrix} CO_2 \\ 44 \end{bmatrix}$$

$$\frac{CO}{O+CO} = \frac{28}{44} = \frac{7}{11}$$

if oxygen were the supporter of combustion. With **air**, however, with 77 parts of nitrogen, the $\frac{16}{28}$ of supplied oxygen must be multiplied by $\frac{100}{23}$ to give equivalent air. Hence

$$\frac{100}{23} \times \frac{16}{28} = 2.5 \text{ lbs. air nearly,}$$

so that the mixture will be

CO.............. 1   pound
O............... 0.56   "
N............... 1.94 pounds
                 ————
                 3.50

made up of 1.56 pounds of CO_2 and 1.94 pounds of N. Hence

$$\frac{y}{x+y} = \frac{1}{3.5} = \frac{2}{7} \text{ instead of } \frac{7}{11}.$$

For CO the value of Q is about 10,000. Then if **the range** T_2-T_1 be found to be

$$T_2 - T_1 = 2000° \text{ F.,}$$

$$Cv_e = \frac{2}{7} \times \frac{10,000}{2000} = \frac{10}{7} = 1.43 \text{ nearly,}$$

wh:ch could be applied for other cases similar to the typical one. The difficulty here is the inaccuracy of the observations of $T_2 - T_1$, especially in engine work, where the changes are very rapid by conduction of the cylinder-walls.

A fourth basis for computing the effective specific heat will be by measuring the increase of volume in a combustion at constant pressure. In this case let v_2 be the greater volume and v_1 :he less. Then will

$$\frac{v_2}{v_1} = \frac{T_2}{T_1},$$

since at constant pressures the volume will be directly as the absolute temperatures. But it is also true that

$$T_2 - T_1 = \frac{yQ}{(x+y)C_p};$$

hence, by dividing by T_1 and transposing,

$$\frac{v_2}{v_1} = \frac{T_2}{T_1} = 1 + \frac{yQ}{(x+y)C_p T_1},$$

and also, by multiplying by v_1,

$$v_2 - v_1 = \frac{y}{x+y} \times \frac{Qv_1}{C_p T_1}.$$

But since

$$\frac{v_1}{T_1} = \frac{R}{p_1},$$

it will be true that the volume range

$$v_2 - v_1 = \frac{R}{p_1} \frac{yQ}{(x+y)C_p},$$

whence the actual or experimental value for the specific heat

$$C_p = \frac{RQy}{(x+y)p_1(v_2 - v_1)}$$

can be computed if p_1, v_2, and v_1 be observed in any case, and R be taken at its value 53.35 for air. If, for example, in a combustion at atmospheric pressure a pound of CO burning to CO_2 gave a value for $v_2 - v_1$ of 40 cubic feet, then the computation as in the preceding example would give

$$C_{ep} = 53.35 \times \frac{2}{7} \times \frac{10,000}{2116.4 \times 40} = 1.83.$$

A fifth method by analogy would be a similar observation or experiment with an increase of pressure caused by a combustion at constant volume, as in an explosive gas-engine. Here obviously the above equation would have the form

$$C_v = 53.35 \frac{Qy}{(x+y)v_1(p_2 - p_1)}.$$

If the pressure range in a closed vessel with CO burning to CO_2 be observed to be 60 pounds per square inch when the volume occupied was 13 cubic feet, then

$$C_{ev} = 53.35 \times \frac{2}{7} \times \frac{10,000}{13 \times 60 \times 144} = 1.35.$$

These values for the effective specific heat deduced from reliable experiment and observation are so much higher than the accepted accurate determinations by the physicists for products of combustion or for air as to confirm the general deduction from all experiment that in the internal heating which occurs in a gas-engine the theoretical temperatures called for with accepted values of the specific heat of air and gas are not attained in practice. The reasons which are most probable for this phenomenon have been already foreshadowed in the discussion of the volume change on chemical combination, due to molecular rearrangement, and possibly to other chemical changes which may occur; the losses in dissociation without a subsequent complete combination within the time allotted to the gas-engine stroke; the reaction between the combining gases and the highly conductive metal walls of the cylinder; the non-instantaneous character of the combustion with imperfect mechanical mixtures of the components; and perhaps also a varying value for the effective specific heat, being greater at high temperatures than at the lower ranges. Internal combustion is also limited in heating effect by the condition that the fuel cannot combine with more oxygen than will chemically unite with it. If an excess of oxygen or air is present, it simply increases the material to be heated with a given calorific power of fuel, and not only does not increase the intensity of the heating, but lowers the resulting temperature. Varying composition of the gaseous mixture due to governing or high speed or other causes affects, therefore, the temperature in the cylinder. With external heating this particular limit is not set.

56. Value of the Exponent in the Equation for Expansion. —When a mixture of gas and air is expanding after ignition, and without a transfer of heat from without to replace the equivalent of the external resistance overcome, the expansion is called adiabatic (par. 50), and the equations for the relations of pressures and volumes will be in the form

$$p_1 v_1^n = p_2 v_2^n.$$

In the gas-engine the exponent n is not unity, but is the ratio between the specific heats of air or the mixture at constant pressure and at constant volume (par. 54). But the preceding treatment has made it clear that neither of these values can be assumed to be those resulting from laboratory determinations and hence, of course, their ratio should not be assumed. To do so is to introduce the likelihood of several serious errors. For example, in the indicator-card in Fig. 22, if

v_1 = the volume of the clearance,
v_2 = the final or terminal volume,
p_0 = atmospheric pressure (14.7 lbs.),

p_v = pressure at compression from card,
l = length of stroke, feet or inches,
x = " . " clearance, feet or inches,

then will

$$\frac{p_v}{p_0} = \left(\frac{v_2}{v_1}\right)^n, \text{ or } \left(\frac{p_v}{p_0}\right)^{\frac{1}{n}} = \frac{v_2}{v_1},$$

which becomes

$$\left(\frac{p_v}{14.7}\right)^{\frac{1}{n}} = \frac{x+l}{x},$$

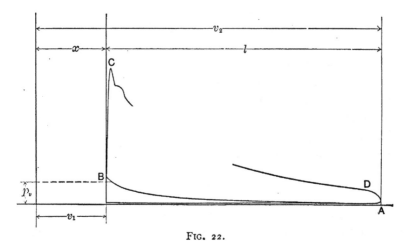

Fig. 22.

from which the third unknown can be calculated if the two others have been observed or the value of n is assumed. If the usual assumption is made that $n = 1.41$ or $\frac{1}{n} = .71$ and p_v be observed, then the value for x will come out too large, and computations for the temperature after ignition from the formula

$$\frac{p_1}{T_1} = \frac{p_0}{T_0},$$

which will hold when the volume does not alter, will give values for T_1 which are too large. Hence the value for n should not be assumed, but from a carefully taken indicator-card, and by careful measurement of clearance and stroke volumes, the experimental or effective value for n should be worked out for different points on the curves of both the expansion and compression lines. In an actual experiment, for example, the expansion lay, as shown in Fig. 23, between the upper dotted or

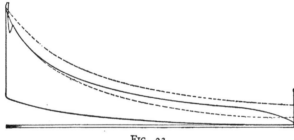

FIG. 23

isothermal line and the lower dotted or adiabatic line, nearer the latter at the beginning. If values for both p and v be taken at

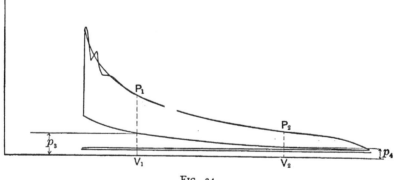

FIG. 24.

various pairs of points in the stroke as indicated in Fig. 24, the calculation for n will take the form

$$p_1 v_1{}^n = p_2 v_2{}^n, \quad \text{or} \quad \left(\frac{p_1}{p_2}\right) = \left(\frac{v_2}{v_1}\right)^n,$$

whence

$$n = \frac{\log \dfrac{p_1}{p_2}}{\log \dfrac{v_2}{v_1}},$$

or

$$n = \frac{\log p_1 - \log p_2}{\log v_2 - \log v_1}.$$

In the experimental case referred to, the values of n from the computations based on the diagram came out 1.10, 1.12, 1.13, 1.14, 1.15, 1.16, with an average of 1.14, and in another test with an average of 1.20. A wider range of generalization has shown that the ends of the real curves and some intermediate points he on a curve whose equations are for compression $pv^{1.35}$ and for expansion $pv^{1.4}$, equal to a constant (see pars. 152, 202). The value for n will often be slightly higher for the compression curve on the diagram than for the expansion curve. Equality of value on the two curves may be tested by the observation whether at the points p_3 and p_4 on the compression curve the ratio is true that $\dfrac{p_1}{p_3} = \dfrac{p_2}{p_4}$. For let x be the exponent for the expansion curve, and y the exponent on compression, then $p_1 v_1^{x} = p_2 v_2^{x}$ will hold for one, and $p_3 v_1^{y} = p_4 v_2^{y}$ for the other. By division,

$$\frac{p_1 v_1^{x}}{p_3 v_1^{y}} = \frac{p_2 v_2^{x}}{p_4 v_2^{y}}.$$

If $x = y$ in both members,

$$\frac{p_1 v_1^{x}}{p_3 v_1^{x}} = \frac{p_2 v_2^{x}}{p_4 v_2^{x}}, \quad \text{or} \quad \frac{p_1}{p_3} = \frac{p_2}{p_4},$$

but not otherwise. In any exhaustive investigation of an internal-combustion engine the value of n should be one of the quantities to be computed for which observations should be made (par. 174).

57. The Continuous Rotative Motor using Pressure, Impulse, or Reaction.—The treatment in the foregoing paragraphs of this chapter has been specially directed towards a statement of the phenomena in motors of the piston class. The operation of media whose pressure or elastic tension is affected by heat is not necessarily restricted to motors of this class. A design of motor similar to the rotary engine in steam-engine practice or similar to the steam turbine can be made available, provided a convenient cycle for the gas can be secured which it will be the purpose of the next chapters to discuss. If the rotary-engine principle is sought for, it will require to receive on rotating vanes the pressure from the expanding gas, and the difficulty will at once be met of securing a satisfactory expansive working of the medium. It is much more likely that motors of the turbine class which utilize the impulse or reaction due to a high velocity of the expanding gas will be found to lie in the direction of success along these lines. The problems are both inherent and structural when air with gas is used as the medium, by reason of the thermal sluggishness of the hot medium acting upon the vanes or plates of the turbine and the difficulties incident to working with the high temperatures required by air. But it is also not certain at the present state of knowledge that the jet-impact method of working a motor would give an economical heat transformation.

CHAPTER IV.

THE HEAT-ENGINE CYCLE.

58. Introductory.—It will be apparent from the considerations in the foregoing discussion that in the operation of gas-engines of the piston class it will be necessary to raise the condition of intrinsic energy of the gas at the beginning of the working stroke, and that the convenient means of doing this is by the combustion within the air of a material having a suitable calorific power. That increased condition of energy makes the gas capable of doing work upon the head of the piston and overcoming external mechanical resistance by its expansion. It is desirable at the end of the expansion that this mass of products of combustion and air shall be as nearly at the state of the atmospheric air surrounding the motor with respect to heat and available energy as is convenient and possible, in order that the least amount of available energy may be thrown away and wasted. That expansion, therefore, should be accompanied by a cooling or lowering of temperature, and connected therewith a reduction of volume. These transformations with respect to the internal energy of the gas will be recurrent or cyclic in their action and will usually repeat themselves in a fixed order of succession. A motor of the class in question, therefore, is said to have a "cycle" of operations, and each step in this cycle may conveniently be called a "phase." The extent of each phase and the period of its recurrence, so far as the medium is concerned, will be affected by the mechanical appliances whereby this succession of phases in the gas or medium are utilized to overcome external resistance.

It is obvious, therefore, that a careful distinction should be made between the cycle as a succession of phases with respect to heat energy in the gas, and the cycle as a succession of events determined by the mechanical construction of the motor, the periodicity of its valve action, and the variation in the supply of heat energy from variations in the load, etc. This distinction between the two types of cycle is believed to be of very considerable importance in a clear analysis of the operations of gas-engines. ·

5). The Cycle of the Steam-engine. — With a view of making clear the meaning of the term cycle as applied to the gas-engine, it may be convenient to refer to the cycle used in the steam-engine.

The media for heat-engine purposes may be roughly divided (par. 5) into those which do work principally by utilizing the expansion which occurs when it changes from the liquid to the gaseous state on the one side, and on the other those that utilize a perfect or permanent gas whose expansion is caused by the absorption of heat (par. 7). In the steam-engine the heat is added to the liquid and its temperature sufficiently raised so that under the conditions of pressure which are fixed upon, the liquid becomes a gas. In the second stage this gas is allowed to expand to as low a pressure as possible or convenient, doing work upon the piston by such expansion in the engine. The gas is then discharged either as a gas or as a liquid. This cycle is not capable of being modified, except in minor details. It compels, as a rule, the presence of three essential elements, of which one shall be the organ concerned with the production of the vapor; the second shall be the organ for the utilization of the pressure of this vapor, and the third, the apparatus for disposing of the vapor discharged. In condensing steam-engines this latter is the condenser, and in non-condensing engines it is the atmosphere. The amount of work which can be done with a given amount of heat in a prime mover of this class is definitely known, within certain limits, when we know how much liquid can be converted into gas by a given amount of heat and the relative specific volumes of the liquid and result-ing gas. The chief data, therefore, are those concerned with the properties of the liquid selected, so far as the medium is con-cerned, and the motor comes in only to affect the mechanical efficiency of the system for the conversion of heat energy into mechanical energy. With the perfect or permanent gases the range of possible methods of heating, expanding, and cooling becomes greatly enlarged since the manner of heating, the method of expansion, and the ultimate disposition of the gas after the work done by the expanding gas and the properties of the medium itself are without significance. This peculiarity was first pre-sented by Carnot, who first called attention to the existence of the cyclic action.

60. The Carnot Cycle.—The cycle of Carnot may serve as a type for all cyclic actions with a permanent gas. The four phases of the cycle are represented by the diagram upon the *PV* plane in Fig. 25. It presents a cylinder in which fits a piston, both of material such that neither have any capacity for heat, nor do they offer any friction. All heat received is to be utilized in the gas which is acting in that cylinder. The end of the cylinder is supposed to be of a material with perfect conductivity, so that the gas may be affected as to its heat condition without loss from the process of transfer. The element A is a source of heat having a great capacity, so that all the heat required for the cycle can be transmitted to the gas at the maximum tempera-ture, which is the condition of maximum efficiency of such transfer and is maintained at the temperature T_1. The element C is a condenser also of great capacity and maintained at the lowest available temperature T_2, so that the cooling of the gas for re-duction of temperature shall be done with the maximum effi-ciency of that process. By having both the heater and cooler of great capacity, no change in T_1 nor T_2 occurs during the cycle. The cover B of a non-conducting material, to be used during the period of expansion and compression, is applied to the cylinder when it is in contact with neither the source of heat nor the cooling apparatus. The relations of pressure and volume for the various stages are given by the subscripts on the diagram. The specific heat at constant pressure will be denoted by C_p, and the ratio

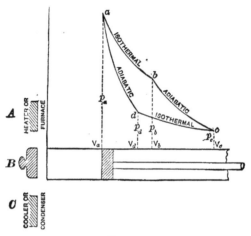

FIG. 25.

between the initial and final volumes $\dfrac{v_c}{v_d}$, which must be the same

as the ratio $\dfrac{v_b}{v_a}$, will be denoted by the factor r, or the ratio of the

expansion.

There will be four steps or stages :

(*a*) Apply the heater A. The pressure rises; the unit weight of gas expands isothermally at T_1. The heat energy taken in is

$$H_1 = CT_1 \text{ hyp. log } r,$$

(see par. 181).

(*b*) Heater A is removed, cover B is applied, and the piston moves out at the expense of the gas temperature, until the temperature falls to T_2 by such expansion against external resistance.

(*c*) Take away cover B and apply cool body or condenser C at T_2. No change will take place, because the expansion is complete, unless the piston be pushed back. But if the piston is retracted, the smallest tendency to an increase of temperature above T_2 is at once met by a flow of energy into the condenser. The gas changes its total heat energy down to the stage represented by T_2, and the amount rejected to the condènser will be the difference between the energy at T_1 and T_2, or

$$H_2 = CT_2 \text{ hyp. log } r.$$

(d) Remove the condenser C and replace B when the point d is reached. The piston is now still further forced in and back until the gas has its initial volume v_a, and if the point d was rightly chosen it has also the temperature T_1 at which it started, because the compression has been adiabatic, and the cycle has been completed. For the relations of v_b and v_c to produce the desired final temperature T_2,

$$\frac{T_1}{T_2} = \left(\frac{v_c}{v_b}\right)^{n-1} \quad \text{or} \quad \frac{T_1}{T_2} = \left(\frac{v_d}{v_a}\right)^{n-1},$$

according as the location of b or d is desired (par. 52).

It will appear, therefore, that the Carnot cycle gives an external work in foot-pounds which will be 778 times the difference between the heat rejected and the heat received (par. 47), or, for the complete cycle,

$$\text{Work} = 778C(T_1 - T_2) \text{ hyp. log } r,$$

which is 778 times the area included in the diagram of curves (Fig. 25), all transfers having been made at maximum efficiency.

The operation of the Carnot cycle is proved to be that of maximum efficiency for the conditions assumed by the expedient of imagining the cycle to be operated in reverse direction by a similar heat-engine. It can be proved, if both engines operate within the same limits of temperature, T_1 and T_2, and one drives the other as a motor, while the other operates as a heat-pump, that such a combination makes both cycles reversible, and that all reversible heat-engines working between the same limits of temperature are equally efficient or that the efficiency, in the thermodynamic sense, is independent of the specific heat or other physical properties of the medium used. The formula for efficiency discussed in paragraph 47 is immediately deducible from the operation of the Carnot cycle.

61. **The Cycle of the Internal-combustion Engine.**—The Carnot cycle and the equation for its condition of maximum efficiency do not apply to the internal-combustion engine if the maximum temperature in the cycle be taken as the temperature after ignition has occurred. It is only true where the maximum compression temperature is taken. It will be apparent that by abandoning the indirect methods which are assumed in the typical Carnot cycle for transferring heat to the gas a much wider range of possible cycles is opened up. In the first place, the heating may occur without a previous compression; in the second place, the compression may be adiabatic and the heating isometric (par. 49–53); in the third place, the compression may be adiabatic, the heating isopiestic; in the fourth place, the compression may be adiabatic and the heating isothermal; in the fifth place, the compression may be adiabatic and the heating may follow any law not reducible to the foregoing standard methods. Each of these may be made to vary again by the method followed in expansion and in cooling. Finally, the heating may be atmospheric without compression or with compression, and where there is compression the cooling may be according to varying forms of the phase. The following table presents in analytic form the possible cycles for such engines:

It will be a matter to be discussed in Chapter XVII which of these cycles are available or unavailable and which of them offer the probabilities for the best efficiency.

It will also suggest itself to determine the effect of the cycle, so far as it requires a larger volume of gas under one method of working than to do the same work in another. The larger volume of gas makes a more bulky motor. Some cycles will operate under higher temperatures than others, and others through wider ranges both of temperature and pressure.

CLASSIFICATION OF CYCLES.

1	2.	3.	4.	5.	6.
Cycle No.	Compression.	Heating.	Expansion.	Cooling.	Cooling.
I	....	Isometric	Adiabatic	Isopiestic	
I A	....	"	"	Isometric	Isopiestic
I B	....	"	"	Isothermal	"
I C	....	"	"	"	
II	Adiabatic	Isometric	Adiabatic	Isopiestic	
II A₁	"	"	"	Isometric	Isopiestic
II A₂	"	"	"	"	
II B	"	"	"	Isothermal	Isopiestic
II C	"	"	"	Isothermal	
III	Adiabatic	Isopiestic	Adiabatic	Isopiestic	
III A	"	"	"	Isometric	Isopiestic
III B	"	"	"	Isothermal	"
III C	"	"	"	"	
IV	Adiabatic	Isothermal	Adiabatic	Isopiestic	
IV A	"	"	"	Isometric	Isopiestic
IV B	"	"	"	Isothermal	"
IV C	"	"	"	"	
V	Adiabatic	Any law	Adiabatic	Isopiestic	
V A	"	"	"	Isometric	Isopiestic
V B	"	"	"	Isothermal	"
V C	"	"	"	"	
VI	....	Atmospheric	....	Isometric	Isothermal
VII	Adiabatic	"	Adiabatic	Isopiestic	
VIII	....	"	"	Isothermal	
IX	Adiabatic	"	"	Isometric	
X	"	"	"	Any law	

A consideration of the relative advantages of the various cycles must be postponed until after a consideration of the mechanisms or motors which they operate.

62. The Otto Cycle with Heating at Constant Volume.— There were several forms of gas-engine proposed and built previous to 1876, to which reference will be made in the historical appendix (Chapter XX), but in that year Dr. Otto brought out, in Germany, the engine which he designated as his "Silent" gas-engine, to distinguish it from his earlier and more noisy type. The Otto cycle was first suggested in a French patent of 1862, by Alphonse Beau de Rochas. He advocated the advantages of a previous compression of the combustible mixture of gas and air, and proposed to do away with a separate compressing pump-cylinder for this purpose by making only one stroke in four to be the working stroke in a single-acting engine. The Beau de Rochas or Otto cycle involves:

1. Aspiration of the mixture of gas and air in proper proportions during an outgoing stroke of the piston (1–2 in Fig. 26).

2. Compression of the mixture by the return of the piston (2–3–4). This compression fills a comparatively large clearance volume behind the piston, which must be so adjusted to the displacement by the piston that there shall be no danger of such

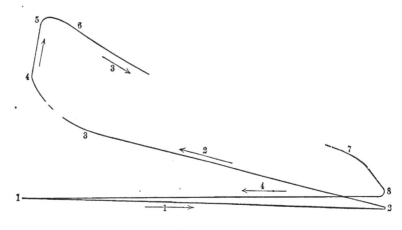

FIG. 26.

elevation of temperature from the compression as to ignite the mixture as the result of compression alone (pars. 152, 202).

3. The piston being at or near its inner dead point (4), the compressed mixture is ignited by some acceptable and reliable device (Chapter XI), at which the pressure rises at once (4–5) and exerts its outward effort to drive the piston forward. Expansion is followed by gradual lowering of pressure during this working stroke (5–6–7). This heating is therefore done at constant volume.

4. The exhaust opens just beyond 7 and the products of combustion are discharged into the open air through the exhaust.

valve by the return of the piston to its inner dead-centre (8–1).
The cycle then repeats itself.

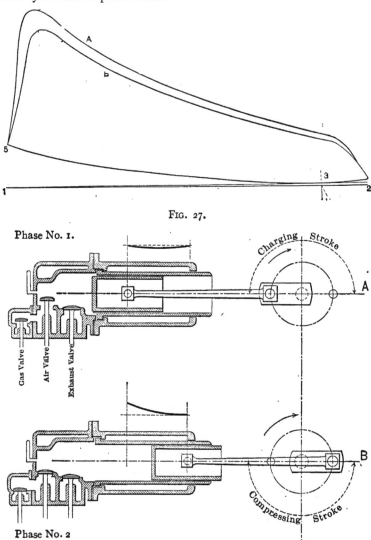

FIG. 27.

Phase No. 1.

Gas Valve

Air Valve

Exhaust Valve

Charging Stroke

A

Compressing Stroke

B

Phase No. 2

FIG. 28.

It is apparent that a heavy fly-wheel must be used to equalize
the motion of the crank-shaft, having energy enough stored in it by

the working stroke to overcome the resistance during the time of the other three strokes, and cause also the piston to perform the acts of the cycle in the cylinder. High rotative speed is therefore an advantage. Furthermore, a high initial pressure and temperature are desired, with a low terminal value for both, so as to secure a high mean value. Rapid inflammation is therefore desired, and the methods of ignition become important. Fig. 26 is a diagrammatic analysis of the succession of events, while Fig. 27 is the normal *PV* diagram from an actual engine. To connect the various steps of the motor mechanism with the effects in the cylinder and upon the gas and mixture, the diagrams of Figs. 28 and 29 will be serviceable. They show the effect of the

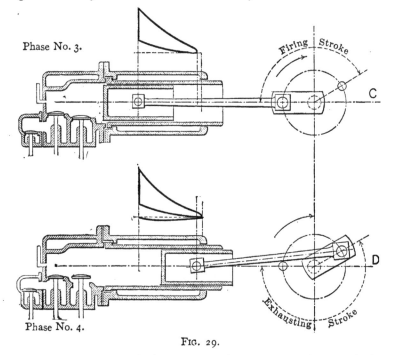

Phase No. 3.

Firing Stroke

C

D

Exhausting Stroke

Phase No. 4.

FIG. 29.

successive traverses of the motor piston, generating each its appropriate line of the diagram as the volume varies and causes the pressure to vary with it. The gradual building up of the typical diagram of pressures is made clear. The same illustrations show the succession of operations performed by the valves. In phase No. 1 the gas and air inlet valves are open, the exhaust is closed; in phases No. 2 and No. 3 all are closed; in phase No. 4 the exhaust is open, and the others are closed. The ignition phenomenon occurs at the beginning of phase No. 3. By the expedient of plotting the observed pressure acting upon the crank-pin at each point of the two strokes of the piston, as is done in Fig. 30, the values of the varying effort on the engine-shaft appear clearly to the eye. It is obvious that in this design only the one working stroke in four traverses of the piston is a working or effective stroke.

63. The Brayton Cycle with Heating at Constant Pressure.—The second important cycle is, historically, a little earlier than that of Otto. Its principle is the compression of a mixture of inflammable gas and air which is introduced into the working cylinder and there ignited so as to burn in such a manner that the pressure shall not increase above a fixed constant value. The power is generated by the increase of volume at constant pressure due to the inflammation of the gas in the air. Such engines are not explosive, but the pressure increases gradually, due to slow combustion. The credit for using this cycle attaches to a Philadelphian named Brayton, who utilized it in an engine in 1873 (see Fig. 218). In England Messrs. Simon used the same

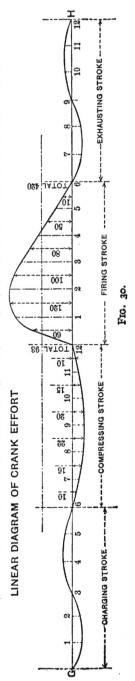

FIG. 30.

cycle in 1878. Credit is also due to Sir William Siemens, who proposed the cycle as far back as 1860, but no engine was built.

In the constant-pressure cycle there are usually two cylinders, one a compressing-pump and the other the working cylinder. The charge of gas and air is drawn into the pump and compressed on the return of the piston into a receiver. The pressure in the receiver may be about 60 or 80 pounds. The mixture flows from the receiver into the working cylinder and is ignited as it enters, receiving, therefore, a supply of hot mixture at constant pressure until the valve cuts off admission. From the point of cut-off to the end of the stroke the volume of gases is expanding, and of course the terminal pressure can be reduced by adjusting the point of cut-off. The work diagram on the PV plane from such a cycle is quite similar to that of the steam-engine.

64. The Cycle with Heating at Constant Temperature.— The third variable in the pressure-volume-temperature series is the temperature, and a cycle in which the addition of the heat to the mixture should be made at a constant temperature would constitute a third class. The nearest actual approach to this cycle is in that which is made use of in the Diesel engine. In this cycle the air is drawn in on the aspirating phase of the cycle and is compressed by the energy in the fly-wheel to a high pressure. It is usually about 500 pounds in small sizes. Into this highly compressed and heated air is introduced the jet of combustible. At first proposed by Diesel, this combustible was kerosene oil, and it was supposed to be completely ignited by the high temperature of the air, so that the air should receive all its heat energy at the temperature prevalent when the ignition was begun. It is more than questionable, in view of the time necessary to ignite the oil and to heat the air, whether this heating of the air is done at a constant temperature throughout its entire mass in actual practice. To the extent to which this result is attained the cycle approaches the Carnot cycle, in which the heating is supposed to be at constant temperature.

65. Advantages of the Internal-combustion Principle.—

In discussing the advantages of the internal-combustion principle as a means of deriving mechanical energy from liberation of heat, it is unavoidable that the comparison be made between this principle and that of the ordinary steam-engine. In the latter there has to be the furnace and the boiler, exterior to the engine proper, both for liberation of heat and for storage of that liberated energy. There should therefore be a distinction drawn between the advantages due to the use of liquid or gaseous fuels, which are practically essential in the internal-combustion engine, and those which belong to the direct utilization of the energy of the fuel by combustion in the cylinder, instead of outside of it, in a furnace. Hence the advantages of the direct internal combustion are:

(1) The energy of the heat liberated by combustion operates directly upon the piston to produce motion, and without intervening appliances.

(2) The economy in fuel per horse-power per hour is greater than with steam or externally heated air, because heat is not wasted in furnace or chimney, or in doing work upon a transferring medium which is not utilized in the engine.

(3) No fuel is consumed wastefully in getting the motor ready to start, nor is any wasted in the furnace after the engine stops. The losses in banking fires under boilers which are run intermittently are avoided, the losses due to blowing steam to waste through safety-valves when the motor is stopped for short periods, and the losses of fuel through the grates from cleaning and when the run is over.

(4) The radiation losses of heat from the boiler-setting or furnace do not occur.

(5) The bulk and weight of the boiler and its setting are eliminated, as well as their cost.

(6) This gives this type of motor a distinctly portable character if desired, even up to considerable sizes, where it may be convenient to have the motor follow to the place where the work is to be done, as in logging and lumbering.

(7) The absence of boiler and chimney eliminates the repair

and maintenance account attaching to them, as well as the labor to operate them and their first cost.

(8) The absence of the boiler and **its** furnace lowers the insurance risk (unless offset by the presence of the producer, the gas-holder, or the stored liquid fuel).

(9) The absence of the boiler avoids a necessity for licensed operators which are required both afloat and ashore where steam-plants are run.

(10) The motor is ready to start on the instant and without previous preparation or delay from starting a fire and getting up pressure.

(11) When the fuel-supply is shut off, the motor stops and there is no attention which the plant requires gradually to shut it down. These two latter considerations are particularly potent with respect to the automobile uses of these motors.

(12) This principle lends itself easily to the condition where storage of energy is required. With gas-burning engines, the producer may be run at high efficiency when convenient, and the gas held in gas-holders till needed, or the energy in liquid fuel may be drawn into the motor through carburetors (Chap. X) as required. This is convenient, for instance, where a plant is to be worked overtime.

(13) Incident to this is the advantage of subdividing power units in a large plant. Each motor may receive its supply of motor energy through pipes as gas without loss, or from fuel tanks, and such motors can be run independently of each other as to capacity, speed, time, and the like, as long as the store of gas or oil holds out.

(14) In compressed gas in tanks under pressure a large amount of fuel energy and power may be stored in small bulk and weight, to be expended through reducing-valves to motors as required. This property is only of moment when the fuel weight must be reduced as in aerodromes. In automobile and yacht practice the liquid fuel does not weigh enough to make gaseous compression worth while.

(15) The rapidity with which flame propagates itself in an explosive mixture of fuel and air renders a high number of rotations of the shaft possible per minute. This makes a high-speed multi-cylinder engine of small weight per horse-power.

(16) The rapid ignition of explosive mixtures gives a high initial pressure at the beginning of a piston-stroke. Where this means also a high mean pressure it gives a powerful engine for a given cylinder volume.

(17) There is no storage of large amounts of energy in the form of pressure in a containing vessel, a rupture of which will cause disaster.

(18) There is no boiler to give difficulties with a water containing salts in solution, and requiring a constant watchfulness to keep properly supplied with water lest an accident result from low water and an overheated boiler.

(19) There is no exposed flame or incandescent fuel-bed requiring care and watchfulness. Such flames in absence of good draft arrangements may blow back or downwards in gusts of wind outdoors, and are sources of danger in accidents, if they can reach the fuel-supply.

(20) The mechanism of the motor is simple in principle and does not involve a great number of parts in motion.

The gas- or oil-engine, furthermore, attaches to itself the advantages of gas-firing and mechanical stoking, in that

(21) The normal and proper combustion is smokeless;

(22) The fuel does not require so great a diluting excess of air for combustion, which lowers the temperature of the latter;

(23) The avoidance of dust, sparks, or ashes;

(24) The liquid or gaseous fuel will be handled mechanically by pumps or pressure organs into the motor apparatus. Hence the labor and cost of such handling are avoided, and the cost of removal of ashes.

It should not be forgotten in a study of the foregoing that in cities where gas is made at central stations and distributed by mains the central generating plant has had to assume many of the disadvantages whose avoidance constitutes the arguments for the internal-combustion motor. If the motor operates its own producer, as an isolated gas-making plant, some of the above arguments disappear. If the motor uses liquid fuel, as in the automobile, then all of the arguments hold which apply to this form of motor.

66. Disadvantages of the Internal-combustion Principle.— On the other hand, there are certain arguments against the internal-combustion motor as now in use, some of which are inherent, and others of which attach more to some types than to others.

. (1) The Otto cycle (pars. 62 and 69) gives only one working stroke in four piston-traverses. In the two-phase cycle (par. 73) there is one working stroke in two traverses. For a given mean pressure the cylinder volume of the gas-engine will be larger than in the double-acting steam-cylinder at the same speed.

(2) In single-cylinder engines the crank effort is irregular (par. 62); hence a heavy fly-wheel is required for steadiness, or, a number of cylinders correcting the unsteadiness adds weight to the engine.

(3) The motor does not start from rest by the simple motion of a lever or valve. It has to be started by an auxiliary apparatus in which the energy required to start it has been previously stored (par. 164) or which may develop enough energy to cause one working stroke to be made.

(4) This entails a clutch or other transmission mechanism between the motor cylinder and the useful resistance, so that auxiliary starting shall not involve overcoming the total external resistance also.

(5) There is no way of increasing the power of the engine beyond a limit set by the cylinder diameter to meet short demands for a greater effort than the normal, except by running normally at less than the full load.

(6) There is no way in Otto engines to increase the period during which the predetermined maximum pressure may be

exerted on the piston. As soon as the piston begins to make its stroke, the pressure begins to fall off at once

(7) There is no storage of energy except in the living force of the fly-wheel to be drawn upon for such temporary emergency of overload. The mass of hot water in the boiler, or the accumulation of pressure in it, is such a reservoir in the other systems.

(8) The inconvenient heat of the combustion in the cylinder makes it necessary to use some system for cooling the metal of walls and valve-chambers, to prevent distortion and rapid oxidation. In small cylinders this cooling may be done by air; with larger motors where the quantity of heat in question is greater this cooling will be best done by circulation of water. This weight or volume of water and the apparatus to circulate it are an objection.

(9) The water in cooling jackets enveloping the cylinder carries off heat unutilized.

(10) The cooling water lowers the mean pressure.

(11) In spite of cooling water the valves become leaky and require attention and renewals.

(12) It is difficult to secure a low final temperature when the exhaust opens. Hence a considerable pressure exists just when the valve releases the cylinder contents, and the escape of these high-pressure products of combustion into the air and their expansion on their escape causes a disagreeable coughing or barking noise.

(13) The high temperature of the cylinder makes lubrication difficult and uncertain.

(14) If combustion is not complete in the cylinder, the odor of the exhaust is offensive.

(15) If rates of propagation of the flame are not adjusted to the speed, or if explosive charges pass unignited into the exhaust passages or pipes, there may occur explosions of some violence in these passages or pipes, with their attendant noise and alarming shock. In many cities the fire laws compel the exhaust-pipes from gas-engines to be caried in pressure-resisting metal pipes

completely to the free air, and do not permit them to be simply introduced into brick chimney-flues.

(16) Imperfect combustion also results in deposits of lampblack or soot, which clog or cake upon the working parts and are not only defiling but presently stop their working. Ignition apparatus is particularly liable to this trouble.

(17) The high initial pressure in the cylinder due to the ignition produces a jar or vibration.

(18) Governing is not easy, since it must effect a phenomenon which is nearly instantaneous in its duration. When the work of the engine is variable, governing may not be close.

(19) If the compression is defective or badly adjusted, the power of the engine suffers.

(20) If the ignition apparatus is defective or out of commission, the motor stops dead.

(21) If the carburetor is out of adjustment, the motor slows down gradually and stops.

(22) The motor does not usually run in both directions, and reversing therefore requires a train of gear to reverse the application of power. Such gears are apt to be noisy.

(23) The normal motor runs at its maximum efficiency only when running at a fixed speed. To get varying speeds, either the reversing train may be made a variable-speed train (with attendant noise and difficulty in shifting speed when motor and resistance are both moving), or the speed may be varied by making the power of the cylinder vary by throttling the mixture. This latter will usually be attended with loss of efficiency; and when carried to a limit of speed, the motor will cease to operate its cycle and will stop.

All of this last group of disadvantages are the result of the one peculiarity of the internal-combustion motor that it generates the power for each independent stroke at the beginning of that stroke, and has no reservoir of stored energy behind it. Hence anything which attacks the reliable action of the processes which culminate in each single stroke will stop the motor. This has given an unreliab lity or tricky character to many forms of motor when the real difficulty was the crudity of the apparatus which was used to conduct the processes. Hence the importance of the subsequent chapters, treating on governing, ignition, carburation, and manipulation, in which these detailed processes receive fuller treatment.

67. Variations in Cycle.—It will be apparent from a consideration of the table in paragraph 61 that minor variations in cycle will be caused by the effects of the appliances for governing, in so far as these operate to vary the initial and final temperatures of the expanding mixture. Where the exhaust-valve of the engine, for example, opens before the expansion is complete, there will be a drop in pressure, resulting from the free expansion in the air without doing work. This free expansion will be the occasion of two steps in the cooling process as indicated in the table, and this action is the occasion for some of the differentiations in cycles of the several groups. The cycles which involve heating at atmospheric pressure compel the use of an engine of such bulky cylinder volume that it is scarcely necessary to give consideration to them.

CHAPTER V.

68. Introductory.—Referring to the distinction drawn heretofore between the cycle used in any engine and the mechanism which is designed to utilize that cycle, the present chapter presents, briefly, the development of the modern engine using gas. The engine using gas as a source of heat is, historically, older than that using the liquid fuels. The types chosen to present the engines which have utilized the cycles treated in the previous chapter are five in number.

69. The Otto Engine.—The Otto Silent engine of 1876, and as since modified, closely resembles a single-acting steam-engine. The cylinder is somewhat longer than required by the stroke as limited by the crank, in order that between the head of the piston and the head of the cylinder on the dead-centre may be a volume of sufficient extent (C) to contain the mass of compressed gas and air which is required for the working stroke (pars. 152, 202). In the usual forms of the European stationary Otto engine the cylinder is horizontal. In it fits a long trunk piston, B. The trunk construction gives a considerable contact area to guide the piston in its traverse of the cylinder, and the piston itself is of the box construction, so that the working face may be kept at a considerable distance from the pin on which the connecting-rod swings. The cylinder is water-jacketed in order that it may be kept cool enough to avoid deformations and to keep the valves tight and to permit of effective lubrication. (Fig. 31.)

The valves which are needed for the Otto cycle are an inlet-valve for gas and an inlet-valve for air (Figs. 28 and 29), which shall be opened upon the aspirating outgoing stroke of the piston. These valves are so proportioned as to give the proper mixture of gas and air when the valves are opened. On the side of the cylinder opposite the inlet openings in the designs copied after Fig. 31 is the exhaust-valve, which is usually a lifting- or poppet-valve similar to the gas-inlet valve. This exhaust-valve opens on the return stroke after the working stroke, and is to allow the products of combustion to escape freely to the

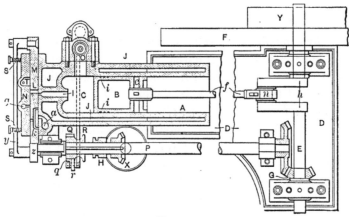

FIG. 31.

outer air. In addition, the mechanism of the engine must provide for the ignition, properly timed, of the mixture which has been compressed in the clearance space behind the piston. In the early form of the Otto engine this ignition was effected by the large slide-valve M operating across the end of the cylinder. This valve was held against the head of the cylinder by a cover-plate N and strong spiral springs. The slide-valve design of Fig. 31 limited the speed or number of revolutions per minute which the engine could make, and the problem of its proper lubrication was always a difficult one. The inlet- and exhaust-valves and this sliding ignition and timing valve were operated from a lateral shaft P at the side of the cylinder, driven by a pair of gears from the main- or crank-shaft. The diameters of these bevelled gears were so adjusted that the valve- or lay- or cam-shaft made one revolution while the main-shaft made two. This made it possible for the valve-shaft to time the actions of its cams so that they would come once in each two revolutions of the main-shaft, as is the requirement of the Otto cycle. The lateral shaft also drives the governing appliance and is a convenient attach-ment for devices to secure mechanical oiling. The feature of the governing of the Otto engine will be referred to in a subsequent chapter, but in brief the governor acted upon the gas-valve to open it more or less, or not to open it at all, while without action upon the air-inlet valve. The effect of this action was to impoverish the mixture when the gas-valve was partly opened, and to admit no combustible whatever when the gas-valve was not open at all.

The subject of ignition also will be treated separately, but in the older standard form the transverse slide-valve, after closing the admission from the gas- and air-passages, presents a cavity in its face which is filled with flame from an exterior flame at T which is burning in the open air. This flaming cavity L in the valve is, by its motion, cut off from connection with the outer air just before it is put into connection with the explosion-port, I, filled with compressed mixture. Through this explosion-port which communicates with the clearance volume, the charge is ignited before the piston has made any considerable move-ment from its dead-centre.

It will be apparent, from this description, that the engine in carrying out the cycle makes one working stroke in two complete revolutions of the fly-wheel shaft, and that each stroke of the piston represents one phase of the cycle. The fly-wheel must, therefore, be massive, in order that during the three auxiliary strokes (Fig. 30) it may have stored up energy from the one working stroke to overcome the resistance external to the engine and to perform the functions of the cycle. It is obvious, further-more, that the engine must be started so as to produce one com-pression and an ignition before it will begin to revolve by its own motor energy. The engine will usually stop, in the case of a single-cylinder motor, with the compression partly begun, but not completed. The advantage of several cylinders operating on one crank-shaft is apparent when this peculiarity of the Otto cycle and the Otto engine is concerned. It is very usual to have a lever whereby a special cam can be thrown into action upon the exhaust-valve, so that in large engines the very great effort to start the engine may be diminished by allowing some of the compression behind the piston to be relieved into the exhaust until the first revolution or two shall have been made.

The more modern forms of the Otto engine use poppet-valves and an electric ignition for the firing of the compressed mixture behind the piston. Of course any of the systems described in the chapter on ignition may be applied to the Otto cycle.

The Clerk engine (Fig. 32) was designed along the lines of the Otto engine and to use its cycle, while securing an impulse from the ignition of a compressed charge at every revolution of the main-shaft. This engine was introduced about 1880. It contains two cylinders, of which one is a charging or displacing cylinder which draws in the combustible charge and transfers it directly or through a receiver to the other, which is the power or motor cylinder. The displacer crank is 90° in advance of the motor crank. The displacer cylinder and receiver volume transfer the charge into the clearance space behind the motor cylinder, where it is compressed by the return stroke of the piston. The exhaust-ports EE' of the motor cylinder are formed in the side of its bore at a point such that as the piston reaches the outer dead-centre it shall have uncovered these ports so as to allow the products of combustion to escape into the exhaust-pipe and so to the open air. The displacer piston sends its charge into the clearance volume at the back end so as to act somewhat to sweep the products of combustion from the preceding stroke before them into the exhaust-openings. Compression resistance in starting can be relieved by a by-pass into the exhaust.

A modification of the Otto and Clerk engine using the same cycle is to close the crank end of the cylinder and so arrange the

valves that the front or crank end of the piston shall discharge
the functions of the displacer piston of the Clerk engine. By
this means, the engine, as in the Clerk engine, gives one impulse

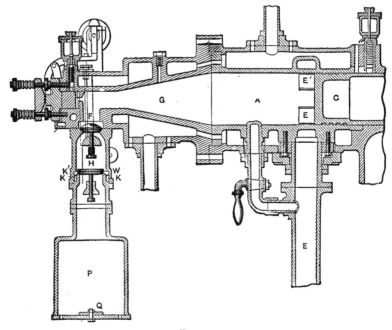

FIG. 32.

or working stroke in every revolution of the crank-shaft. Such
engines are called "two-cycle" engines (par. 73).

70. The Nash Engine.—The Nash engine is an American
design using two or more vertical cylinders each operating upon
the Otto cycle (Fig. 33). By using two cylinders side by side,
there will be two working strokes in two revolutions, which tend
to give a more equal turning movement to the fly-wheel shaft.
The inlet-valves for gas and air are mechanically operated, but
the action of the governor is either to cut off the supply of gas
alone, without cutting off also the corresponding supply of air;
that is, the governing impoverishes the mixture; or to throttle
the mixture of gas and air. By the vertical arrangement

the crank-shaft can be operated in an oil-bath, securing
perfect lubrication of the crank-pins, and a certain amount of

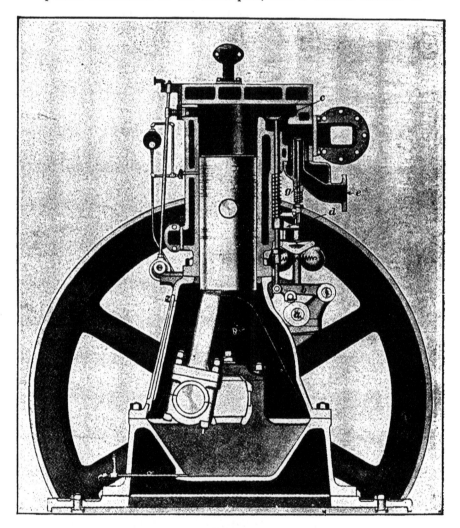

Fig. 33.

spattering of the oil into the trunk end of the cylinder secures a
lubrication at this point. The valves are operated by cams on

FIG. 34.

(*To face page* 177.)

an independent shaft *a* outside of the crank-case, driven by spur-gears at half the speed of the main-shaft. The cam lifts a roll upon a pivoted lever *b* whereby the large poppet admission-valve *c* is opened at each charging stroke. The admission valve-stem carries an arm *d* by which its rise will lift the gas-valve stem *g* and admit gas to the mixing-chamber through the gas-inlet pipe *e*. Air enters around the gas-valve whose stem is *g* in an annular passage, and when *c* is open the mixture passes to the cylinder on the charging stroke. The action of the governor is to throw out a short link attached to *g* so that the rise of the arm *d* will fail to hit the end of this link. The gas-valve does not open in this case, but air only enters through the admission-valve; hence the mixture is not ignited in the following stroke. The exhaust-valve is behind the admission-valve, and is also a cam-operated poppet. The ignition is by an exterior flame or hot tubes in the older and smaller forms and by electricity in the recent and larger ones.

71. The Korting Engine.—In the Korting engine (Fig. 34) an effort is made to secure a proper proportioning of the air and gas mixture by means of mechanical aspiration of each constituent in a separate cylinder (*a* and *b*) whose volume is proportioned to the desired proportions of gas and air. This pair of cylinders may be single- or double-acting and is on a common rod, and the pistons are to act as the displacing pistons in the Clark engine, with a positive proportioning instead of the automatic proportioning of the Clerk design. The motor cylinder in the single-acting type receives an impulse at each revolution, since it is only a compressing and working cylinder and is not compelled to draw in the charge. The governing is effected by diminishing the proportion of time during which the displacing pistons are open to the working cylinder or to the admission of mixture. When open during a full stroke the working cylinder receives the maximum charge. When open less than this the charge is proportionally diminished, while its relative proportions remain unaffected.

72. The Westinghouse Engine.—In the vertical gas-engine designed by the Westinghouse Machine Co., and shown in section in Fig. 25, there are three vertical cylinders, each operating upon the Otto cycle. By the use of three cylinders there will be three impulse or working strokes in two revolutions of the crank-shaft. Hence there is only one less working or impulse stroke than would occur in the single-cylinder double-acting steam-engine. The valves of the poppet type are mechanically operated by cams on driven shafts which are geared to the crank-shaft. Air and gas are admitted in desired proportions to the mixing-chamber *N*, and when the admission-valve *J* is opened by the cam *B* and lever *C* the charge flows into the cylinder on the down stroke. On the upward stroke the charge is compressed, and at the upper dead-centre a second cam on the shaft *B* closes and breaks an electric circuit at the bottom of the spark or igniting-plug *F*. The passage of this spark ignites the charge and this produces the working stroke. The exhaust-valve *E* is lifted by the cam on the shaft *A* by the roller on the lever *G*, and the products of combustion are exhausted through *O*. Through *H* and *K* circulates the cooling water. The construction illustrates the use of removable plugs or bonnets for convenient access to the valves. The governing in this engine is effected by a fly-ball governor at the side of the engine, shown at *B* in Fig. 36, which controls the areas of the ports through which gas and air are admitted to the mixing-chamber. The lever *H* above the chamber turns a cylindrical valve or shell by which the area of the gas-port is adjusted by varying the length of the port uncovered to the passage *G*. The lower lever *H* similarly controls the air-port *D*, so that any desired ratio of gas to air can be fixed upon and will be permanent until readjusted by hand. Pointers over graduated arcs indicate the relative positions. Inside these cylindrical shells the governing valve *A* is adjusted for position by the balls of the governor so that the *width* of the two ports uncovered is made to vary with the speed of the engine, so that a throttling action occurs without affecting the proportions of the mixture. This uniform mixture of varying volume is fired at every working stroke. The amount of energy is determined by the volume which fills the clearance and therefore by the compression of the charge, and not by its constitution or the

Fig. 35.

(To face page 178.)

proportion of combustible in it. When the gas changes in quality or richness, an adjustment of the relative areas of gas and air introduced in the mixing-valve makes the necessary adjustment for such change. A convenient provision on the cam-shaft enables one of the three cylinders to be disconnected from the

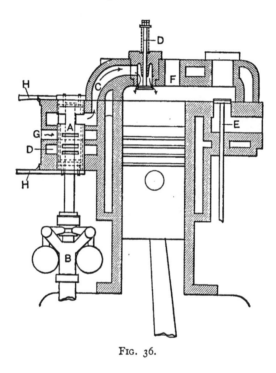

FIG. 36.

inlet service of gas and air, so that this cylinder can be connected with a reservoir of compressed air to operate as an air-engine and turn the engine over until the other two cylinders have begun their normal function of compression and ignition. The sliding of the starting-cam on the shaft can then throw the air-inlets from the air-reservoir out of action and the gas-inlets come into service.

For the larger sizes of engine developing several hundred horse-power the horizontal arrangement of cylinder is preferred (Fig. 37), and in this design the desired smooth and regular action can be secured by using the two cylinders in tandem, making a double-acting engine.

73. The Two-cycle Engine.—The term "two cycle" has been applied to those forms of engine in which, as in the derivatives of the Clerk design, an impulse or working stroke takes place

FIG. 37.

once in each revolution. This is nearly always secured by the expedient of closing the front or crank end of the cylinder or the crank-case, so as to make this end serve to draw in the charge and displace it into the back end, where it is finally compressed and ignited. This construction compels the exhaust-ports to be located as in the Clerk engine, in the side of the cylinder-bore, so as to be uncovered by the piston just before it reaches its outermost position. The pressure of the gases from the front or displacer end will help to carry the exhaust gases out, but care must be taken to prevent the escape of fresh, unused mixture through the exhaust-port by having this latter uncovered too long. Any products of combustion which remain behind the working face of the piston act to heat the incoming mixture and to increase its volume and pressure while diminishing its weight and density, and also serving to dilute the composition of this new mixture. There has been an opinion that the mixture and the products of combustion have a tendency to stratify behind the piston, but the existence of this action is decidedly questionable.

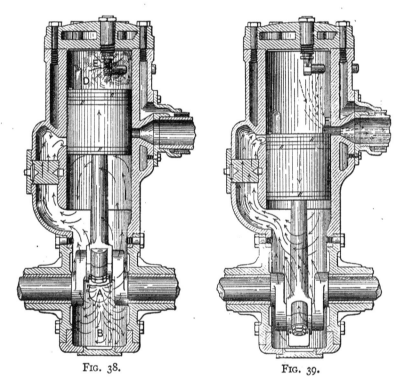

FIG. 38.   FIG. 39.

In a successful form of two-cycle engine (Figs. 38, 39, and 40) the mixture of carbureted air enters from the carburetor (see Chapter X) through the opening *A* into the crank-case on the upward stroke of the piston. The downward or working stroke (Fig. 38) after the ignition of the working charge above the piston slightly compresses this mixture in the case so that it tends to escape through the channel at the left of the cylinder as soon as the descent of the piston shall uncover the upper end of it. The descent of the piston first uncovers the exhaust-port *F* at the right (Fig. 39), and the burnt gases flow out as shown by dotted arrows. The enlarging volume of the exhaust-pipe lowers the velocity of the escaping gases and lessens the noise. The further descent of the piston, now nearing its lower dead-centre, uncovers the inlet-port _C_ (Fig. 40) when the compression in the crank-case is the greatest, whereupon the fresh mixture flows into the volume above the pis-

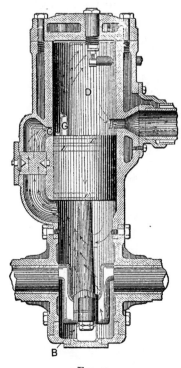

FIG. 40.

FIG. 253

ton, now at its greatest value, filling this with the new charge. The deflector _G_ (shown in more modern form in Fig. 253 and in better detail) throws the charge of fresh mixture away from the open exhaust-port and up against the top head, so that it acts to force out from above downward any burnt gases still left behind in the cylinder, and this action continues until the ascent of the piston closes first the inlet-port _C_ (Fig. 39) and then the exhaust. Compression now ensues above the piston after both ports are closed by the piston (Fig. 38) until the upper dead-centre is reached, when the compressed charge is fired, and the working stroke is made again, repeating the cycle.

Fig. 252 shows a form of well designed engine in which the piston is made to act as a valve on the inlet passage from the carburetor. The piston of this engine (Fig. 253) shows modern aspects of the deflector.

The simplicity of this cycle has attracted many designers and users, since it avoids a cam-shaft and has only a valve on the inlet connection which enters at *A* from the carburetor, if even this. Hence many parts are dispensed with, in some cases running over 100 in a multi-cylinder engine. Its

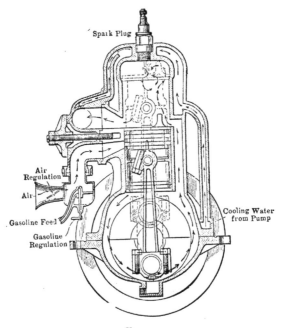

FIG. 252.

principal objections are the difficulty of regulating it closely to speed, and the trouble which arises when under variable resistance it happens that the combustion of the charge has not been complete when the inlet opens (Fig. 40). When the space *D* above the piston has flame in it, and the port *C* is opened, the flame will run back and ignite the mixture stored in the crank-case, and it may not be renewed again sufficiently soon to keep the engine turning against its load. Premature ignitions result sometimes from the fact that the compression is necessarily invariable, and particularly if there are any projections which gradually become red-hot. The deflector G is kept cool enough not to give trouble from this cause by the blowing action upon it of the relatively cool current of fresh charge. Fig. 252 shows the arrangement which results when both inlet and exhaust-ports come in the plane of the rotation of the cranks as in multi-cylinder engines.

74. Comparison of Types.—For a full mathematical and analytical treatment of cycles independent of the mechanisms which utilize them the reader is referred to Chapter XVII. But for the present purpose a summary consideration of the cycles in connection with the engines which utilize them would bring out the following results:

$$\text{Given} \begin{cases} \text{a certain mass of gas,} \\ \text{the same compression,} \\ \text{the same heat-supply after compression,} \end{cases}$$

there will be the same work done and hence the same efficiency in the cycles of

> Carnot,
> Otto,
> Brayton.

For the reasons for this conclusion reference should be had to the full treatment later in this volume (Chapter XVII).

If, further, a comparison be made to ascertain the highest, lowest, and intermediate values, the following table results:

Item.	Lowest.	Intermediate	Highest.
Maximum temperature...................	Carnot	Brayton	Otto
Pressure range........................	Brayton	Carnot	Otto
Volume range.........................	Otto	Brayton	Carnot
Temperature range....................	Carnot	Brayton	Otto
Mean effective pressure..............	Carnot	Brayton	Otto
Mean effective temperature	Carnot	Brayton	Otto
$\dfrac{\text{Pressure range}}{\text{Mean effective pressure}}$	Brayton	Carnot	Otto

The Diesel, being regarded as a modified Carnot, can be brought into this grouping.

It does not necessarily follow that the maxima of theory are the most convenient or practicable in practice. For example, while the Carnot holds the first place so far as maximum temperature is concerned, its impracticability gives the place to Brayton. Since neither pressure range nor large work area per stroke is wanted by itself, but only the ratio between them, Brayton holds the most favorable place, since it is to this ratio that the weight of the engine will be approximately proportional. The volume range should be low, which is the great advantage of the Otto. The mean effective temperature should be low, and Carnot is the only one which exceeds the Brayton in this matter. The low mean effective pressure of the Carnot and all other isothermal combustion cycles puts them out of consideration in comparison with Otto and Brayton.

It is not, however, a matter of indifference as to the means used to get the heat into the working medium. When the air contains varying degrees of moisture, so that the fuel becomes not only carbonic acid upon burning in the air, but there is also a proportion of steam present, what value should be used for the specific heat in such a combustion? (par. 55). In the second place. the chemical change is accompanied by a change in the intrinsic volume (par. 14). It is, furthermore, likely, in the third place, that a fuel may give out more heat when burned in one way than when burned in another.

75. Other Forms of Gas-engine.—In the foregoing enumeration of types of gas-engine motor a certain limited number only have been referred to. A modification from these designs has been much favored in England and is coming into use in automobile practice in this country, in which the two cylinders operating on a common crank-shaft are put on opposite sides of the crank-shaft with the crank revolving between them. Both cylinders may take hold upon a common crank-pin, or they may be connected to separate cranks. When they take hold upon a common pin and lie in the same axial line, the two cylinders can be tied together conveniently so as to take off some of the shock or jar due to the liberation of forces inside the cylinder. This arrangement is known as the opposed or, sometimes, the double-opposed system. The use of four cylinders is coming increasingly into use for automobile practice by reason of the action of this system in causing four impulses in every two revolutions of the shaft. This gives the same quality of turning effort as is given by the steam-engine, provided the ignitions are suitably timed. This four-cylinder system has also the advantage that in nearly all circumstances one of the four cylinders will have in it a compressed mixture ready to be ignited, so that if a spark can be fired in all cylinders at once, with the transmission-gear detached from the motor-shaft, the engine becomes self-starting if the pistons are tight enough to have held the mixture from escape by leakage, and the motor was stopped with the igniting device either out of action or set considerably behind (pars. 140, 164).

There has been proposed by certain of the English designers a plan to have the exhaust-gases which remained in the combustion space swept out or to have the cylinder and combustion-chamber " scavenged " by pure air so that the combustible charge should be a mixture of gas and air without exhaust-gases as diluents. In this system advantage is taken of the oscillations or waves of pressure which occur in the exhaust-pipe, due to the inertia of the discharge from the cylinder. The high-pressure discharge being succeeded by a less pressure, it is possible to make the period of this diminished pressure coincide with the approach of the piston to the end of its exhaust-stroke. If the exhaust-valve is kept open, in communication with this space of diminished pressure and the charge or air-inlet valve is held open, while the exhaust-valve is also open, a charge of pure air comes in through the combustion space and sweeps the burned gases from before it. There have also been designs proposed with unusual arrangements of the mechanisms so that the volumes swept through by the piston should be different for the different phases of the cycle so as to secure a maximum increase of volume in the expansion stroke and lower the terminal pressure when exhaust opens. These designs are due to Mr. Atkinson of England. The advantages offered by the theory of these designs have been more than offset by the inconvenience and complication of the mechanism with which they were carried out. Combinations are also in use with the Clerk arrangement of exhaust-port opened by the piston, with the ordinary cam-driven exhaust-valve. The arrangement of cylinders tandem on a single piston-rod (Fig. 37) forms another modification.

76. The Compound Gas-engine.—The compound gas-engine is also a further modification in which the expansion of the gas, after ignition, keeps on by continuous action through two cylinders of successively increasing volume, instead of being completed in one cylinder only. The purpose of compounding is to diminish the terminal pressure at which the expanded mixture leaves the engine and thus utilize the heat energy of the charge more completely. It also acts to diminish exhaust noise. The difficulties in the way are those which attach to the loss of pressure and heat incident to a free expansion between the two cylinders, which causes a heat loss greater than is regained by the longer range of the expansion. The second cylinder increases the engine friction, and the additional work which it gives out is small compared to the work done in the first one. If the attempt is made to get more out of the second cylinder in the way of crank-pin effort, it becomes of larger volume, with the friction proportionately increased. The second cylinder will always be colder than the first, and the passage of the hot gases into it causes a loss or drop of pressure or volume by this chilling action which is not regained in work. The general opinion concerning the compound engine to date has been that the gain was not worth the sacrifices made to secure it.

CHAPTER VI.

GAS-ENGINES USING KEROSENE OIL.

77. Introductory.—The only difference between an internal-combustion engine using kerosene oil and the gas-engine proper is that the oil-engine requires a device whereby the liquid fuel may be atomized or pulverized so as to be introduced into the mixture in a state of such fine division that the liquid fuel in a condition analogous to a mist shall be distributed all through the mixture of oil and air in such a condition that the propagation of flame shall be instantaneous or practically so, as it is in a mixture of air and gas. The difficulty of governing in the oil-engine is somewhat greater than in the gas-engine, since a drop of liquid oil makes a considerable volume of gas when vaporized If a slight excess of liquid fuel is injected into the mixture above that which the mixture can handle with complete combustion, the liquid fuel is broken up by the heat and is either oxidized or dissociated. If simply oxidized, it burns as a liquid in the cylinder and exhaust-passages, making an unpleasant odor and depositing soot. If dissociated, the volatile elements burn off and leave behind a carbon residue which coats the surfaces and clogs the passages. It is the variation in composition of crude petroleums which makes it practically impossible to use them directly in oil-engines. The volatile parts will vaporize and the less volatile will be deposited forming coatings and clogging cylinders, valves, and passages

78. The Priestman Engine.—The Priestman engine was one of the earliest to use kerosene in liquid form It was a four-phase

188

or Otto cycle engine with electric ignition. A jet of kerosene is forced by air-pressure maintained by a pump in a reservoir into a jet or current of air from that same reservoir. The kerosene meets the air, which attacks it in annular form and atomizes or pulverizes it. The atomized oil in a vehicle of air enters a vessel called a vaporizer, which is kept hot by the exhaust-gases which are in a jacket surrounding the vaporizing-chamber On the out-stroke of the piston the mixture from the vaporizer passes into the cylinder behind the piston with the necessary supply of additional air to make an explosive mixture. The mixture is compressed by the return of the piston, and is fired by the spark from an induction-coil.

To start the engine a hand-pump had to be operated to get pressure to force the oil through the spraying-nozzle, and by means of an external lamp the vaporizer was heated. When the vaporizer was hot, the engine was started in the usual way.

In this engine governing was effected by throttling the oil and air-supply and the effort made to maintain the proportions by weight of oil and volume of air. The compression pressure of the mixture before ignition is, however, steadily reduced as the load is reduced, so that at very light loads the engine would run almost as a non-compression engine. The vaporizer was liable to become flooded with oil, which lowered its temperature on the one hand, and if anything happened to make the vaporizer too hot, the oil would decompose with a deposit of carbon as the result.

79. The Hornsby-Akroyd Engine.—The more successful British form of the kerosene engine is known as the Hornsby-Akroyd. The kerosene is carried in a chamber from which it is drawn by an oil-pump driven from the valve-shaft. This pump sends the oil to a water-jacketed chamber at the side of the cylinder having two outlets. One of these is a by-pass which is operated by the governor permitting the return of excess of oil. When the by-pass is open by the speed of the governor, the entire capacity of the oil-pump is returned into the oil-reservoir.

When it is closed by the slowing of the engine and the increase of the load, the entire capacity of the pump is delivered through the other opening, which is a small needle-hole, into the hot chamber behind the piston at the end of the cylinder, which is the clearance or combustion volume. This chamber is first heated in order to start the engine by a lamp until it is at a good red heat. After the engine is started the heat of compression and of the ignition of the oil keeps the chamber hot enough so that no ignition apparatus is needed. The oil, being injected into the atmosphere of air in the chamber and heated by the compression and the hot walls, becomes a gas, and the compression and the heat of the walls fire it as the compression reaches its maximum at the end of the stroke. If the load on the engine falls off, so that too little oil is delivered to the combustion-chamber, it cools, and little by little the engine will slow down until it will finally stop. The engine is therefore at its best under a constant and adequate load which will keep the pump in normal discharge and the vaporizing-chamber at normal heat. The troubles in this engine are due to a deposit of carbon in the chamber due to dissociation of the oil at high temperatures and the clogging of the needle-hole in the jet either from carbon or from some impurity in the oil.

80. The Secor Kerosene-engine.—The Secor engine (Fig. 41) is an American design in which the external vaporizer is discarded, and in which the liquid oil is drawn in at atmospheric pressure with the necessary air by the aspirating or charging stroke of the piston. The proportion of liquid to air is proportioned by a micrometric adjustment of the inlet-valve controlled by the governor. The lack of precision incident to a forced oil-supply and an inhaled air-current is thus avoided. The fuel and air come together in a mixing-chamber, which is only warmed by conduction from the motor cylinder. Cam-operated poppet-valves are used. The charge is ignited electrically, using the hammer-break system described later in Chapter XI. If too much oil should collect in the mixing-chamber, it will make the

FIG. 41.

(*To face page* 190)

governing sluggish under varying loads, since several revolutions must take place before the governor can effect the passage of oil into the mixing-chamber.

81. The Mietz and Weiss Engine.—The Mietz and Weiss kerosene-engine represents the two-phase type. It requires, therefore, that the crank-chamber should be enclosed in order that a moderate compression of air may be effected on the outgoing

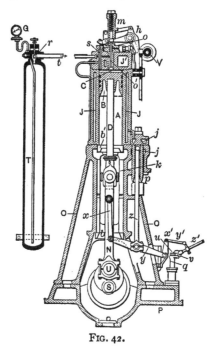

FIG. 42.

or working stroke of the piston. An eccentric on the shaft of the engine operates a small plunger by which the oil is injected into the cylinder. This oil is delivered upon a conical vaporizer which is preheated by a lamp in starting the engine, but which is kept hot by the ignitions after the engine is moving. The air charge is received from the crank-chamber through a port which is opened at the end of the impulse stroke after the larger exhaust-port has been first opened through which the previous charge escapes. A projection or deflector on the piston directs the in-coming charge towards the head of the cylinder and away from the exhaust-ports as in Figs. 38–40. A small valve at the end

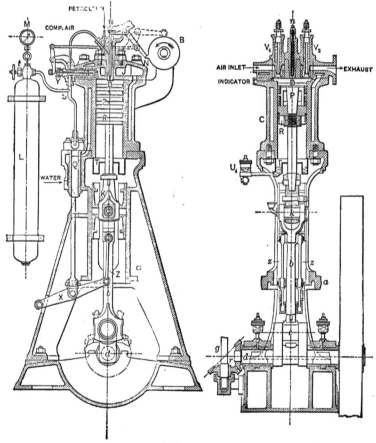

Fig. 43.

of the oil-pump cylinder limits the amount of oil injected, and the governing is done by a push-blade which is lifted to miss contact with the plunger if no charge of oil is desired. This engine is also made into a two-phase gas-engine by a very simple conver-sion involving only the omission of the oil system.

82. The Diesel Engine. The Hirsch Engine.—-The Diesel engine is an effort to secure the realization of the Carnot cycle by having the heating of the gas take place at constant temperature. It was presented in 1897 by Mr. Rudolph Diesel. The

FIG. 44.

kerosene-oil fuel is injected by a pump into the cylinder at the end of an adiabatic compression of the air drawn into the cylinder on the aspirating stroke. Figs. 42 to 44 show sections of the Diesel engine as variously designed, and Fig. 45 is its characteristic indicator-card. To secure a high value for the initial temperature, and a high range of pressure the compression of the

.air is made to approximate 500 pounds per square inch. The oil entering into this heated atmosphere is at once raised above the temperature of ignition, so that it burns with a utilization of the necessary weight of air. In order, however, to secure a further addition of heat to make the addition approximate an isothermal addition, a further quantity of oil may be admitted so as to maintain a constant temperature as the gas is expanded up to the point at which the governor should cut off a supply of fuel. After the cut-off of the fuel-supply an attempt at adiabatic expansion sets in and is continued to the end of the stroke. The cylinder is, therefore, necessarily made longer and of smaller cross-section. The air is delivered into the cylinder against the compressed air within it by pressure from a separate pump. The

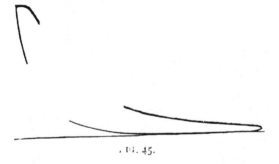

. Fig. 45.

injection of oil is once in each two complete revolutions as in the Otto cycle.

The F. C. Hirsch motor illustrated in Fig. 46 operates on essentially the same principle as the Diesel. The jet of oil enters the hot bulb at the top of the cylinder at the moment when this is filled with hot compressed air from the compression or upward stroke. The heat of the walls and of the compression gives an initial heat and pressure to the air sufficient to raise the oil to ignition-point and still further raise the initial pressure of the working stroke. In the card in Fig. 47 this initial pressure is 180 pounds at 320 revolutions per minute, giving a mean pressure of 74.2. The oil-supply is regulated by a governor, or can be controlled in marine use by a thumb-screw operating a needle-valve. The engine is started by warming the bulb by a Primus lamp for six to ten minutes; after one or two turns of the starting-crank the engine should take care of itself, and the heating-lamps

FIG. 46.

may thereafter be put out. In this motor, as in the Diesel, there can be no pre-ignition, since the compression is on pure air without fuel. The latter is introduced only close to the dead-centre. There can be no condensation of liquid fuel in the cylinder, with consequent irregular action and offensive odor to the exhaust.

83. The Ver Planck - Lucke Kerosene-engine. — Messrs. Wm. E. Ver Planck and Charles E. Lucke have designed and operated a kerosene-engine in which the liquid fuel is heated so that it gives off a vapor from the liquid until the tension in the closed chamber containing the kerosene amounts to about 10

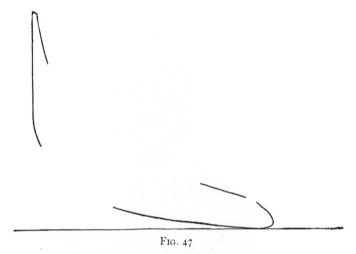

Fig. 47

pounds per square inch. The heating is done at first by an exterior lamp like a plumber's torch, but afterward the pressure is maintained by the heat from the exhaust-gases. The kerosene vapor from this closed reservoir is delivered to a mixing and proportioning valve so located that the vapor cannot condense in transit. The oil vaporized at this valve meets and mixes with a similarly controlled air-current. As cold air is used, fairly high compressions are possible, much higher than with the hot-bulb systems. All vapor that may have partly condensed to form a cloud on the suction is re-evaporated the next instant on the compression when the mixture has the lowest possible temperature for the compression pressure. Ignition is by electric spark.

Should a sudden load through a demand for a large amount of vapor result in a lowered pressure in the boiling-chamber, there would result instantly a great evolution of vapor to meet the supply from the heat in the kerosene which has a temperature due to the higher pressure, and is therefore superheated for any lower pressure. As there is always a mass of liquid oil present in the kerosene boiler, its temperature can never rise enough to cause decomposition and deposit of coke. The boiling and feeding is continuou ; all excess vapor passes into a water-jacketed, liquid-kerosene feed-pipe, where it is condensed and returned for re-evaporation.

84. Comparison of Types.—The introduction of the successful atomizing and vaporizing carburetor has been the most notable step in putting the various types of kerosene-engine upon practically the same footing. It has become obvious that for the satisfactory working of kerosene it must be pulverized or broken up by the atomizing method and vaporized before entering the cylinder, in order to give the best results, particularly where the load varies (see par. 32). Where the carbureted air enters the cylinder in gaseous form under all variations, the efficiency of the various types will approach each other. The superior efficiency of the Diesel type over the others results from the high compression and from the fact that the combination of atomizing and vaporizing takes place at such a high temperature as to make the average temperature of the absorption of heat higher than in the other types, and because it occurs in the presence of an excess of air, so that the combustion of the oil is practically complete in an atmosphere of air supporting combustion. Previous to the introduction of the carburetor, kerosene-engines only worked well under conditions of constant load.

CHAPTER VII.

GAS-ENGINES USING GASOLINE.

AUTOMOBILE ENGINES.

85. Introductory.—When it became necessary to furnish motors of light weight for mechanically dirven bicycles and horseless vehicles, attention was at once directed to the use of air carbureted by liquid gasoline to make a gas to use in the cylinder. For this class of service, where light weight was the prime requisite, the high-speed engine was at once decided on. It was further often of advantage to use both ends of the cylinder, so that the type called the two-cycle type has been much used. Gasoline was preferred to kerosene, since it carburets the required air more rapidly and certainly, without the application of an exterior source of heat. It could be obtained readily in ordinary stores in small villages and towns, and became the accepted source of heat in spite of the elements of danger resulting from its volatile character. It is usually carried in tanks, either under atmospheric pressure or under a very slight air-pressure sufficient only to insure its displacement to the motor as required. The advantages incident to its use have brought it forward for bicycles, automobiles, and launches. The inconvenience of carrying water for the cooling of the cylinders and valves of such engines has introduced what is called the air-cooled motor, in which the motion of the cylinder itself through the air should be depended on to cool the cylinder to the necessary point. Where water is used for cooling, it will usually be a limited weight of it which will be cooled by circulating through a radiator,—the ıadiator being air-cooled for land practice In launch practice the water in which the boat moves can be used for cooling directly or for cooling the radiator. It may be undesirable to circulate salt water containing both acids and mineral salts through the water-jacket, where it may become of a high enough temperature to be vaporized, when it will concentrate the acid and precipitate the mineral matter with n the jackets, from which it is removed with difficulty.

86. The Air-cooled Bicycle Motor.—Fcr operating a motor bicycle or light tricycle the air-cooled system is the only convenient one. In order to keep the weight of the motor and the load on the tires to their lowest terms, the crank-shaft of the motor will revolve at 2500 revolutions per minute, and in order to secure frequent working strokes the engine will be of the two-cycle type, giving 2500 explosions per minute. If the arrangement shown in Fig. 50 be selected, for example, the motor cylinder is attached to the rear member of the frame, so as to bring the crank-shaft bearing at the point where the two members of the frame join. The reduction of speed is effected by a belt transmission to a flat surface forming part of the tire. To take up stretch in the belt caused by dampness and use, a tightener pulley is adjustable by link and nut. Chain transmissions are of course not open to this trouble, but are less silent. The gasoline is carried in the tank hung to the upper member of the frame and delivers by gravity to the carburetor under the tank. The battery and coil for the ignition are under the saddle. The exhaust passes into a muffler under the lower frame, at the front. It will be observed that the cylinder is cast with deep external ribs, so as to expose a large radiating and contact surface to the air as the motor moves through it, and cause an effective air-cooling.

To compel the motor to make the first few strokes necessary to start the machine. the bicycle is fitted with the ordinary pedal equipment with coaster brake, so that by starting the machine as an ordinary bicycle the first few strokes are made by the movement of the machine itself. The presence of the pedals makes it possible to operate the machine by foot-power when desirable. Uusally there is an arrangement whereby the compression on the return of the piston in starting can be relieved until the motor starts its cycle regularly. The ignition of the charge is by electric spark, which can be advanced or retarded or prevented by a lever on the handle-bar or between the knees of the operator,

FIG. 50.

and this is used as a speed-control as well as the throttle-valve, whereby the quantity of fuel admitted to the carburetor may be varied. The bicycle conditions favor the use of the closed crank-case, so that the aspiration stroke draws upon the mixture which is enclosed around the crank under slight tension. A bicycle with a motor of this class will weigh about 125 pounds, and the nominal horse-power of the motor will be about $1\frac{1}{4}$.

87. The Air-cooled Automobile Motor.—The most prominent and successful American air-cooled automobile engine is that used to drive the Knox automobile. There is nothing distinctive, however, except the construction of the cylinder. which, instead of a water-jacket. has a large number of metallic spines screwed radially into the walls. A fan driven by a belt keeps up a good circulation of air through and around these spines, which readily conduct the heat from the cylinder walls outward along their own surfaces, where by the increased surface it is dissipated to the air more rapidly. These spines are made of $\frac{1}{4}''$ round iron threaded their entire length, and are screwed into the walls. If the surfaces of these iron rods were smooth and l inches long and d inches in diameter, the surface for radiation from such a cylinder would be increased by the cylindrical surface of the (n) rods or by πdln square inches. If also by threading the entire length the surface were increased y times. so that the threaded surface$=y\times$plain surface, then the radiating surface by the addition of n threaded spines would be increased $\pi dlny$ square inches

Example.—500 $\frac{1}{4}''$ spines 2″ long in which the threaded surface is 1.4 t.mes the plain surface would add

$$\frac{22}{7}\times\tfrac{1}{4}\times 2\times 500\times 1.4 = 1100 \text{ square inches.}$$

These engines cannot be built in large sizes. however, the limit being about .7 H.P. in one cylinder, by reason of the impossibility of dissipating large quantities of heat by the use of a medium having so low a specific heat as air.

88. The Water-cooled Automobile Motor.—In choosing a representative type of water-cooled gasoline automobile engine it would seem appropriate to select for such type the Daimler engine. It was in 1885 that Mr. Gottlieb Daimler patented his high-speed gasoline engine, and in the same year Carl Benz of Mannheim, Germany, constructed and patented his first gasoline tricycles. The next period of progress brought to the front the French designers Peugeot, Panhard, De Dion, and Mors. The American introduction of these same types is to be credited to Haynes & Apperson of Indiana, and Winton of Cleveland.

A typical Daimler engine would be such a one as is shown in Figs. 51 and 52, having four cylinders arranged in pairs of a cylinder diameter a little less than $3\frac{1}{2}$ and of stroke a little less than 4 inches. The crank shaft revolves at 930 revolutions. It operates on the Otto cycle, with the inlet-valves opening automatically on the suction stroke and the exhaust-valves mechani-

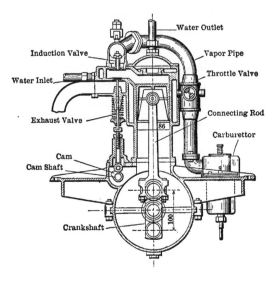

FIG 51

cally driven from a cam-shaft turning at half speed, driven from the motor-shaft by gears. The governing is effected by the outward movement of weights on a horizontal shaft whose centrifugal tendency is counteracted by springs. This governor throttles the inlet of charge into the cylinder by means of a rod, and this same rod can be controlled from the operator's seat for varying speed at will. The cylinders are arranged vertically, which is the

most convenient arrangement where there are ɛs many cylinders
as four in se ies, and the cylinders are water-cooled by the ciren-
lation in jackets of water from a tank by means of a centrifugal
pump driven from the motor-shaft. The ignition ma;y be by a hot
tube, or, in order to secure variations in the point of ignition and
avoid the limits set by the hot tube, electric ignition can also be
used. For convenience of access for inspection and repair, the

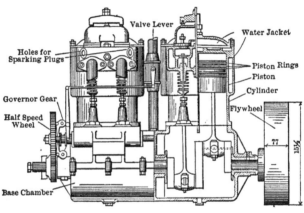

FIG. 52.

ι alve-bonnets are held down by covers, seating on ground joints,
so that by loosening a bolt which holds down a dog bearing upon
these detachable covers, the latter can be removed and the valves
inspected. The details of transmitting the motion of the motor-
shaft to the propelling wheels of the vehicle are aside from the
present purpose.

89. Variations in the Automobile Motor.—The variations
from the foregoing typical form which are to be met in the successful forms of the present day cover a wide range. These variations are often of commercial origin in the matter of affecting
the price at which the motor and its vehicle can be sold, as well
as the results of diffe ence in selection of type. The engine may
have three vertical cylinders or two. The motor cylinders may
be arranged horizontally instead ɾf vertically. They may be

placed on opposite sides of the shaft, if arranged horizontally, or on the same side. They may be of the two-phase or the four-phase system.

They may d ffer in their methods of ignition, although the electric system in one or the other of its forms (par. 128). by reason of the c nvenience in varying the time of ignition has practically displaced all other forms. The control of speed by the hit-or-miss method of governing has been superseded by one of the other forms of control, and the throttle system controlling both air and gasoline is the one which is in most frequent use. This control of the motor action by throttle and by spark gives a wide range of power and of speed in th motor tself without calling for readjustment in the transmission machinery between the motor and the propelling wheels of the vehicle. This double control is usually attached to the steering column of the car. or it may be operated by the feet. Probably 1200 revolutions per minute may be considered as the normal speed of motors of this class. Their horse-power ranges from six in the single-cylinder designs to thirty-five and forty, and even sixty or eighty horse-power in machines intended exclusively for racing upon prepared tracks. The condition in motor trucks introduces no considerable differences in the motor design. but mainly in the gearing whereby the speed of the motor is reduced to that of the propelling wheels with a corresponding gain in leverage In trucks for heavy loads the motor requires to have a considerable torque in order to start the vehicle from rest on grades or on a rough and resistant road-way.

The internal-combustion engine is year by year finding a widespread application to the demands of heavy motor trucks.

90. The Launch Engine.—The conditions in the launch engine resemble those in the motor vehicle except that the load for resistance is not so likely to vary within the same range of limits. The variation in resistance will be a variation in speed. For this reason the two-cycle design has been a favorite for launch practice, and in the Lozier type selected, this feature is

embodied in the cycle illustrated in paragraph 73, Figs. 38, 39, and 40. The speed and power are varied by throttling the mixture. The mixed charge slightly compressed in the crank-case passes into the working cylinder at the end of the working stroke and is there compressed   In common with all engines of this design, if the mixture is impoverished to a point at which its combustion is retarded and is not completed by the time the exhaust opens, and the inlet-val e immediately thereafter, it is easily possible for a flame to pass through the inlet-valve and ignite the mixture in the crank-case with a somewhat disconcerting report, and of course a stoppage of the motor until the crank-case shall be filled again with a fresh unburned mixture,   The regularity of the resistance in launch practice is favorable also to the employment of kerosene-engines.   An increasing development of recent years has been the introduction of auxiliary gasoline-motors into sailing yachts and catboats.   Under ordinary circumstances the sail power is greatly in excess of the power of the auxiliary engine, but the latter can be used when the wind has failed or as a means of manœuvring the boat in starting and in landing without reference to the sail-power   The requirement of such an engine, so far as its propelling features are concerned, is that when the screw is not turning it should not oppose an undue resistance to the motion of the boat.   For this reason variable-pitch screws or blades whose pitch can be reversed are quite usual   By reversing the pitch the engine can be made to back the boat without changing the direction of the rotation of the motor-shaft.

91 **Converted Gas-engines.**—It will be apparent that any of the gas-engines discussed in Chapter V can be made into gasoline engines when they are to be used in places where gas is not natural or easily manufactured by a very simple conversion.   All that needs to be done is to introduce a carbureting device of acceptable form (Chapter X). so that the suction stroke shall draw in carbureted air instead of distinct and separate supplies of gas and air through two inlet-openings.   It may possibly

happen, where the compression volume relatively to the piston displacement in the design of the engine was small, that the mixture of carbureted air rich in gasoline will give trouble from pre-igniting, due to the richness of the mixture. This can be corrected, of course, by impoverishing the mixture, but the engine will not be as efficient as it would be in running upon a fuel for which the compression volume was more correctly designed (pars. 152, 202).

CHAPTER VIII.

ALCOHOL-ENGINES.

92. Introductory. — It is possible to carburet the charge of air entering the motor cylinder by the use of alcohol, as well as by either kerosene or gasoline. The greatest development of the alcohol-motor has been in Germany and France, rather than in America, for reasons of a purely economic rather than technical sort. The high revenue tax which has prevailed hitherto upon alcohol as a feature of American practice has made it less attractive than the petroleum derivatives on which there is no such tax. In France and Germany there has been a governmental encouragement towards the production of alcohol by distillation which has been lacking in America. It is not usual to use the ethyl alcohol or spirits of wine, but more usual to use the methyl or wood alcohol in spite of its pungent and disagreeable odor. It is, again, often impossible to get alcohol free from water, and in automobile practice it is usual to render the alcohol non-potable by introducing some form of hydrocarbon to a degree which makes it unpalatable. In France, on the other hand, the ethyl alcohol, from the prevalence of the vineyard, is more used than the wood-alcohol. Recapitulating, the statements in paragraph 34, it should be noted that a very usual engine mixture in use quite extensively is as follows:

Ethyl 90 per cent.................... 100   vols.
Methyl 90 per cent.................. 10   "
Hydrocarbon..................... 0.50 "
                                  _____
                                  110.50 "

What is designated as denatured alcohol, sometimes called by the trade name of electrine, takes the above mixture of alcohols and adds an equal volume of benzol. This mixture has a specific gravity of 0.835, as compared with water, and has a calorific power of 13,150 B.T.U. The hydrocarbon referred to in the alcohol mixture is usually not defined beyond that it should have a boiling-point between 350° and 440° Fahrenheit.

The reader is referred to parag. 34 for the general conclusions of recent researches on the use of alcohol for power. The important differences are in the use of a greater compression and in the carburetor. A carburetor to handle alcohol requires to be hot and to be kept at a higher temperature than works satisfactorily with the more volatile gasoline. A very usual method is to start the motor with gasoline, using a gasoline carburetor, which will usually perform its functions while the engine is cold. After the engine has become well heated the gasoline is shut off and the other part of the carburetor is turned upon alcohol. Fig. 69 illustrates the Marienfelde form of duplex carburetor for gasoline and alcohol.

93. Alcohol-automobile Motor. The Gobron-Brillié. Since the use of alcohol as fuel has not been extensive as yet in America, the student has to turn to French sources for an example of an alcohol-motor. The type selected might be any of the gasoline types, fitted with such form of carburetor (Chapter X) as should be adapted to work with alcohol. The engine part would require no modification. For a type specially designed for the use of alcohol probably one of the most interesting is the Gobron-Brillié, illustrated in Fig. 48.

This engine is fitted with two pistons in one cylinder, one through a direct and the other through a back-acting connecting-rod acting on the same crank-shaft. The two cranks are set at 180°. The arrangement selected is for a two-cylinder engine carrying, therefore, four pistons; the two lower pistons act on a single crank. The carburetor (Chapter X) is designed especially to keep the mixture constant in proportion by mechanical means. A spindle is rotated by the machine through a small angle by a ratchet. Corresponding to each ratchet tooth is a small bucket on a wheel inside the carburetor casing. These buckets pick up the alcohol from the bottom of the feed-chamber and deposit the measured amount in the aspiration-pipe where it meets the inrushing stream of heated air. Of course this measured amount must be just so much as can be carried in one bucket, and every stroke of the engine causes just exactly the same amount of alcohol

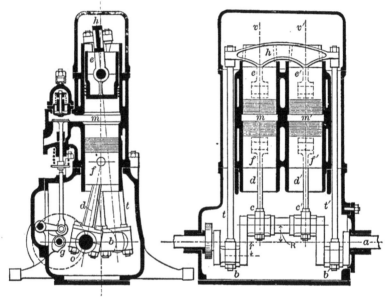

FIG. 48.

to be fed, i.e., the contents of one bucket. Ignition is electric, and governing effected by spark variation (Chapters X to XII). The double-piston mechanism offers some interesting peculiarities and advantages as respects balancing on opposite sides of the crankshaft, and permits a construction whereby variable volumes of clearance and compression become possible.

94. Alcohol-launch Engine.—It has long been desired to use alcohol instead of naphtha or gasoline in pleasure-launches, on account of the avoidance of the odor and the unquestionably greater safety of alcohol as a fuel. Any automobile motor can be applied to launch uses, as the problem of propulsion on the water is usually less complicated than on the land, by reason of the practically uniform resistance offered by water. The alcohol-launches in America most usually have condensing engines on account of the cost of alcohol, and when the apparatus must operate condensing it is usually more convenient to use the alcohol as a heat medium, and by the usual steam-engine cycle with the heat applied from without or externally, and abandon the internal-combustion principle. In the alco-vapor launches, for example, alcohol is not burned, but the heat is furnished by burning kerosene under a retort in which the alcohol is vaporized, and its vapor tension drives the piston. In the ordinary naphtha-launches part of the naphtha is used in a burner to vaporize and give tension to that part which drives the engine, operating in a closed circuit or cycle.

CHAPTER IX.

PROPORTIONING OF MIXTURES.

95. Introductory. — Since the internal-combustion engine operates by the oxidation of the fuel in the cylinder, it is of vital importance that the proportion of fuel and air should be properly adjusted to each other and to the work to be done. As the resistance on the crank-shaft may increase or decrease the supply of fuel should increase or decrease, and the system of governing should be adjusted so as to keep this mixture and the proportioning of it at the point of highest efficiency. In liquid fuel this matter is of special consequence, since a drop of the liquid makes a considerable volume of gas when vaporized, and a very small variation in the supply of liquid will make a wide variation in the supply of gas. It will be seen in the discussion on governing that the proportioning of the mixture requires as careful attention as the amount.

In the engine which draws its supply of fuel from a gas-main, and particularly from one which has been divided and has ramified through a building or a plant, it may easily happen that the pressure in the pipe supplying the engine will vary from day to day or from hour to hour. In engines supplied through carburetors the speed of the engine may easily produce a considerable difference in the flow of fuel and the proportions of the mixture due to inertia of inlet-valves or any circumstance which causes them to open sluggishly or reluctantly for the inspiration stroke.

Where the engine receives gas from a house-pipe which supplies illuminating fixtures, a further difficulty is to be guarded against in the fluctuation of these lights when the engine makes its draught upon the volume of gas in the pipe. Such fluctuation is not only bad for the working of the engine, but the flickering of the lights is disagreeable and must be prevented. The manner of doing this in most frequent practice is the introduction of a chamber of variable volume close to the engine on the pipe which supplies the gas to it. This variable chamber is most frequently a bag of flexible rubber, which fills during the three strokes during which no gas is withdrawn and collapses partially when the inspiration occurs. If a collapsible rubber bag is inconvenient, somewhat the same effect can be produced by an enlargement of the pipe so as to form a volume or storage space in which the elasticity of the gas itself shall act somewhat as the flexible rubber of the bag. These enlargements are best arranged so that the gas in flowing into them comes in at one end and passes out at the end opposite with some distance between the inlet and outlet. The enlarged cross-section at once reduces the linear velocity of the gas at that end which is towards the lines to be affected, and the withdrawal by the engine from the other end does not produce a perceptible pulsation, where the waves of such pulsation are broken by so considerable a change of cross-section.

96. Automatic Mixing by Suction.—The most frequent method of securing the desired proportions of gas and air in the gas-engine is by means of a separate valve for air and for gas (See Figs. 28, 29), the areas of whose openings are adjusted to the desired proportion. When the pressure on the working side of the piston is reduced by the outgoing stroke, these two valves will lift automatically by the excess of pressure without, and through each of them will stream the proper volume by reason of the difference of pressure below them and above them. It will be apparent, however, that if either pressure varies (it is usually the gas-pressure which varies), the mixture will not be that for which their areas have been adjusted and a different proportion of mixture will be the result. There will be, usually, a cock in the gas-supply pipe which is supposed to be wide open when the engine is at work. It is obvious that if it is partly closed a very considerable variation in the proportions of the mixture will prevail, constantly, as long as that condition lasts. If the gas changes in quality or richness, a desirable change in mixture can only be reached by adjustment of that gas-cock, if the areas of the valves are themselves unalterable. The air-valve, as a rule, is not capable of adjustment. It opens to the outer air or to the air of the room in which the engine operates, and there is no way in which it can be increased. It may be diminished if its lift is controlled by a spring or by a stop, but, in general, the adjustment of the air-valve is not practised in the automatic system. The inlet of air is often an occasion for noise in the gas-engine, since the cross-section of the opening will of necessity be usually much less than that of the piston, so that the velocity of the air through the opening is many times greater than the linear velocity of the piston on its inspiration stroke. In some forms of engine a muffler is introduced on the inlet so as to quiet the sound of the air through a constricted opening.

The system of mixing by automatic action is the cheapest to construct. In its simple form it attaches itself to governing by the hit-or-miss system (par. 136). It has the objections at high speeds which apply to any automatically operated valve resulting from the delay in opening to admit the incoming charge when these valves have any weight or mass or must overcome the action of the spring which holds them shut. This system is particularly unreliable when there is no pressure in the tank or vessel containing the fuel. As the speeds increase the mixture drawn in becomes less and less rich, due to the inertia of the fuel as well as the inertia of the valve.

97. Proportioning by Adjustable Valves. — A better system of proportioning the mixture makes the inlet area both of fuel and of air controllable. This system appears in both the Nash and Westinghouse engines (Figs. 33. 36). In the Nash engine the lift of the valve is controlled both for gas and for air to a proper adjustment of proportions as revealed by the indicator-card, and the governor of the hit-or-miss type never varies the proportions after they are once adjusted. In the Westinghouse engine the proportions, when once adjusted, are not varied, but the governor acting on a second valve draws in more or less volume of the uniform mixture as the demands of the resistance may require (Fig 36). The proportioning valve is adjusted experimentally with the indicator for the best effect with the fuel and pressure prevailing at the engine. If any change takes place in either quality or pressure of the fuel, the adjustment of the proportioning valve must be altered accordingly. When once correct, however, the adjustment need not be changed. In the engines which draw their supply of air from out of doors, as in automobile practice the proportion of air and fuel is more likely to vary widely. There is a greater weight belonging to a given volume in cold weather when the air is dry, while in warm weather and when the air is full of moisture there is a corresponding difference in the amount of oxygen in a given volume, which will mar the proper working of the engine as the result of variations in the character of the mixture. This system leaves much to be desired. The logical outcome of this system is governing by throttling the mixture and not by varying or impoverishing it.

98. **Proportioning by Mechanically Operated Valves.**—By reason of the difficulty referred to above from the inertia or sluggishness in action of automatic valves the tendency in high-speed practice in the automobile has been distinctly towards mechanically operated inlet-valves which shall open positively at a definite point of the stroke by means of cams driven from the half-time shaft of the motor (see Fig. 49). With mechanically operated valves governing must be done by throttling the amount of mixture which reaches the valve. The objection to the single mechanically operated inlet-valve which has been urged is that it admits a charge of fuel to the cylinder at every stroke whether the engine requires it or not. This is avoided in some engines by the use of two inlet-valves, one for gas alone,

and the other for the mixture (see Fig. 33). The effect, however, of overcoming the resistance of the valves by mechanical means in enabling the cylinder to receive its full charge of fuel at high speeds overbalances the objection from this point of view. It may be serviceable to show quantitatively how considerable the suction-throttling may easily become. At 32° F.

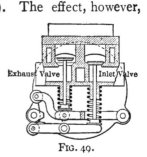

Exhaust Valve   Inlet Valve

FIG. 49.

and at one atmosphere pressure the volume of a pound of air is 12.387 cubic feet and it weighs .0808 pound per cubic foot. At any other pressure p_1 its volume v_1 will be

$$v_1 = \frac{p_0 v_0}{p_1} = \frac{12.387 \times 14.7}{p_1},$$

and the weight per cubic foot the reciprocal of this, or

$$w_1 = \frac{p_1}{12.387 \times 14.7} = .0055 p_1 \text{ pound.}$$

If it be assumed that by the sluggishness of the valves and air due to inertia and friction the pressure of the aspiration stroke is only 10 pounds absolute instead of 14.7, then the weight w_1 will be

$$w_1 = .0055 \times 10 = .055 \text{ pound.}$$

But at atmospheric pressure there should have entered .0808 pound, hence .0808 − .055 = .0253 pound less went in to fill the volume generated, or a loss of $\frac{.0253}{.0808} = 31$ per cent as compared with that weight of mixture that would have gone in if the speed had been lower, the ports large enough, and the valves effectively opened (see par. 178).

99. Proportioning by Volumes of Pump Cylinders.—A system of proportioning the charge by means of separate pumps for air and fuel secures a positive and continuous adjustment of their relative proportions when the volume displaced by the piston of each pump is once fixed. This system is independent of variations of pressure in the gas-mains and to a great extent independent of barometric pressure and atmospheric temperature. This principle is carried out in the Korting engine and is discussed under the method of governing that engine (par. 71).

100. Proportioning by Control of the Carburetor.—The design of carburetor can be made to determine the proportions of the mixture. This can be done either by varying the proportion of pure air which meets the carbureted air in order to furnish the necessary amount of oxygen for complete combustion, or the entire amount of air may pass through the carbureting appliance and the amount of fuel be regulated by restricting its flow (see Chapter X). It is more convenient to have the proportioning done by the former process, since under these circumstances the only necessary adjustment will be for variation in the quality of the fuel due to changes in its calorific value and by changes in the barometric or hygrometric state of the air. When a proportion for constant conditions has once been established, the governing appliance will take care of the quantity of mixture, while the adjustment of the carburetor takes care of the quality. It is inconvenient to make the carburetor vary both quality and quantity and to saddle the combined functions upon the governing apparatus. As will be seen in the treatment of carburetors in the next chapter, it is quite easy to adjust the proportion of liquid fuel when the suction effect is practically constant so as to make the desired mixture.

101. Effect of Scavenging.—It has already been foreshadowed in a previous paragraph (No. 75) that methods have been designed to cleanse the cylinder from burnt products of combustion by providing for a scavenging effect by means of pure air. The effect of this scavenging stroke is to relieve the cylinder and the combustion-chamber of gases which are not supporters of combustion, so that when the fresh mixture came in it should not become diluted by being mixed with exhaust-gases which were not combustible.

It is apparent, therefore, that any device in the design of the engine which shall produce an effect the reverse of scavenging will produce a material variation in the composition of the mixture whose ignition performs the working stroke. Those methods of governing which preclose the exhaust-valve before the stroke is ended leave a residue of such incombustible gases in the cylinder. This residue not only prevents the inlet of the same quantity of new mixture on the inspiration stroke as would enter if the combustion-chamber had been completely emptied, but by their presence in the mixture they retard the rapidity with which the combustion takes place in the fresh mixture; their heat lowers the density of the charge and therefore diminishes the intensity of the initial pressure and the average or mean pressure throughout the working stroke.

It will be apparent, therefore, that the effect of these gases when the governor uses them as a means of controlling speed is twofold. They act to diminish the quantity of combustible and to modify its normal behavior after ignition. It is the latter effect which is the element of uncertainty in the proportioning of the mixture with this system.

102. Effect of Variations in the Mixture.—The principal effect of variations in composition of the explosive mixture is upon the rapidity with which the flame propagates itself throughout the combustion-chamber. Experiment has shown that there is a proportion at which the pressure at the beginning of the stroke caused by the inflammation of the mixture rises most rapidly and produces the greatest effect (see Chapter XIX). To impoverish this mixture by diminishing the proportion of fuel in it retards the ignition process, diminishes the initial pressure, lowers the average or mean pressure through the forward stroke, and may, perhaps, be carried to a point at which ignition will not occur at all. At or near the limit of such impoverishment it will be apparent that variations in the amount of the compression on the return stroke will vary the readiness of the mixture to ignite. Where the impoverished mixture is also throttled, it may result that on the return or compression stroke the compression may not reach a point at which that particular mixture would ignite at all, whereas a richer mixture or a higher compression would both of them favor such ignition and cause it to take place. An impoverished mixture, furthermore, and particularly one which is diluted with products of combustion, may ignite slowly enough so that it is not completely burned at the time when the exhaust-valves should open at the beginning of the return stroke. This state of affairs is particularly annoying with the two-cycle type of engine, since the incoming charge of fresh combustible mixture is expected to follow the discharge of the products of combustion. If the latter are flaming after the exhaust-valve is opened, they will ignite the incoming mixture, and usually that ignition will run back into the crank-case or other end of the cylinder, setting fire to the charges of mixture in that space, which will result, of course, in the stoppage of the engine. Retarded ignition which continues into the exhaust-pipe will obviously make the latter excessively hot.

It is apparent, furthermore, that indeterminate variations of the mixture render it difficult or impossible to regulate the engine closely to a predetermined speed. If the mechanical appliances for regulation, acting according to law, produce their effects upon a mixture which is not determined by law, an uncertainty in regulation is at once unavoidable. If the mixture is varied, it should be varied, in a determinate way.

103. Effect of Speed Variations in Varying the Mixture.— With engines working upon gaseous fuel, the effects of variation in speed, and particularly the effects of high speed, are not apparent in producing wide variations in the proportions of their mixture. In the engines which carburate the air by a liquid fuel, the effect of speed variations and high speed in varying the mixture will usually be considerable. If the liquid fuel is inspirated into a current of air, the pressure which causes that inspiration of liquid will be greater when the speed of such inspiration is higher. There will result, therefore, that the mixture will be richer from the action of this cause when the speed is high than when the speed is low.

On the other hand, if the inlet-valves of the engine are operated automatically by differences of pressure inside the cylinder on the inspiration stroke, as compared with the pressure of the external air, the inertia of these valves, and of the column of air in the pipe and of the liquid to be inspirated into the column of air have to be overcome by that difference of pressure; and if the time of the inspiration stroke varies as the speed of the engine varies, it will be obvious that carburated air will flow into the cylinder through a less proportion of the inspiration stroke at high speed than when the period of that stroke is longer. If, on the other hand, the valves are mechanically operated, the inertia of the valves is eliminated from the problem, but only the inertia of the air and liquid remain. For these reasons the mechanically operated inlet-valves cause a less wide variation in the mixture than is certain to occur with the automatic system of inlet-valves.

It may easily occur with the automatic system that a speed should be reached at which the inertia of the flow of mixture, together with friction in pipes, bends, and valves, may result in a relatively small proportion of mixture reaching the cylinder (par. 98). On the return of the piston the compression will be less, and the less volume of fuel will make the next working stroke a weaker one as the result of both effects. Barometric pressure of the external air will obviously influence the response of the air in the pipes to the differences of pressure inside the cylinder and out. If the air is cool and dry, it weighs more to the cubic foot than when the air is warm and moist. These causes produce as their effect the curious phenomenon of a diminishing horse-power in the motor with increase of its speed of revolution. This is a condition which is practically unknown in engines of the constant-pressure class, such as the steam-engine receiving energy in the form of a **gas** under pressure from a reservoir. As in the previous case, this variation of speed may be both in quantity and in quality, with carburated mixtures, since the inertia of the liquid will be different from that of the air, and the effects of speed on such inertia will be different.

If, from any circumstance, the mixture becomes too rich in fuel, the combustion will be probably incomplete within the cylinder, and the exhaust will have an offensive odor from partly burned and partly carbonized fuel. This state of affairs will reveal itself also by the presence of visible vapor resembling smoke in the otherwise colorless exhaust-gases.

Obviously, also, defective proportioning of this sort consumes an unnecessary or wasteful amount of fuel.

CHAPTER X.

CARBURATION AND CARBURETORS.

105. Introductory.—In cities and elsewhere the stationary internal-combustion engine may receive its supply of hydrocarbon for use as fuel in the form of gas from a central generating station. This gas distributed through mains and pipes is ready for use as received. When the plant using gas-engines is a large one the necessary gas supply can be more cheaply supplied from an independent producer (pars. 24–28).

In small isolated plants, such as the automobile and the launch for marine purposes, it is convenient to make use of the hydrocarbon in liquid form. It can be carried conveniently in tanks and supplied to the engine as required, and is consumed in the form in which it is bought and sold in the market.

The use of the liquid hydrocarbon, however, will necessitate an apparatus whereby the gas can be manufactured from the liquid fuel as required by the engine. The most convenient form of gas for engines, perhaps, will be that which is made by carburating atmospheric air as described in paragraph 22. One of the great steps in the development of the modern internal-combustion engine has been the design of satisfactory apparatus to carburate air just before it enters into the combustion-chamber. The idea of carburation is not a new one, but the improvement in the forms which have been produced for the purpose has drawn a distinct line between the early and the more modern forms. In fact it is not too much to say that the successful working of the automobile engine and of all other engines of the same class is principally dependent upon the certainty, reliability, and satisfactory working of the carburating device.

The carburating apparatus will serve to saturate atmospheric air with any liquid hydrocarbon. There will, therefore, be carburetors for gasoline, for kerosene and for alcohol, divided only as required by the varying characteristics of the liquid. In general the process of carburation is to saturate the atmospheric air with the liquid fuel in a finely divided or atomized state like a mist. This general principle of atomization has long been used in medicine and surgery and is familiar in the form of the apparatus used in spraying perfumes. The air saturated with a mist of hydrocarbon will subsequently undergo a further mixture with an additional supply of air such as may be required for its full and complete combustion in the working cylinder. With the less volatile hydrocarbons the process of carburating the air cannot be satisfactorily carried on at the ordinary temperatures of the external air. The carburetor for such liquids will have both the principle of atomization and the subsequent vaporization by heat. When the engine is working, the vaporization can be effected by waste heat from the hot exhaust-gas. In starting the motor, however, when all is cold, the vaporization requires an outside source of heat in lamp or torch or otherwise.

The first principle in carburation, historically, is the evaporation of the volatile hydrocarbon at atmospheric temperatures, from the surface of its own liquid. Such carburation may be called surface carburation, and the evaporation may then be from the cool surface, or the volatility of the liquid may be increased by heating. This system requires that a current of air to be carburated moves over the surface of the liquid.

The second system may be called the principle of mechanical ebullition. The current of air to be saturated is made to pass through the liquid mass, so that it bubbles up through the liquid and escapes at the surface. By this bubbling the liquid is mechanically agitated and a certain proportion of it is entrained with the air in a finely divided state or mist.

The third principle is that of the spray carburetor. These are true atomizers in which the jet of liquid fuel is thrown up into the current of moving air by the fact that the air on its way to the cylinder on the aspirating stroke of the engine has a pressure less than atmosphere. A small orifice or nozzle opening into the suction-pipe delivers the liquid fuel into that moving current, and by the mechanical action of this current the mist or cloud of liquid particles is distributed through the moving current which it saturates.

It will be seen in the later treatment that the form of the apparatus utilizing this third principle for the less volatile hydrocarbons will require that the spray be made into a gas by heat. With gasoline, as a rule, it is not necessary to vaporize the mist. The first two principles are practically out of competition with the third, which is the modern form.

106. The Surface Carburetor. The De Dion Motor-cycle Type.—One of the earliest forms of the surface carbureter was

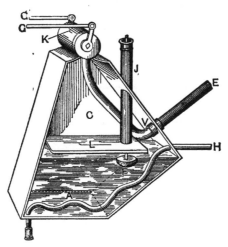

Fig. 53.

brought out for the early motor cycles and is illustrated in Figs. 53 and 54. The liquid gasoline is poured into the containing vessel

and lies in the lower part. The current of external air is drawn
down through the inlet *J* so that underneath the plate *L* it spreads
itself over the surface of the gasoline and picks up the vapor
which rises to the surface. The level of the plate *L* can be adjusted as the level of the liquid varies. It acts both as a spatter-
plate and to discharge the air in an even volume over the surface.
The volatility of the gasoline may be increased by passing hot
exhaust-gas through the tube *A* and out into the exhaust-pipe
at *H*. The carburated air rises at the top of the carburetor
through the opening *B* into the chamber *K*, which is known as

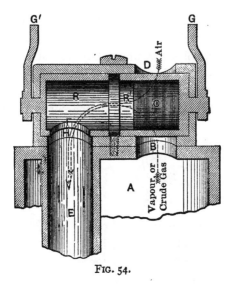

FIG. 54.

the twin tap from its construction as shown in Fig. 54. The
carburated air from the carburetor meets an additional supply
of air from outside through *D*, which is protected by a wire
cage and can be controlled in area by means of the lever *G*.
This control can be made to vary the proportion of fuel and
air which passes through the passage *R* into the pipe *E*, which
delivers the mixture in explosive proportions to the engine cylinder. The lever *G'* is, therefore, a throttle lever varying the

amount of mixture delivered to the cylinder, while the lever **G**
varies the proportions of air and fuel in the mixture. That is,
G regulates the quality and *G'* regulates the quantity.

107. Wick or Flannel Carburetors.—Belonging to this same
type and form, the second class are the carburetors which are
known as the felt or flannel type, of which Fig. 55 will serve as

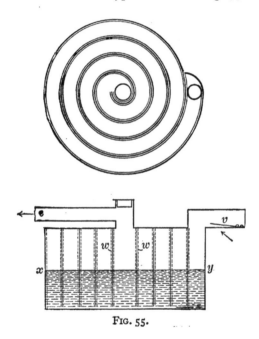

FIG. 55.

an illustration. The air is drawn in by suction of the engine
stroke through the pipe *e* and enters at *v*, so that a pressure less
than the atmosphere is created in the carburating-chamber. The
carburetor is made of thin metal and is divided into a spiral by a
thin metal coil which is fast to the top of the carbureter, but does
not extend all the way to the bottom. It will be plain, therefore,
that when the carbureter is filled half full of gasoline, as to the
level of the line *xy*, for example, the atmospheric air which enters
through the opening of the valve *v* will be compelled by the spiral
to pass around the coil in order to reach the central outlet and

be discharged through the pipe *e*. The surface of the spiral coil is covered with felt or flannel loosely stretched on the thin metal by basting it through holes made for the purpose. This felt or flannel reaching down into the gasoline draws up the liquid by capillary action, and the passage of the air-current over the wick surface evaporates off the liquid and thus saturates the air.

In surface or wick carburetors of this type experience shows that the best results are secured when about four inches of liquid lie in the bottom of a carburetor about eight inches deep.

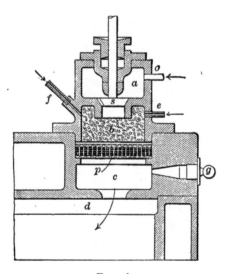

Fig. 56.

In another form of wick or surface type (the Brayton, Fig. 56) the space *b* is filled with sponge or felt or some material of similar sort. The liquid hydrocarbon enters at the top through the pipe *e*, while a jet of air is forced through the pipe *f*, serving to atomize or spray the liquid. The additional air necessary for the complete combustion enters through the pipe *o* and passes through the porous bed *b*, when the valve *s* is opened. The cut shows this form of carburetor applied directly to the engine cyl·

inder. The opening closed by the plug *g* is provided to receive
a taper to effect the ignition in starting the engine.*

The objection to the flannel or wick carburetor for out-of-door
use has been the gradual fouling of the fibres of the wick with
dust, so that on becoming clogged they would no longer serve
as an evaporating surface. The objections to the De Dion form
of evaporation directly from the surface of the mass were that
the process of vaporization requires a certain amount of heat

FIG. 57.

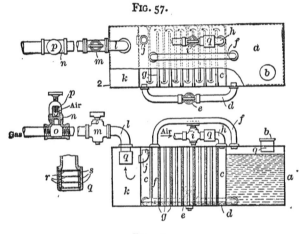

FIG. 58.

which was naturally absent from the liquid mass. It gradually,
therefore, became chilled and frozen, or in any case lost its
readiness to give off its volatile components as the temperature
lowered. Furthermore, as the volatile elements were naturally
drawn off first, the mixture became less and less volatile, until
finally its capacity for saturating air at atmospheric temperatures
disappears entirely by the continued process of fractional distillation at atmospheric temperatures.

Figs. 57 and 58 show a form of wick or flannel carburetor in

---

* For the privilege of using Figs. 55 and 56 in their present form, as reproduced from patent drawings, the author is indebted to the International Text-
book Co. of Scranton, Pa.

which the current of air is made to traverse a considerable length of porous surface by the construction of baffle-plates attached alternately to the opposite sides of the vessel and reaching nearly across. The air enters into the first compartment *c* and, after passing from side to side over the felt surface which is moistened by the gasoline, it passes outward through the pipe *l* through *q*.

108. Carburation from a Gauze Surface. Olds Type.—To avoid the inconvenience from a fibrous or porous material and yet secure the convenient vaporization from the gasoline surface

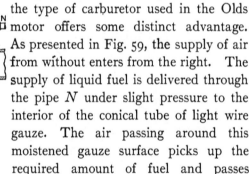

the type of carburetor used in the Olds motor offers some distinct advantage. As presented in Fig. 59, the supply of air from without enters from the right. The supply of liquid fuel is delivered through the pipe *N* under slight pressure to the interior of the conical tube of light wire gauze. The air passing around this moistened gauze surface picks up the required amount of fuel and passes

Fig. 59.

through the throttle-valve *V* to the engine. Any liquid which the air does not absorb runs down through the conical tube and is delivered back to the supply-tank. Fig. 60 shows the con-

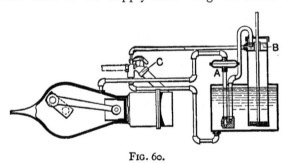

Fig. 60.

struction of the carburetor system complete. A small leather diaphragm at *A* has upon its surface a varying pressure resulting from the pulsations caused by the trunk of the engine in the closed crank-case. It therefore acts as an air-pressure pump, lifting fuel to *B*. From here it circulates through the carburetor proper *C*, and any unused excess goes back to the bottom of the fuel-tank.

109. Carburation by Mechanical Ebullition.—In the early Daimler cycles and motor cars the form of carburetor devised by Gottlieb Daimler was used, which is shown in Fig. 61. In the cylindrical vessel containing gasoline was placed a hollow float. The entering air came down through the central tube, which was borne by the float so that there should be a constant immersion of the lower end of that tube below the surface of the liquid. As the engine made its aspirating stroke air was drawn in both through the top of the smaller cylinder and through the central tube. The air which passed through the gasoline became carburated and, uniting with the air from without, passed to the engine cylinder in explosive proportions.

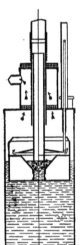

The objections to this system were the same as those attached to the De Dion type as far as the lowering of temperature and the fractional distillation are concerned.

FIG. 61.

110. Spray Carburetors. Float-feed Type. Maybach's.—The third principle in carburation which involves the spraying action of the liquid fuel into the current of air is the modern system. It appears in two general forms. In the one the level of the gasoline in the tank or chamber which supplies the spraying jet is kept at a constant level a little below that of the nozzle, so that the reduction of pressure causes the flow of liquid. When the aspiration ceases the flow ceases without the intervention of a valve whose closure shuts off the delivery of fuel. In the second type the flow of fuel is checked by the closure of a valve, and therefore no float is required to maintain a constant level with respect to the orifice of the jet.

One of the earliest of the float carburetors was that of William Maybach, a colleague of Daimler, which is shown in Fig. 62. The gasoline is delivered by gravity or pressure into the top of the chamber at the right through the opening which is controlled by a needle-valve attached to the float *A*. As the float falls the supply of liquid is permitted to rise, and as it rises the opening is closed. The bottom of the float-chamber is connected to the carburetor proper through the pipe *B*. The air enters upon the aspirating stroke when the valve *D* is opened, whereby the pressure in the mixing-chamber is made less than atmosphere, so that the liquid fuel rises through the capillary orifice by excess of pressure and mixes with the air in *C*. It will be apparent that no valve is necessary except the one

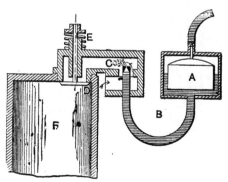

FIG. 62.

controlled by the float, and the fuel will only enter the mixing-chamber *C* as required, with the pressure variation upon the suction stroke of the motor. The more complicated forms of float carburetors are really all derivatives of the simple Maybach type. Some illustrative types may be useful.

III. Float Carburetor Constant Level. Distributing Cone. The Phœnix-Daimler, and Longuemare.—The float in the chamber with the needle-valve attached directly to it was found to offer some inconvenience when applied to the motor vehicle exposed to jolts. The needle-valve would be opened by the inertia of the float, even when the chamber was full enough to close the valve when the carburetor stood still. It was, therefore, a simple modification to separate the float from the needle-valve and to cause the latter to be held shut by counterweight levers, whose action should be overcome by the float when the

level fell. The form of carburetor shown in Fig. 63 illustrates
the counterweighted spindle, and in addition the plan of making

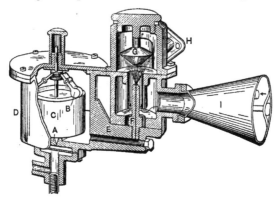

FIG. 63.

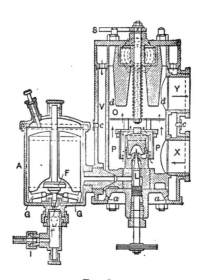

FIG. 64.

the jet of gasoline to impinge upon a conical surface where it
should be spread in a thin film over which the incoming air must
pass.

The Longuemare carburetor, shown in Fig. 64, illustrates the same type of float and counterweighted levers for the needle-valve and the same principle of baffling the flow of liquid fuel. The gasoline enters at the inlet I at the lower left hand and is discharged through the nozzle which is controlled by the valve L. The air enters from without at X, and the mixture passes to the engine through the connection Y. The valve L controls the size of the fuel-jet, and by means of the handle S the proportion of air which the gasoline saturates is controlled by means of the lift of the check-valve. The additional supply of air which does not undergo saturation passes around through the space P to form the explosive mixture. This form of carburetor is fitted with the heat-jacket V within which the hot products of combustion may circulate if desired. For starting the carburetor with the more reluctant liquids the hollow jacket can be packed with cotton or similar material soaked with liquid fuel and then ignited. The openings c permit the flow of air for combustion until there shall be a flow of hot gas. At d are wings of thin metal which are intended to grow hot and to serve as vaporizers in addition to the carburetor effect below.

112. Float Carburetor. Constant Level with Baffle-plates. —Fig. 65 shows a type of carburetor intended to compel the intimate mixture of the mist of fuel with the incoming air. At the right is the float-chamber which supplies the liquid fuel to a diffusing orifice in the carburetor proper. The air entering from below is forced by baffle-plates to take a circuitous course over the surface of these plates, from which it takes up any liquid which may have run down by gravity from the diffusing orifice. The initial supply of air comes in with the gasoline through the ball valve between the two chambers, and the passage between these chambers can be closed by the needle-valve from without. This form of carbureter has an interesting detail by using the glass front through which the operation of the diffusing appliance can be observed.

113. Carburetors without Floats.—In the third class of carburetors, in which the jet of gasoline enters the incoming air through the valve or other appliance actuated by the air, its apparent simplicity is secured by doing away with the float and its attachments. It will be apparent by the study of the Longue-mare design of carburetor that it consists of a number of parts

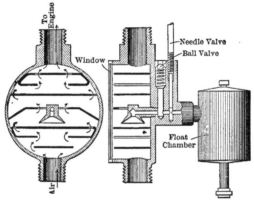

FIG. 65.

which make it costly. The float principle, furthermore, is liable to derangement by jolts or jars in a moving vehicle on a rough road.

In the form of carburetor shown in Fig. 66, which is known

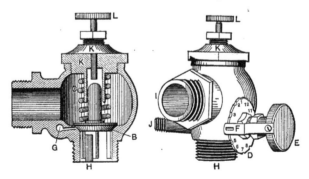

FIG. 66.

as the James-Lunkenheimer design, the air enters from below through the inlet *H*. The cylinder is connected to the side

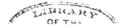

inlet *I*, so that when the piston make its aspirating stroke the pressure from without overcomes the pressure of the spring which holds down the valve *B*. The gasoline enters through the pipe *J*, which supplies the channel *G* in the casing of the valve, from which the necessary number of outlets open into the seat, which is closed by the valve *B*. It will be apparent, therefore, that when the valve lifts by the lowering of the pressure upon it, it opens the gasoline passages and the liquid fuel enters at various points in the annular air-current moving past the valve. When the valve shuts, both the gasoline and the air supply are shut off at once. The opening from the gasoline-pipe *J* is controlled at will by the needle-valve, which has a milled head *E* and an indicator and locking device whereby it can be set once for all for any desired fuel-supply. The lift of the valve is also controllable by the stop which is adjustable by the milled head *L*.

Another form involving much the same principle is illustrated in Fig. 67. The air enters from above through the openings

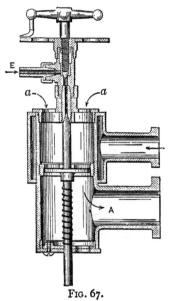

Fig. 67.

a, and the gasoline is supplied through the pipe *E*. The valve in the gasoline-pipe is held upward by the spring so that its normal position is closed. When the motor aspirates, the piston on the valve-spindle is lowered, compressing the spring and opening the gasoline-valve at the same time that the passage *A* to the cylinder is opened by the air-pressure on the top of the carburetor piston. At the end of the charging stroke the spring forces the piston up, closing the valve and shutting off the access of air. This design shows also the jacketing of the carbureter by the hot products of combustion surrounding this upper part and entering through the upper nozzle at the right hand. The carburetors in use on the majority of the American automobile motors belong to this third class and are made by the builders of the engine themselves under license from the basal patent illustrated in Fig. 66. The objections to the principle are met in the high-speed types of motor which operate with wide variations in the load. The difficulty with the high-speed requirement is due to the inertia of the inlet-valve and the resistance offered by the spring. At high speeds the actuating pressure caused by the motor piston is applied so rapidly that the interval occupied by the entire charging stroke becomes so short that the inertia of the valve and the resistance of the spring retard the opening of the valve until the motor piston has traversed a considerable fraction of its stroke. In consequence the volume of the motor cylinder is not filled with the weight of combustible charge which would enter at atmospheric pressure if the engine were moving slowly (see par. 98). The diminished weight of charge or the less mass in the motor cylinder results in a diminished compression and in the presence of a less amount of explosive energy and in a less initial pressure over the working stroke. It follows, therefore, that the horsepower of the motor supplied with a carburetor of this class may not necessarily increase with the number of revolutions as computation would require. The horse-power will increase with the speed up to a point which may be called the critical speed of the motor; and beyond that the increase of speed is followed by decrease of mean pressure propelling the piston, so that the motor has a limit of its capacity set by this condition and it does not become more powerful by increasing its speed. The difficulty set by variable resistance results from the fact that the flow of gasoline is determined by the adjustment of the valve which corresponds to E in Fig. 66 and which it is not convenient to adjust for the variations of the load. This is the case if the gasoline is supplied by gravity or under a constant head through the opening G. Too much fuel will come in when the valve is open for a considerable interval when an engine is moving slowly, and not enough may pass during the very short interval when the engine is working rapidly. On the other hand, with engines of high speed and a certain adjustment of the spring it may easily happen that the valve hardly ever closes down tight upon its seat, but hangs suspended in the continual flow of air which in a multiple-cylinder engine is practically continuous with the pulsations hardly noticeable. In such a case the adjustment of the valve E if correctly made may cause the motor to work satisfactorily even under variations of resistance.

Belonging in the same general group as to operation to which the floatless carbureter belongs may be grouped certain other

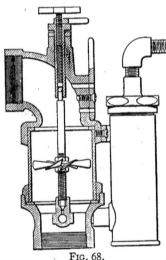

FIG. 68.

forms in which the float is used. That is, they may be operated either with or without the float. Fig. 68 illustrates the type of a large number in which the effort is made to subdivide the liquid fuel by causing eddies or spiral currents in the air. The supply of air coming in from the bottom by atmospheric pressure will cause the spindle to rise which carries the vanes or propeller-shaped blades. The ascending currents will twist the spindle and give a spiral motion to the air in the chamber which will help to complete the mixing, Sometimes these helical areas are doubled. The rise of the needle-valve spindle is controlled by the stop at the top of the mixer, so that the supply of fuel can be varied. This carbureter could work without the float-chamber if desired.

114. Carburetors for Motor-Vehicles. Automatic Carburetors. — While all of the modern forms of float and floatless carburetors have been applied experimentally to motor-vehicles, the recent practice of builders has developed some principles of design, which are the outcome of the special difficulties as to variable fuel supply which are there present.

The requirements as to fuel supply in these motors may be summarized as follows:

1. The adjustment of fuel and air to each other should be as automatic as possible, under all varying conditions. The operator of the motor is likely to be unskilled in making this adjustment himself, and it is therefore perhaps not desirable that he be encouraged to alter it. But the mixture is a combination of the liquid fuel (gasoline, alcohol or kerosene) with the gaseous air. The latter will be subject to a control of its volume while the former to a control of its weight in a given time. The mass of the liquid is different from that of the air, by reason of the great density difference between them. Temperature of the air will have great effect on the weight occupying a given bulk or space, and the hygrometric state of the air affects both weight and the amount of oxygen in a cubic foot, as well as the temperature of the resulting mixture after ignition. It has been frequently noted how smoothly and "sweetly" (to use a colloquialism) the motor runs in the evening and night, as compared to the day. The greater density without inconvenient lowering of the temperature which occurs in winter is doubtless one explanation of this. To lessen the variation of air quality, the modern method is to draw the combustion air from a heated zone, such as that surrounding the exhaust pipe of the motor by means of a flaring mouth-piece. Or, the carburetor may be heat-jacketed by products of combustion, or hot-water-jacketed from the outflow of warm water from the motor jackets. The carburetor works best with air at about 80° F. at entry. To make it warmer is to diminish the oxygen weight inconveniently; to have it cooler is to retard the quick ignition of the fuel and air mixture when it reaches the combustion chamber. To keep the air and fuel at constant temperature under all outside temperature conditions of weather is to secure best results when the mixture is once correct.

2. The motor-car presents a more difficult problem in service than the stationary engine under uniform load, or than the motor-boat. In the latter the variation in power is a variation in speed, and the resistance only varies outside of that as respects a few factors which do not vary directly as the speed. Even in buffeting waves and against a head-wind and tide, the motor-speed can be kept up. With the motor-car, however, the following range of demand must be met:

(*a*) A minimum speed of pistons and against minimum resistance, which occurs when the car is stopped or standing.

(*b*) A minimum speed of pistons but a maximum turning resistance, as on heavy hills, deep and heavy road surface, and in starting or working amid crowded and irregular traffic.

(*c*) A maximum piston speed, and maximum power in car speed as in racing or in speeding with full loads on levels.

(*d*) Combinations of intermediate piston speeds with all car speeds and intermediate resistances, as in hill and level work in open country.

(*e*) The requirement of minimum fuel waste at exhaust in form of smoke, and of maximum efficiency of fuel in vehicles of large power capacity as in commercial vehicles. These latter are in a special class where fuel expense is significant, and which have usually a time schedule to maintain. They must not be stalled on the road because at any time the excess consumption as emptied the fuel tank.

(*f*) Changes of level of the fuel tank relatively to the carburetor or the changing level of the surface of the liquid fuel in the tank must not vary the fuel supply. Grade changes must not also change the ratio of the fuel level to the jet-nozzle in the carburetor itself. The jet should be about $\frac{1}{16}$ of an inch above the fuel level, and this should not change materially.

(*g*) The varying piston speed brings with it a variation in the compression pressure and hence in the mean effective pressure of the succeeding expansion stroke when the water-cooling system is effective. The compression pressure when it takes place at high speed is nearly adiabatic (pars. 50, 203), because the cooling water in the jackets or the direct air-cooling effect has scarcely time to lower the temperature of more than the film of mixture which is nearest to the cooling surfaces. At low speeds, the mixture is exposed longer to such action of withdrawal of the heat due to compression; the heating process approaches, therefore, more nearly to an isothermal (par. 49) having, therefore, less energy at the end of the process, and giving a lower mean effective pressure in the next stroke. This is the same in effect as calling for more fuel or heat energy per cubic foot of cylinder volume at low speeds than at the higher.

(*h*) The high piston speed with a given area of exhaust passage or through the exhaust valve, keeps more heat in the cylinder by retarded flow of the products of the combustion of the last stroke. The hotter cylinder is filled with a less weight of incoming charge when its volume is filled, as compared to one which is more effectively cooled by the longer action of the cooled and cooling walls. This is the same as a throttling effect of partial closure of the controlling or accelerating valve, and is the opposite in effect to the action treated in the previous paragraph. Every individual motor will differ from all its alleged duplicates in the sum of these two influences. Constant-speed motors are not subject to these.

(*i*) Constant-speed motors are not subject to the variations of effect of inertia or demand for acceleration of the air in the inlet passages. The elasticity of the air causes it to expand or compress before the column moves as a whole under the pressure of the external atmosphere. This latter is the force which sends the air into the carburetor (there is no such thing as suction, regarded as a force) and as the air flow induces the fuel flow, variations in the former cause corresponding variations in the latter.

Modern carburetors will therefore be found to seek after these results:

1. Constant temperature and moisture conditions for the air by drawing it from a warmed zone close to hot surfaces about the motor.

2. Constant temperature and preheated conditions for the fuel, by jacketing the fuel chamber of the carburetor, and using waste heat of jackets or exhaust to secure this. (Fig. 258.)

3. Constant level adjustment relative to the fuel jet by using an annular float which surrounds the jet chamber (as in Fig. 256) and by seeking to neutralize inertia effects on float and automatic fuel valve, by counterbalancing the former over a fulcrum by a weight (Figs. 256 and 259) on the valve-spindle, or by a

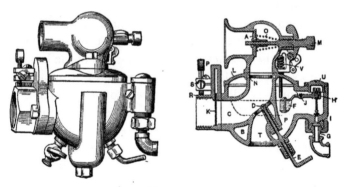

FIG. 254.

spring (Fig. 255). The spring has no inertia of its own as respects its response to a float motion due to a jolt or jar, and its effect can be varied by adjusting its initial compression by a screw and nut.

4. But the feature which has brought automatic action nearer for high-speed work has been the introduction of the automatic or compensating air valve. This is A in Fig. 254 and D in Fig. 255. It is held shut when the air column is moving slowly towards the motor outlet, or when the motor is at rest. In hand-starting, therefore, the flow of fuel relatively to the air volume is considerable, and the mixture rich in fuel, making the first ignitions easy and sure. Fig. 254 shows a detail at V whereby

the depression of that spindle from without, forces down the annular float *F*, raising the level of gasoline around the jet, and flooding the carburetor with excess of fuel. As the motor piston speeds up, the aspiration strokes become more frequent and the tension in the air passage grows less, increasing the flow of fuel through the jet opening due to the atmospheric or other pressure behind it, and the pressure against the spring which holds the automatic shut. The spring yields more and more as the difference of pressure increases more and more with speed, opening a larger and larger area for air, until the maximum is reached at the highest practical speed. While the fuel flow

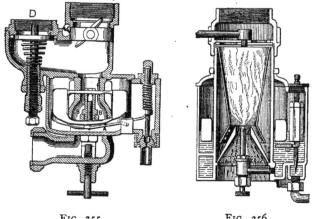

FIG. 255.   FIG. 256.

increases, therefore, the air flow increases faster by reason of the increasing area, producing, therefore, a mixture relatively leaner in fuel at high speeds, while the fuel flow has intrinsically or actually increased with the speed. Danger of pre-ignitions from undue compression of a rich mixture is prevented, and an excess of fuel which might not find oxygen enough to burn in the limited time of the high-speed stroke. The high speed favors also an intimacy of mixture of the fuel and air due to the high spraying velocity of the jet, or special pains in design may be taken, as in Fig. 256, to favor mixing by the shape of the

nozzles before and after combination has occurred. The throttle
acts upon the completed mixture in all cases, but a hand adjust-

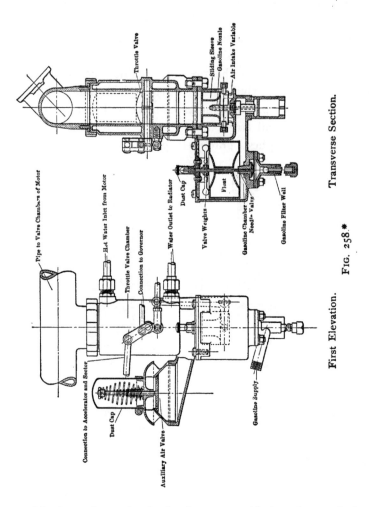

Fig. 258.*

Transverse Section.

First Elevation.

ment of fuel supply to the jet is always provided, to be used also
as a shut-off of fuel if desired.

* The author is indebted to the courtesy of Mr. Benj. R. Tillson for permission to use Figs. 258 and 259 in the clear form here presented.

5. Another solution looking to the same or better results is to place two carburetors in series, or tandem. The first or smaller one is for running the engine at low speeds or under lighter demands for power. It has its own throttle, and delivers into the intake piping between the motor and the throttle of the second or larger carburetor. This latter has the automatic air-valve provision, fitted with a dash-pot detail to prevent inconvenient fluctuations. As the engine will be speeded up, or the

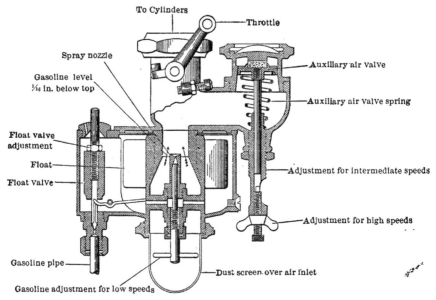

To Cylinders

Throttle

Spray nozzle

Gasoline level
⅟₁₆ in. below top

Auxiliary air valve

Auxiliary air valve spring

Float valve
adjustment

Float

Adjustment for intermediate speeds

Float valve

Gasoline pipe

Adjustment for high speeds

Dust screen over air inlet

Gasoline adjustment for low speeds

FIG. 259.*

throttle is opened wider, with increased demand for power, the second carburetor comes into play, and provides a full capacity up to the maximum of the motor with that fuel and with a greater fuel efficiency. A copper coil of pipe brings hot exhaust gases into a space below the main throttle to secure constant temperature and necessary heat for certain vaporization. The fuel jet plays against the outside surface of this hot coil at all times.

6. To secure positive and increasing flow of the liquid fuel to the carburetor with increasing speed, the practice of putting the fuel tank under pressure has been introduced. This pressure is brought from the exhaust connections by a by-pass pipe on top of the liquid in the tank, and being a gas without oxygen it renders the tank free from danger of an ignition of an explosive mixture of fuel vapor and air. By increasing the difference of pressure on the two sides of the liquid fuel, the flow of the latter is hastened through a small orifice, which is thereby more quickly controlled; by using exhaust pressure, the fuel pressure is greater as the speed increases; the flow is not made variable by differences of level of tank, full or empty, or by the inclination of the car-frame on grades up or down. Mountain-work is troublesome without the pressure feed. It is inconvenient to have the tank high enough to secure gravity feed, if the tank is large; and gravity feed is variable with the amount of liquid in the tank, and can never be equal to more than a few ounces per square inch. With the pressure feed, however, an apparatus to get up pressure is required for starting, since the working of the motor is required to produce the pressure feed to the carburetor. Or, a small fuel tank with gravity feed may be used to start the carburetor when the motor is at rest, and to be shut off when pressure is established.

7. Or again, instead of a disk-valve opening by air-pressure from without, a flexible diaphragm may be used, actuated either by differences of air and gas pressure, or by a spring. By the motion or collapse of the diaphragm more air openings are made available, and an increased fuel supply may be effected by the same motion. Or a sleeve or thimble with holes or slots may be used, whose motion under spring resistance to flow may open larger areas as the thimble moves in front of openings in the casing which surrounds it.

In motor-car practice to date, no attempt has been made to supply the air to the carburetor under pressure by mechanical means such as a fan or blower. The size, weight, and speed of such a fan would make an inconvenient addition to the weight of the equipment, and the power to operate it would offset the power which would be saved. But in theory, and in plants of considerable size, the serving of several units from one independently driven pump, particularly if motors of the two-cycle type were to be operated, would offer considerable advantage in securing full weight of charge per stroke, and could relieve the motor of the function of supplying mixture of fuel and air for the following power stroke.

115. Alcohol Carburetors. Martha, Japy, Richard, Brouhot, Marienfelde.—The only difference which requires to be made when alcohol is to be used as the fuel and it is to be atomized and vaporized in a carburetor is that the carburetor must be

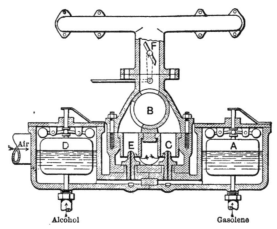

FIG. 69.

kept hot so as to secure the vaporization in addition to the simple atomizing which is required for the gasoline carburetor previously discussed. In some cases the conducted heat from the working cylinder will keep the pipe hot enough to vaporize the alcohol after the engine is once working, so that it is only necessary to get the engine started and well warmed. After this the same equipment will work indifferently on alcohol or on gasoline. A convenient plan which has been much used in Germany and in France is to make the carbureter double, as is shown in Fig. 69, which illustrates the Marienfelde design.   This is a constant-level or float-feed carburetor into which air enters through the orifice at the left hand and surrounds the two jets C and E in an annular current.   The gasoline enters at the right-hand chamber and, with its level controlled by the float A, passes into the air-current through the nozzle at C.   Above the jets is the shell valve B, which is shown in the position in the cut for the working of the carburetor with alcohol through the left-hand inlet-float D and jet E.   The engine is started cold with the gasoline side in action, which will require no heat in the connections to start the motor.   After the motor is running at speed, the shell valve is turned over in the position shown in the cut, when the alcohol begins to act and the gasoline supply is cut off.   The valve F above the shell valve B acts as a throttle-valve to vary the amount of air and fuel which passes to the cylinder from the carbureter.

The more usual forms do not use the double principle, but depend upon heating the atomized alcohol by passing it in a circuitous passage around a hot exhaust-pipe.   In the Martha carburetor, Fig. 70, the alcohol enters from below into the spray-

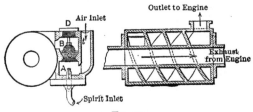

FIG. 70.

ing part of the carburetor and is aspirated by the charging stroke with the air which enters through t.   The alcohol is atomized by contact with the corrugated surface and the netting in the part B, from which it enters at the side of the horizontal chamber, which is the vaporizer.   The alcohol vapor and air move in the spiral channel in contact with the exhaust-conduit and thus out to the engine.

In the Brouhot carburetor, which is more properly called a vaporizer, the exhaust-gas enters at the bottom, as shown by the inlet arrow (Fig. 71). If the valve which regulates the mixture is shut tight, the exhaust-gas passes spirally around the central tube in one direction, while the alcohol passes in a reverse spiral in the other. The alcohol spiral ends at the top where the vaporized alcohol meets the pure air and they pass together to the motor. The exhaust-gases descending from the top passes out at the bottom as they would if the regulating-valve were open. Of course a partial opening of the regulating-valve splits, the exhaust-current so that only a part of it passes through the vaporizer.

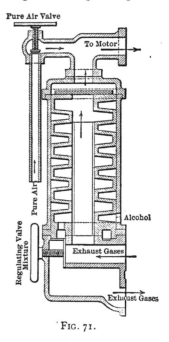

Fig. 71.

In the Richard form the float principle is used (Fig. 72), and the jacketing of the impact surface by the hot products of combustion entering at e and leaving at e' causes the atomized alcohol, spread in a thin film on the hot cone by the deflector d, to become a gas and to move to the engine through the outlet m. In the Japy carburetor, illustrated in Fig. 73, the ribbing of the passage L, where it is surrounded with hot gas, makes that surface act as vaporizer for the alcohol which passes through the float-chamber A and the controlling-valve C. The inlet of air at E is controlled by the valve V.

116. Kerosene Carburetors.—In the carburation of air by kerosene it is particularly necessary to pay close attention to the vaporization process. In the discussion of the kerosene-oil engine in its earlier forms (par. 79) the difficulties were referred to which result when the attempt is made to inject the liquid oil without atomizing. The motor cylinder either becomes coated with a hard carbon coat resulting from decomposition of the oil by a cracking process, or the cylinder is flooded with liquid oil which it will not vaporize completely. If, however, the ato- mized mixture of carburated air is drawn through a vaporizer at nearly a red heat, the carbon deposit disappears, apparently carried away by the next succeeding rush of air, and when this carburated hot air meets the main supply of air required for combustion and is thoroughly mixed with it, the combustion appears to be practically complete without deposit of carbon. Before the engine starts, the vaporization has to be effected by a separate source of heat or, as discussed in the paragraphs above; or the engine may be started on a more volatile liquid, such as gasoline, and be changed over to kerosene when the motor has

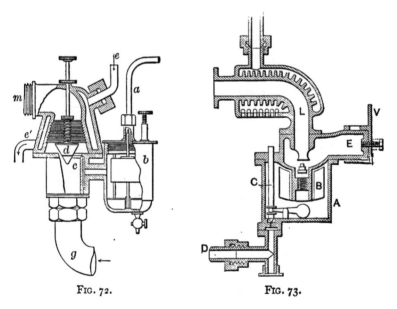

FIG. 72.   FIG. 73.

become well heated. It will be apparent, however, that where the heat from the exhaust-gases is used as a supply to meet the vaporizer, the regularity of its action must be affected by every condition which varies the discharge of heat in the exhaust, so

that every stroke without explosion, every slow-down with diminished fuel energy in the charge, and every stop of the motor
permitting a cooling of the vaporizer, will interfere with its regularity. Reference should be made to paragraphs 32 and 83.

117. Some Principles of Design of Carburetors.—In the
design of gasoline carburetors, experiment seems to show that
a good spraying effect at the jet is best secured by a velocity of
the incoming air past the nozzle between 75 and 80 feet per
second. Some experiments made by Mr. L. Berger have indicated that when the suction-pipe between the carburetor and the
motor is sufficiently hot to cause the globules of liquid gasoline
to form a gas on striking the hot surface, 35 square inches of
surface per horse-power at a temperature of 180° F. will secure
complete vaporization at atmospheric pressure. The vapor of
gasoline, according to these same experiments, diffuses in the air
with a velocity of 0.2 of an inch per second. So that if the velocity
of flow of the incoming charge is known, the length of the suction
pipe can be calculated so that the gasoline vapor may have time
to permeate completely the air by coming laterally from the
walls of the pipe before the mixture is admitted to the cylinder.

With kerosene, on the other hand, about 31 square inches
of vaporization surface are required per horse-power, heated to
a temperature of 390° F. at atmospheric pressure. When this
principle is carried out in a kerosene motor the volume between
the atomizer and the suction-valve of the motor will become
nearly equal to the volume of the piston displacement with a
high-speed motor.

CHAPTER XI.

IGNITION.

120. Introductory.—It has been already considered that the problem of increasing the heat energy of the mixture of gas and air behind the working piston demanded that after the mixture had been compressed it should be ignited so that the gas should burn in the oxygen of the mixture and impart the increased pressure due to this heat. This ignition should be so timed as to occur at the proper point of the cycle so far as the gas is concerned and at the proper point of the stroke of the piston so far as the motor is concerned. In the Otto cycle this ignition is to take place at such a point that the combustion shall be complete or nearly so when expansion begins. There have been many methods proposed for the accomplishment of this purpose, each of which offers certain features.

121. Ignition by an Auxiliary Flame.—The plan of igniting the mixture by an auxiliary flame was early tried. In its simplest form it consisted of having two jets or burners connected to the gas-supply by flexible tubes. These jets were alternately presented to the explosive mixture and ignited it. The ignition of the explosive mixture was necessarily followed by the extinction of the auxiliary jet, so that this required a secondary or free jet burning in the open air by which the igniting jet could be lighted after each succeeding extinction. One of the earliest forms of this combination of igniting jet and continuous lighting jet was the system of Barnet, whereby the igniting jet burned within a shell or casing which rotated like a valve, presenting the opening in the casing alternately to the explosive mixture in the cylinder and the lighting jet which burned outside of it. This system is open to the very serious objection of the escape of gas with its attendant odor, and it is not available for use in the open air. If, at any time, both flames were extinguished, the engine ceased to operate. It could also give only about 40 ignitions per minute. Very ingenious combinations of the flame-ignition, and the slide-valve have been made in the design of Otto engines, with a view to increasing the number of ignitions to over 100 per minute and to secure continuity of the igniting flame and overcome the flow of the flame in the wrong direction when the pressure due to the compression was greater than the pressure which the jet would resist.

122. Ignition by Internal Flame.—In the Brayton cycle the igniting flame was kept burning continuously within the working cylinder itself. This arrangement was possible with the continuous heating process of the Brayton cycle as long as the supply of mixture flowed through the open valve. Since the pressure in the working cylinder was enough greater than that in the reservoir which supplied the continuous burner, there was a tendency for the flame to be blown out by a reversal of the direction of flow in the jet. A safeguard had to be abundantly provided, lest the flame should blow back into the reservoir within which it would, of course, be propagated and would result in an explosion. This was secured by a provision of wire-gauze safety attachment, but in case of the deterioration of this gauze the danger was always present. This gauze was an element of weakness in the Brayton engine, and in spite of care the flame would become occasionally extinguished, when, of course, the operation of the engine stopped.

123. Ignition by Heated Metal from External Jet.—A system of ignition which avoided bringing a flame or jet into the cylinder was to ignite the mixture by bringing into it by the action of a slide-valve, a surface which had been heated outside the cylinder by the action of an auxiliary jet or flame in which the surface stood at rest during the phases of the cycle in which there was no necessity for ignition. The difficulty connected with this system was the uncertainty that the metal surface could be heated sufficiently to insure ignition when the mixture was such as to ignite reluctantly by reason of its impoverishment, and the difficulty connected with the deterioration of metal exposed to oxidation at high heat. This metal igniter was often made in cage form, so that a large metal surface should be exposed to the gas as soon as it was intruded into the combustion-chamber. (Clerk.)

124. Ignition by Catalysis —It has long been known that one of the properties of spongy platinum (known as catalysis) is that the impact upon it of a jet of combustible gas would so raise its temperature that it would act as an igniter. This property was thought of and applied to a limited extent for gas-engines, but had the objection that with a mixture of varying richness the ignition was not positive nor reliable.

125. Ignition by Incandescent Wire or Cage Electrically Heated.—In order to avoid the objections connected with the flame-heating of a metal to be introduced into the combustion-chamber, the plan has been used of heating a coil of platinum or other resistant wire or thin strips of platinum by means of the resistance which they offer to the passage of an electric current. This incandescent platinum is carried on a slide from which it is electrically insulated and is introduced white-hot into the explosive mixture. By this plan any danger of blowing out of flame is avoided, the temperature of the incandescent metal can be made high enough to insure ignition, and several of the objections to the previous systems are avoided. The objections to the plan are those incident to the presence and unreliability of the electrical apparatus to produce the incandescence. If electrical methods are to be used, the sparking plan presently to be discussed is more convenient and cheap.

126. Ignition by Hot Tube.—A plan of ignition which offers some advantages is to pass the gas mixture into a tube of platinum, porcelain, nickel-steel, or similar fire-resisting material, which is kept incandescent by an external flame (Fig. 76). The entry of a small portion of the mixture into this tube brings it to the ignition-point, and the fire is propagated through the entire mass from this one point. The objection to this system is the fragility of the incandescent tube if made of porcelain, which is liable to break under jars or as the result of accidental injury. When platinum or steel is used the tube is not so fragile, but the system

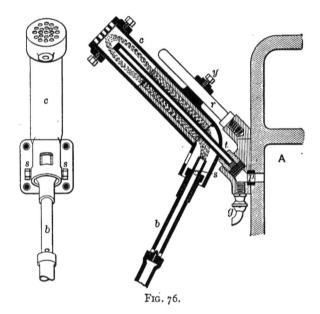

FIG. 76.

is more costly and the tube oxidizes or deteriorates. With the gasoline motors a special burner for burning the gasoline is required which shall keep the tube hot.

In the handling of this system of igniting many designers have used a timing-valve, which should open the hot tube to the explosive mixture compressed in the cylinder just at the right moment to have the ignition propagate itself from the tube to the full volume of mixture. The American designers have not used the timing-valve, finding that by adjusting the flame which heats the tube to different points in its length it is possible to vary the time at which with varying compressions and characters of mixture the full ignition shall occur. The time of ignition with the hot tube will depend upon:

1. The length of the tube.
2. The size or volume of the passage leading to the tube.
3. The amount or degree of compression of the mixture by the piston.
4. The temperature of the tube; the hotter the tube, the earlier the ignition; the cooler the tube, the later.
5. The fact whether it was hottest near the open or the closed end; if heated near the open end, the earlier the ignition.
6. The temperature of the mixing- and ignition-chambers.
7. The temperature of the jacket-water outlet.
8. The speed of the engine.
9. The quality or proportions of the air and fuel admitted.
10. The pressure of the intake or suction stroke.
11. The governing action and the system of governing.
12. Leakages: at piston, at exhaust, past valves.
13. The state of the surfaces of the tube, outside and in.
14. The location of the tube, with respect to receiving and acting on new or fresh mixtures, or mixtures containing burnt gases.

It will be apparent, therefore, that where an engine is to run at variable speeds and powers so that the governing process is to vary Nos. 8, 9, 10, and 11 from stroke to stroke, and particularly in automobile practice, where it may be desirable to retard the ignition sometimes until after the piston stroke has begun and been partly made, the hot tube has given way before the electric ignition methods. The hot-tube ignition cannot be retarded without a timing-valve, and even with it it is uncertain. A nickel-steel tube about 4 inches long of $\frac{1}{8}''$ thickness of walls and with a $\frac{3}{16}''$ hole through it will last about three years when heated continuously to a good red heat.

127. Ignition by High Temperature of Compression.—In the Hornsby and in the Diesel oil-engines, ignition is secured by the expedient of having the compression space of the cylinder kept hot by having no water-jacket at this point of the cylinder, so that when the piston returns and compresses the air behind it the temperature of that air shall be so raised by compression that a jet of combustible oil entering that air will be at once raised above the firing-point and will ignite without flame or spark. This system requires that in order to start the engine the combustion-chamber shall be heated from without by some form of lamp or heater until it is hot enough to produce the first ignition. After that it will be kept at ignition temperature as long as the engine is working, provided the governing action does not so impoverish the mixture that it will not ignite within the compression limits at which the engine is working. These engines, therefore, must always receive a charge of combustible so that there may be a source of heat to keep up the temperature even when the engine is lightly loaded. If the weight of fuel is so reduced by governing under light or variable loads as to furnish too little heat, the strokes will gradually become less powerful as the compression-chamber cools, until finally the engine stops.

With heavy loads and high compressions, where any fuel may have remained unburned from a previous charge, this system gives trouble from back-firing or pre-ignitions. Pre-ignitions of this character often result from the presence in the combustion-chamber of any projection or small isolated mass from which the conduction of heat will be slow, so that from the compressions and ignitions this projecting part may become highly heated, and heated faster than it can be cooled by conduction. Such a projection will ignite the compressed mixture, even while the general surface of the cylinder and combustion-chamber may be too effectively cooled to do so. A piece of an asbestos gasket sticking out of its joint; or a bolt-head; or the heated points of the terminals of an intended electric system, will act in this way. Even in a water-jacketed cylinder with unusually heavy loads the cylinder metal may itself become hot enough to ignite the charges for a while, even when the regular ignition system is out of action. The air-cooled motor will often prove reluctant to stop from this same action. The objection to the system is its uncertainty with low and variable compressions and widely varying loads. Its advantage is its avoidance of all subsidiary apparatus to cause the desired ignition, and its dependence upon a fundamental law.

128. Ignition by Electrodes and Electric Spark. The Jump-spark System.—The exceeding convenience and compactness of an electric system of ignition of the explosive mixture in a gas-engine cylinder early directed attention to this method. It was

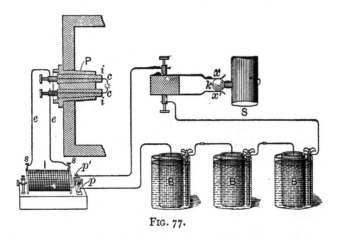

Fig. 77.

used by Lenoir in his historic motor. The principle is to cause a spark of sufficient intensity to pass between two terminals on an electric circuit, the spark to be in the mixture which its heat is to ignite (Fig. 77). It is only necessary, therefore, to insert into the compression volume behind the piston, a plug P insulated from the metal of the cylinder walls and carrying into the cylinder and its mixture the two points c, c, of some resisting metal with a spark-gap between them, and then at the proper instant. as determined by k on the shaft S of the motor, to make the electric current jump the gap and fire the mixture. Fig. 78 shows the general appearance of the plug with the cylinder terminals projecting at the bottom. The plug usually fits a standard half-inch pipe-thread, and may easily be inserted or renewed. The inner point in the form illustrated is insulated electrically by a porcelain or mica or lava or soapstone lining or core from the metal of the engine, while the outer terminal is in electrical or metallic

FIG. 78.

contact with the engine or its frame or mounting. The spark therefore passes the gap in contact with the mixture when the exterior circuit is excited. The failing cases for these plugs are the closing of the gap by oil or lampblack deposit or by water, so that the spark does not form, because there is no gap for it to jump; or the breakage of the insulating material whereby the current is short-circuited in the plug and does not reach the points. If the gap is filled up, it can be formed anew by washing the points in liquid gasoline; if the gap has become too wide by the erosion of the points by heat, the spark will not jump across, and they should be brought together. About a sixteenth of an inch is the working distance with ordinary electrical currents used in motors. Fig. 78 also shows the double-gap arrangement which has been found to contribute to the certainty of the passage of the spark. The sec nd gap is external to the cylinder and does not become fouled; it shows plainly to the eye whether the spark is passing; its resistance acts to intensify the igniting spark.

There are two systems or principles of electric ignition. One is called the jump-spark system, and depends upon the principle which is utilized in the Ruhmkorff or Faradaic coil, whereby a secondary current of high intensity flows through a coil of fine wire which surrounds a primary coil of coarser wire when the primary current is made or broken, so that the flow of electrical energy is intermittent in that primary circuit. In Fig. 79, for example, which shows a typical arrangement, the primary circuit starts from the primary or storage battery at the left, and its wires *A* and *E* are shown in heavy lines. The motor is at the lower centre, and the primary circuit is made and broken by the contact end *K* on the spring *L*, which bears against a commutator surface cn the motor shaft. This surface carries a conducting arc *H*, so that as the shaft revolves the primary circuit will be made when *H* passes under *K* and *L*. This primary circuit ends in a few turns of large wire (perhaps No. 14) around the

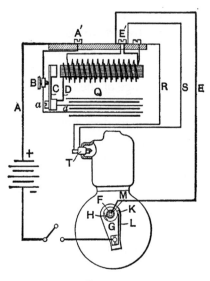

FIG. 79.

core of the coil above. Outside of this is the secondary coil (perhaps of No. 36 wire) making a large number of turns and shown in Fig. 79 by the finer lines. This secondary circuit has the sparking-plug *T* in it, projecting into the motor cylinder. When, therefore, the main switch on the primary circuit is closed (it is shown open at the lower left hand of Fig 79), and the motor revolves, the primary circuit will be closed at an adjusted angle of the motor-crank, whereupon a stream of sparks of high intensity will cross the gap at the plug and fire the charge. In the arrangement shown in Fig. 79, the vibrator or trembler *B* is made to vibrate very rapidly by the make-and-break of primary circuit which occurs at *B*, so that while *H* and *K* are in contact mechanically several electric makes-and-breaks occur at the coil. The other way to effect this same flow is the mechanical vibrator shown in Fig. 80. Here the shaft of the motor *C* is recessed, so

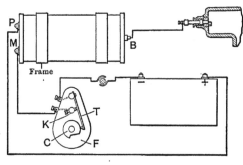

FIG. 80.

that the vibrator *T* is held away from the contact-point *K* except when the end of *T* falls into the recess. The primary circuit is then made through *T* and *K* to the primary terminals *M* and *P* and causes the secondary to flow through *B* to the spark-plug. By giving a certain mass to the contact end of *T* the latter will vibrate against and away from *K*, while the shaft-gap is passing, making and breaking the primary circuit mechanically, which is followed, of course, by similar breaks in the secondary, and a flow of sparks.

It will be apparent from either of the two preceding illustrations or from Fig. 81 that this system of making contact by a commutator surface on the motor shaft and a conducting arc *B* makes it very easy to advance or retard the moment of the passage of the igniting-spark relatively to the dead-centre of the piston stroke. By arranging that the centre of the vibrator *H* in Fig. 81 may have an adjustment angularly around *A* by hand or by governor, the angular position of the sparking instant is obviously varied. The limit to this adjustment which is wise may easily be set by stops which shall prevent too wide a range of variation of the ignition. Too early ignition may cause the engine to start backward; too late ignition may cause the flaming of the mixture to be still in progress when the exhaust opens.

The advantages of the jump-spark system are the avoiding of any moving parts inside the cylinder, and the strong spark

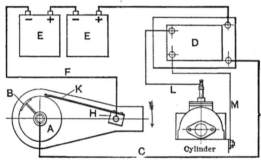

FIG. 81.

which results even from low battery power in the primary circuit. A battery of four, six, or eight cells, giving a voltage of from $4\frac{1}{2}$ to 6 volts with $\frac{3}{4}$ or 1 volt per cell, and an amperage of 8 to 16 amperes is about the usual standard. Storage batteries giving a capacity of from 100 to 300 ampere-hours are much in use for automobile motors. Dry batteries with carbon and zinc elements and depending on the usual sal-ammoniac paste are apt to grow weak as the paste dries with time.

129. Ignition by Electric Arc. Hammer-break System.— The other system of electric ignition does not use the secondary circuit, but depends upon the fact that a break in the flow of a primary electric current will reveal an arc or spark passing between the broken ends, until the distance between them becomes too great for the arc to jump. If such a break can be made inside the cylinder, and so that the arc when formed is surrounded by the explosive mixture, its heat will ignite it and the stroke will be made. Fig. 82 shows a typical arrangement. The cam at the bottom is so timed that it shall lift the toe *D* outside the cylinder and carry inside the angular motion which *D* transmits to *C* at the proper instant. The angular motion of *D* and *C* causes the arm or lever inside the cylinder to be torn away from the contact surface of the insulated pillar against which it rests. The severing of this contact causes the arc to pass and fires the charge. The spring *E* causes the contact to be remade when the push-rod *G* is forced away from *D* by the pin *L,*

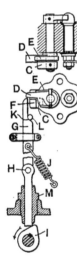

and the inclined surface *K.* From the fact that the contact terminals must waste away by the electric oxidation caused by the passage of the spark, it is usual to give some enlargement to the contact end of the internal lever, so that it receives something the shape of a hammer on a handle. This has caused this arc method to be often known as the hammer-break system. It is also called the contact system. To give greater intensity and duration to the arc a sparking-coil is usually inserted in the circuit, consisting of a coil of about No. 14 insulated wire surrounding a core made of a bundle of soft Swedish-iron wires of No. 20 gauge and from 6 to 8 inches long. Such a coil acts as a sort of condenser, and increases the tendency of the current to flow across the gap after

FIG. 82.

the break is made. If its resistance is excessive it takes too much battery power and current, which will deteriorate the contacts unnecessarily. From $1\frac{1}{2}$ to 2 amperes is usually enough battery capacity.

The failure of this system comes about either from the interposition of non-conductive deposit on the terminals, so that no current passes, or from a conductive deposit which short-circuits the current and prevents the arc, so that there is no arc formed across a gap. Deposits have been mitigated in ill effect by making the contact of the terminals a wiping or sliding one, so that the abrasion of the moving element kept the fixed one clean. It is also an objection to the system that it compels moving parts to be inserted into the hot combustion-chamber and to pass through stuffing-boxes in the walls of that chamber. To give a spark of a given intensity, this system requires more battery power than the other. On the other hand, there are fewer points where failure of the system may lurk, since there is but one circuit and not two, and no vibrator or trembler is required.

130. Dynamo- or **Magneto-electrical Ignition. General.—** In the foregoing discussion the source of electrical energy has been some form of battery. It is obvious that an electrical current can be generated by the revolving armature of a dynamo or magneto-electrical machine, from whose action the necessary spark action can be secured, and the cost and deterioration incident to battery action can be avoided. Hence a tendency to use generators is a feature of many modern designs, particularlly of automobile motors. The dynamo can be quite small and light in weight, having its armature driven at from 1500 to 2000 revolutions per minute by belt or friction drive from the motor-shaft, and giving a current of 10 volts. The objection is obviously that a battery either of primary or storage type is required to start the motor from rest. The storage battery is more usual, since the dynamo can be wired to charge it while the motor is running. Such dynamo and battery can also be used for lighting the lamps for night use. The power consumed is small, and its cost is offset by the convenience of having fewer battery cells.

The general argument concerning electrical ignition in either form urges against the system the troubles from defective wiring, chafing of the wires in moving motors, short circuits from water, and leakage of current from poor insulation. The deterioration of the batteries, and the defective contacts from dirt, oxidation, loosened connections, and the like, can be avoided by careful attention and inspection. The electric systems also fail when the quality of the mixture to be ignited or its temperature falls off, so that a "fat" spark of normal intensity will not insure a satisfactory ignition. In old weather and with a cold motor it will easily happen that ignition will fail at the start. With battery circuits this can be avoided by using more calls in starting than will be required after the motor is warmed to its work, and gradually cutting cells out of the circuit until only those are in use which are needed. The spark must plainly be powerful enough to raise the mixture at the igniting-point to the necessary temperature for a propagation of the flame. This temperature will be higher for a lean than for a rich mixture, and variation of quality may make variation of the spark intensity necessary if governing of the motor acts to impoverish the working mixture.

It will be apparent, of course, that a failure of the ignition system must cause the motor to stop, and an opening of the switch on an electric ignition circuit is an effective means of stopping the motor for short periods in automobile practice.

CHAPTER XII.

GOVERNING.

135. Introductory.—The gas-engine differs from the steam-engine in the method to be used to vary the effort as the resistance varies, by reason of the fact that it does not draw a supply of energy from a reservoir, but that the energy is generated by combustion in the cylinder for each working stroke. The capacity of the governor to increase the power of the cylinder at need is much more limited than in the steam-engine.

It follows, therefore, that the methods which are usual for the steam-engine require not only modification but reconstruction when applied to the problem of governing speed under variations of load. The same principles which apply in the steam-engine apply to the gas-engine, concerning the desirability o. making t' e speed of the engine as nearly automatically controlled by the governor action as is possible, and that the governor should be isochronous, in the sense that it shall make the engine perform its cycle in equal times under all variations of load. Referring back to the fundamental formula,

$$\text{H.P.} = \frac{PALN}{33,000},$$

it will be apparent that this can be written

$$\text{H.P.} = PNK$$

after the engine has been actually constructed, since A and L are not variables and 33,000 is a constant factor, so that the fraction

$\dfrac{AL}{33,000}$ can be denoted by K. If it be desirable to keep the
number of revolutions invariable as the horse-power varies, the
quantity to be varied will be the pressure P, and the methods
to be used will be directed to produce that variation. But the
engine will be at its best when working with the maximum value
for P for its normal condition; hence the governor as a rule acts
mainly to diminish P as the resistance diminishes. The governor will usually be of the shaft type with revolving weights
which are moved outward by the acceleration due to centrifugal
force, while this tendency to move outward is resisted by springs
which draw the weights inward as the speed falls. By having
an initial tension upon the springs, there will be a tendency to
equilibrium between the centrifugal action and the springs only
at that speed for which the governor is adjusted. The governor
may either be on the principal shaft of the motor or on a sub-

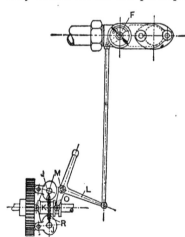

sidiary shaft, driven by gears.
By this latter arrangement the
governor may be placed where
convenient. In many forms the
operator of the engine can control the action of the governor
by hand as a means of varying
the speed of the engine, where it
may be desirable to do so. In
Fig. 83, for example, which shows
an enlarged detail from Figs. 51
and 52, J and J are the masses
thrown radially outward by the
centrifugal acceleration due to
the rotation of the motor-shaft.

FIG. 83.

The springs R tend to draw them in. As the balls or weights fly
outward, their tendency is to slide the collar K to the right, while
the springs R will slide the collar to the left when they pre-
ponderate. The bent-lever arms O and L cause this motion of

K and its equilibrium under action of the centrifugal and spring forces to adjust the amount of opening of the throttle-valve *F* in the pipe through which energy is supplied or controlled with respect to the motor cylinder. The long upper arm *M* can be connected by a rod to a throttle-lever at the operator's hand, so that his will can add either to the action of centrifugal forces to close the throttle-valve, or to the action of the springs *R* to open it. The sketch of course applies in detail to only one system of governing.

136. Governing by Missing a Charge. The Hit-or-miss Governor.—The first and simplest system of governing was to arrange the cam by which the inlet-valve for gas was raised for the suction stroke of the piston, so that when the engine was above speed this cam did not meet the lever which it was to lift. The engine, therefore, made the aspirating stroke without drawing in any charge of gas, and of course, when the ignition occurred, there was nothing to ignite. In some early forms of governor the principle of inertia was applied by causing a reciprocating catch which should normally meet the end of the valve-spindle, to be lifted or lowered out of the plane of that spindle by the inertia of a weight attached to the lever. When the engine was above speed, the reciprocating element, moving faster than the inertia rate of the weighted ball, caused the weight and attached lever to lag behind and miss connection with the valve-stem.

The obvious objection to this system, when close regulation is demanded, is that in the Otto cycle and engine the missing of a working stroke results in an inoperative complete cycle. If the load suddenly increases just after the charge has been missed, there will be a notable diminution of speed, since the last working stroke took place two revolutions previous, and even with a large weight of fly-wheel the variation in speed could not fail to be detected. Where the gas-engine was to be applied to incandescent electric lighting, or to other purposes where close regulation of speed was a vital matter, the hit-or-miss system proved unsatisfactory. It has still been retained on some automobile practice, in order that when the motor is disconnected from the transmission machinery, and the motor-shaft with its fly-wheel is revolving idly, there shall be no unnecessary consumption of fuel from having an impulse in every cycle when the engine is thus running light.

137. Governing by Impoverishing the Charge.—A method of governing in some respects analogous to the foregoing, and derived from it, is to have the governor act to reduce the proportion of gas in the charge relatively to the amount of air on the aspiration stroke, but not to cut off the supply of fuel completely. This will result in having an inflammable mixture in the cylinder on compression, but one which is so low in fuel that the stroke is a comparatively feeble one when the charge is ignited. This governing is effected by having a cam of variable section adjusted on the shaft by the position of the governor-weights, so that the gas-valve is held open a less proportion of the suction stroke. The difficulty with this system is that the inflammability of the mixture is so widely varied by the proportion of gas to air that it may easily occur that with a given compression the mixture will fail to ignite at the beginning of the working stroke, whereupon a charge of combustible mixture is expelled through the exhaust with a waste of fuel and a possible danger of its being ignited in some undesired place. This system has been a favorite one with gasoline machines in which the air was carbureted by aspirating it through a mixing device. If the gasoline vapor was not ignited in the cylinder, it might be ignited by the flame in the next subsequent somewhat violent exhaust, giving rise to explosions in the exhaust-pipes which are noisy and alarming.

138. Governing by Throttling the Normal Charge.—A more judicious system than the preceding is to cause the governor to act upon both the air- and the gas-inlet, so that a less quantity of the normal mixture is drawn into the cylinder, but the proportions of that mixture are not altered. This is a feature of the Westinghouse engine for gas (see Fig. 36), and of the great major-

ity of the newer automobile engines which receive carbureted air
from a carburetor.   The system figured in Fig. 83 has this system
in view, the throttle-valve being between the carburetor and the
motor.   See also Figs. 51 and 52.   It has the advantage that there
is an ignition and a working stroke in every cycle, but that the
pressure in the cylinder is less by reason both of the diminished
compression and the diminished amount of fuel.   By keeping
the mixture in constant proportions the danger of failure in the
ignition is lessened.   It is safe to say that the advantages offered by
this system are so great that the tendency in design is to make use of
it more and more, either exclusively or in connection with the methods of governing by cut-off (par. 142) and by ignition shortly to
be discussed.   It avoids the difficulties of the other systems and
brings the governing of the gas-engine more closely into parallel
with the systems used in the steam-engine.   An ingenious system
of throttle-control by speed has been applied to an American automobile motor (Winton, Fig. 84).   A rotary air-pump driven by the

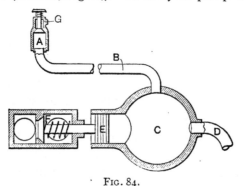

FIG. 84.

motor supplies a moderate air-pressure through D into a reservoir
C.   Opening from this is a cylinder in which fits the piston E.
On the rod of the latter is the throttle-valve F.   As the motor
speeds up, the pressure rises in C and, overcoming the spring,
closes the throttle opening.   To give the operator control over
this action, an outlet from C is controlled by a push-button and
valve at A and G under the heel of his foot.   When A is opened

wide the governing action is practically suspended, since the air-pressure cannot accumulate in *C*. When the motor is stopped *G* will be opened, or in any event leakage out of *C* will gradually cause the throttle to open wide, so that when the start is to be made there is no annoyance from reduction of pressure.

139. Governing by Throttling the Exhaust.—It will be apparent that the net work of the working stroke in a multi-cylinder engine can be reduced and a braking effect in a single-cylinder engine can be produced if the free discharge of the products of combustion into the open air can be restricted by a governing action upon the opening of the exhaust-valve. This throttling of the exhaust operates not only directly to diminish the net forward or driving effort on the crank-pin, but it acts to leave in the cylinder at the end of the exhaust-stroke a certain proportion of products of combustion which are confined therein and which must expand by the aspirating action of the piston down to and below atmospheric pressure before the inlet-valves for gas and air for the new charge will open. The aspiration stroke, therefore, when completed, finds the cylinder filled in part with neutrals resulting from the throttled exhaust which act to dilute the new charge of fresh mixture. This action is, therefore, in effect the same as that of throttling the normal mixture discussed in the previous paragraph. It makes a hot cylinder, also, from the compression of the hot exhaust-gases. This same action results if the exhaust-valve is opened late or closed early in the exhaust part of the cycle and during the exhaust-stroke.

140. Governing by Retarding the Ignition.—If the passage of the spark between the points of the sparking-plug does not occur at the time when the return of the piston for its compressing stroke has produced the greatest pressure of the mixture in the compression space, but takes place a little later, after the piston has begun to move forward, it will be obvious that the igniting of that mixture will not produce the same forward effect. The area of the work diagram is diminished by having the mixture at the beginning of the working stroke retrace the curve of its compression before the ignition raises its pressure by increasing its temperature. The electrical methods of ignition are particularly favorable to this method of governing, which is not possible with the hot-tube systems nor the compression plan.

It is the common practice to adjust the normal engine to fire its charge when the crank is about 15° below or in advance of its dead-centre. This lead of the ignition at high speed particularly gives a chance for the propagation of the flame at constant volume and the complete establishment of pressure. The best mechanical efficiency, however, favors the establishment of maximum pressure, after the piston has begun to move forward for its working stroke, so that the maximum effect should not be so entirely taken up on the shaft-bearings, and before a turning moment has been created for the crank. That is, referring to Fig. 85, if the angle of the ignition line *ab* is inclined to the ver-

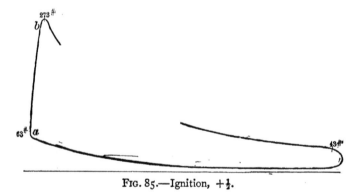

FIG. 85.—Ignition, $+\frac{1}{4}$.

tical by about 5° or 6°. If the ignition be retarded by delaying the spark-instant, the ignition line becomes more inclined, and the maximum pressure comes not only later, but by the expansion of the compressed mixture before ignition the value of that maximum is less. (Fig. 86.) If the spark is still further retarded, the diagram of effort takes the shape of Fig. 87, which is from a $6\frac{3}{4}$ by 12 engine at 240 revolutions, with the spark timed to come when the piston is $\frac{1}{4}$ of an inch past the

dead-centre. The diagrams are scaled to show the fall in pressure.

A difficulty with this system of governing is that the time of
the stroke after ignition may not be long enough for the complete
combustion of the mixture before the stroke is completed and
the exhaust-valve opens. Hence the combustion continues into
the exhaust passages and pipe, with waste of heat, objectionable

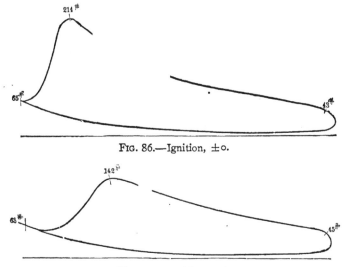

FIG. 86.—Ignition, ±o.

FIG. 87.—Ignition, −¼.

noise due to the pressure, and possible exhaust explosions. If
this system is used in connection with the plan of throttling the
mixture of the normal charge, a wide range of speed and power
control may be easily attained.

The retarded-ignition principle is also of advantage in the
manipulation of multiple-cylinder engines for convenient starting. If the spark period be delayed quite late or the circuit
broken when the motor is being brought to rest, it may result
that when the engine stops there is a compressed charge in one
of the cylinders ready to ignite, but which has not been fired.
When the spark is passed on drawing the retarding arrangement

back toward the dead-centre period, this compressed charge will be fired and the motor start without the necessity for hand starting or the use of auxiliary apparatus.

141. Governing by Advancing the Spark. Pre-igniting the Mixture.—If, instead of timing the·ignition-spark after the working stroke has begun, the charge be fired before the compression stroke is completed, and more than normally before the piston has reached its dead-centre, it will be apparent that the effect is not only to act as a brake upon the compression stroke and retard the engine, but also to diminish the effective energy of the working stroke, since the ignition takes place before the compression is complete. This action can be carried to a point at which the tendency of the fly-wheel and crank to produce compression shall be made by the ignition of the charge early enough to have a leverage sufficient to start the crank revolving in the opposite direction, which would cause the motor to make a backward stroke. This is, of course, the limit in such pre-ignition and should not, under ordinary circumstances, be allowed to occur. Usually, in automobile

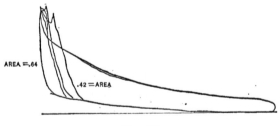

AREA =.64

.42 = AREA

Fig. 88.

practice, the control over the point of ignition is in both directions and on each side of the normal point, so that at the will of the operator the ignition may be retarded or put forward. Fig. 88 shows the work diagram reduced in area from 64 to 42 by advancing the spark, and shows the form of diagrams which result.

142. Governing by Cutting Off Admission.—A system of governing has been used by Mr. Chas. E. Sargent whereby the governing effect shall be to cut off the admission of mixture when the intake stroke is only partly completed. The mixture will be rarefied during the rest of the stroke, but this entails no loss, as the work is restored upon the return of the piston, and the pressure is restored when the cut-off is reached on the compression stroke. Then compression begins and runs through the remainder of the stroke only. Hence the compression is varied according to the work to be done, and the area of the work diagram

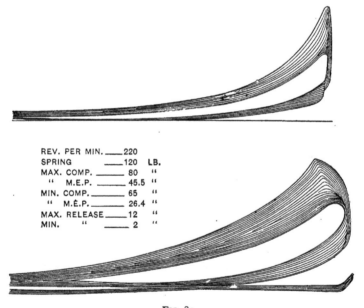

REV. PER MIN. ____220
SPRING ____120 LB.
MAX. COMP. _____ 80 "
  " M.E.P. _____ 45.5 "
MIN. COMP._____ 65 "
  " M.E.P._____ 26.4 "
MAX. RELEASE____12 "
MIN. " ____ 2 "

FIG. 89.

varies with it. Fig. 89 shows a number of superposed strokes, and the governor action in varying the cut-off of admission and the mean effective pressure. Usually the ignition is made to take place earlier as the speed tends to increase, but there is no loss by wire-drawing due to throttling effect. The terminal pressure goes down with this system, which is an obvious advantage as respects noise from the exhaust. This cycle in modified form has been proposed by Clerk of England, Forest of France, and Kohler of Germany. It bears the same relation to the throttling system (par. 138) as the automatic cut-off engine bears to the throttling-engine in steam practice. It is a system which will doubtless prevail more and more.

143. Governing in the Two-cycle System.—In the two-cycle system the governing by throttling the exhaust and by varying the points of ignition is not as simple as in the four-phase cycle. The fact that the exhaust-ports are opened by the motion of the piston itself at the completion of the working stroke and are closed after a part of its return stroke has been completed nearly always leaves a proportion of the products of combustion entrapped in the cylinder, so that exhaust throttling does not apply and the opening of the exhaust-ports before the end of the stroke makes the retarded-ignition method unsatisfactory. Most two-cycle engines throttle the mixture. The method adopted in the Korting engine of retarding the admission of the constant mixture until part of the compression stroke has been completed is, in principle, the same as that of throttling the normal charge in combination with an action to dilute that charge with products of combustion which are not completely expelled when the admission of fresh mixture is retarded. It is apparent that combinations of the methods discussed above can be made other than those which have been described as actually applied.

144. Limitations of the Gas-Engine by the Problem of Governing. — The internal combustion engine, whether applied to motor-cars, launches or to station work and isolated plants, has the same problems to meet respecting its uniform motion as are to be met when power is taken from the steam engine. Except in the launch and motor-car the requirement is usually rigorous that the motion of the crank shall be so nearly uniform per stroke of the piston that all revolutions shall be made in equal times, or with a permitted variation therefrom of a very small percentage. The function of controlling such uniform rotative speed is divided between the fly-wheel on the crank or motor shaft and the governor.

It will be plain that absolutely uniform speed of rotation under all variations of resistance and effort, and under the variations of the forces at play upon the crank-shaft itself, will be very difficult of practical attainment. A governing system which would do this would be called *isochronous*, or equal timed, making the motor make not only the same number of strokes per second or per minute but making each stroke in the same time.

The forces at play upon the crank-shaft and which tend to make its revolutions non-isochronous are those which tend to turn it in its normal direction or revolution, and those which retard such turning, and may therefore be considered to act as forces in the opposite direction. On the positive side are:

1. The effort of the expanding gas mixture upon the piston, acting on the crank-pin.

2. The inertia or stored energy in the reciprocating masses attached to the piston.

3. The inertia or stored energy in the revolving crank-arm, fly-wheel, and part of the connecting rod.

Retarding these energies, and therefore acting in opposite sense are:

4. The useful resistances being overcome reduced to the path of the crank-pin.

5. The resistances in the cylinder valves and passages due to the compression exhaust and aspiration stroke.

6. The mechanical friction of the engines at bearings and contact surfaces, and resistance of auxiliaries.

7. Inertia of reciprocating parts acting to retard the crank, and possible inertia of revolving parts.

These have been more exhaustively analyzed in paragraph 7a, and its accompanying diagram. These forces cannot be constant in any case, and their variations will be due to:

1*a.* The varying effective leverage of the crank-arm, from a zero at the inner dead-centre to a maximum at the 90° point. Some designers have aimed to secure some advantage by not putting the centre of the crank-shaft in the prolongation of the cylinder axis, but have "offset" the cylinder to get better working angles for the crank on the power stroke.

1*b.* The pressure of the expanding gas is varying all through the traverse.

2*a.* The inertia of the reciprocating masses is only acting positively in the second half of the stroke. In the first half during the acceleration period they are retarding.

3*a.* The stored energy in revolving masses is only liberated to do positive work when the crank-shaft begins to slow down from excessive resistance.

The resistant forces vary because:

4*a.* No load is ever constant. Some only vary more than others do.

5*a.* These vary with condition of the air, power being used, condition of motor parts, etc.

6*a.* Vary with care in lubrication, quality of the lubricant, conditions of the feeding apparatus.

7*a.* When reciprocating masses must be accelerated or the fly-wheel speeded up, the inertia of these must be overcome.

When there are several cylinders some of these masses may be acting in opposite ways upon the crank, and these influences will go through cycles of increase and decrease.

It may be said to be the function of the fly-wheel to take care of variations per stroke or revolution in Nos. 1, 2, 3, 5, and 7. It is the function of the governor to take care of Nos. 4 and 6, and to supply sufficient turning force to meet the variations in the load. A massive fly-wheel may also act as an accumulator for turning effort under sudden variations of load, but when such accumulation is used up, or its capacity to accumulate is reached, the motor slows down in the first case, and speeds up in the second. The governor meets variation in mean effort over several strokes; the fly-wheel takes care of maxima and minima of effort per stroke or cycle. The governor is to affect the area or the shape of the indicator diagram from which mean effective pressure is deduced; the fly-wheel must provide for cyclic irregularities, inertia effects, and part of the sudden variation in resistance for a very short time. The governor itself as a problem in design must be quickly sensitive to the retarding or accelerating effect of increase or decrease in the external resistance, and must quickly thereafter adjust the supply of energy so as to modify the area of the indicator diagram and vary the mean effective pressure. Governors which are massive beyond the needs of the power they must exert to affect the readjustment necessary, will be sluggish from their inertia, and will have unnecessary friction. Both will make the governor unsensitive. The problem of governor design is aside from the present purpose, but it may be said in general that governors which use springs to oppose the effect of speed increases are more quick-. acting than those which depend on gravity.

In recent applications of the gas-engine to the driving of electric-generators, and notably of alternators in parallel, the demand for regulation of turning effort upon the fly-wheel shaft has not only been for an isochronism as respects numbers of revolutions per minute, but further than this, that the angular variations in each stroke or revolution shall be kept inside of definite limits. This specification demands a close study of the inertia forces in the reciprocating masses and the securing of balance of effort by the arrangement of the cylinders relative to the shaft, and the timing of the impulse strokes which together shall reduce irregular actions per stroke to a minimum. This may be done graph-ically from a combination of the ordinates of gas pressure, with the ordinates of the inertia forces, and a resultant curve drawn, giving a force curve. On the latter is superposed the resistance curve, supposed constant for one revolution (or for two as may be required) and from the two is derived a coefficient of fluctuation of energy per period, which can be used in any of the standard deductions for fly-wheel mass in accepted formulæ. The study of balance is specially important in design of motor-car engines which are not fastened to foundation masses as in stationary engines, but it is not to be neglected in any case. Students are referred to other treatises for exhaustive discussion.*

It is obviously difficult in the single-cylinder motor operating with the four-phase cycle to secure a uniform torque of the engine-shaft, even with a massive fly-wheel, since the working impulse comes but once in each two revolutions, even when the resistance is constant. This difficulty is mitigated by increasing the number of the cylinders up to four, so that there shall be an impulse at each half-revolution, as occurs with the double-acting single-cylinder steam-engine. It is also made of less consequence by greatly increasing the speed of the fly-wheel shaft so that the interval of time between impulse strokes shall be correspondingly lessened and the resistance attacked by the impulse stroke with correspondingly greater frequency. For these reasons the limitations set to the use of the single-cylinder four-phase cycle where uniform speed of rotation was demanded have been to a great extent removed. It remains the fact, however, that even the best methods of governing have not yet produced such nicety of adjustment of the working pressure to the resistance as is possible with the automatic cut-off steam-engine in view of the practical impossibility of adjusting the release of energy in the stroke within the narrower limits imposed by the rate at which this energy is released when the charge is ignited. The steam-engine draws from a reservoir of accumulated pressure, and may draw more or less as required per stroke. The internal-combustion system must receive all the energy resident in the mixture received, and can only govern by varying this amount of energy before it is released.

---

* For example, " Gas Engine Design," pp. 61–146. C. E. Lucke. Van Nostrand Co., 1905.

CHAPTER XIII.

THE COOLING OF THE CYLINDER.

145. Introductory.—Since the effective utilization of the heat energy in an internal-combustion motor depends upon the difference between the initial and final temperatures of the mixture which drives the piston, it will be apparent that the ideal and economic method for cooling the gas would be the effective transfer of all of its sensible heat into mechanical work. It is desirable that the gas should thus be cooled in order that it may carry out to waste the minimum amount of available energy. On the other hand the metal of the cylinder should be cooled not only as a matter of comfort to those who are about it, but to prevent deformations, leaky valves, defective alignment, and oxidation resulting from high heat. It will be apparent, furthermore, that the cooling of the metal must be done with convenient means and without requiring too great bulk or weight of the medium used for cooling.

It may be said, in general terms, that there are·two methods of cooling the metal. One is by the use of water and the other by the use of air. For the cooling of the gas one method will be by water injection and the other will be by jacket-cooling through the metal walls of the cylinder. The water may be most effectively used upon the mixture by injecting it. The water or air cooling of the metal will be done by circulation.

146. Cooling by Injection into the Air, into the Expanded Gases, into the Products of Combustion.—It has been proposed to inject into the air which is drawn into the cylinder on the aspiration stroke a certain quantity of water in the form of mist. The compression of the mixture which raises its temperature will convert this finely divided water into steam, which will partake of the heating when the gas ignites and by its higher specific heat shall tend to keep the temperature of the mixture lower than it would be in the absence of such water. The water partakes of the cooling due to the adiabatic expansion, but the heat to vaporize it when the pressure is lowered at the opening of the exhaust will be absorbed from the products of combustion, whereby their temperature will be lowered.

A variant upon this plan is to inject the water in spray into the expanded gases after the ignition has taken place, with the idea that the steam thus formed should partake of the expansion in the cylinder and, by the absorption of the heat for the vaporization, cool the mixture as in the foregoing method.

The difficulty, so far as both systems are concerned, is that the injection of this spray or mist of water with its higher specific heat tends to lower the temperature in the cylinder and thus to diminish the net forward effect upon the piston; and to be effective enough to render a water-jacket unnecessary the quantity of injection water required would render combustion impossible, by reason of the great dilution by water-vapor and spray. If little water is injected, so that this effect is not produced, the water does little cooling, but in any case its effect is to diminish the area of the work diagram in the cylinder.

Injection has also been practised as respects the products of combustion after they have left the cylinder and are in the exhaust-passage. This lowers the tension of the exhaust-gases by lowering their temperature, but produces no effect upon the medium which is working in the cylinder itself. It is not, therefore, usually worth while to pay much attention to the heat in these products of combustion, since it is inconvenient to utilize it in any practical way.

147. Cooling of Metal by a Water-jacket, the Steam to be Utilized or Wasted.—Since the specific heat of water is unity, it is the most convenient medium to use for withdrawing excess of heat from the metal of the cylinder and its valves. This is usually accomplished by casting the cylinder with double walls or by surrounding it with a brazed copper-jacket and circulating the cooling water through the hollow spaces. The cool water enters at the bottom and, becoming warmed, it flows off at the top, carrying with it any bubbles of steam which may form in the process, and which would have a tendency to rise. If there is an abundance of water, it is usually convenient to waste the heated water without attempting to apply it to any useful purpose. If water is limited, as in the case of the automobile engine, it will be necessary to use some means for cooling it in some form of radiator, whereby the heat which it absorbs in circulating around the cylinder shall be withdrawn and the same water used over and over again. It is possible to utilize the heat which the water-jacket will carry away, but ordinarily this is more trouble than the economy which it represents.

In the automobile water-cooled engine the radiator for cooling the heated water is made of a coil of pipes, each pipe being armed with a very large external radiating surface so that the movement of the vehicle through the air shall give to the latter a great surface upon which to act for the withdrawal of the heat of the pipe (see par. 87). Some forms of radiator have transverse cooling pipes through the main body of the water, and in addition to the motion of the vehicle through the air a further current of air is stimulated by a propeller fan driven from the engine, so that the velocity of the motor-shaft shall determine the volume of cooling air and not alone the velocity of the car.

It is usual to let the cooling water reach nearly the boiling-point, or about 180° Fahr., so as to keep the metal below the point of deformations, and yet not cool the cylinder-walls unduly. If the water boils, of course it gradually dissipates as vapor and must be replaced. About one gallon of water per horse-power seems to be a convenient proportion for the tanks which carry the cooling water. In cold climates care has to be taken in the cooling of out-of-door motors lest the cooling water freeze when the motor is at rest. This has resulted in the mixing of glycerine with the water in equal parts, or the adding of chloride of calcium ($CaCl_2$) till the solution has a specific gravity of 1.20. Such a solution will not freeze at 15° below zero Fahrenheit. Trouble has often been experienced with water-cooled motors which use water containing mineral matter, from the deposit of such scale-forming material in the jackets. The narrow spaces become clogged with the deposits, and the water cannot get access for cooling the metal.

148. Water-cooling of the Piston.—While the surrounding of the cylinder-walls by circulating water in the water-jacket has a tendency to keep their metal cool, it does not affect the reciprocating piston which touches these walls over a comparatively narrow surface while exposed to the high temperature of the charge over the large area of its head. This has necessitated, in engines of large size, that provision be made for cooling the piston by circulating water through hollows cast in its structure. This is shown to be necessary by the fact that even when the sides are water-jacketed the pistons of large gas-engines will appear red to the eye in the dark, in the absence of such inner circulation. The water is introduced either by means of a flexible connection or by means of two hollow tubes, finished on the outside, which slide through a stuffing-box, entering into corresponding chambers within one of which cool water is maintained under pressure and through the other of which the heated water is discharged.

149. Cooling by Air-jacket. — The inconvenience of carrying the necessary weight of cooling water in the automobile, and the annoyance which is offered in winter by the presence of this water, and the danger of its freezing in cold weather, when the engine is not in operation, have brought about the design called the air-cooled motor. The cylinders of such engines are cast with deep corrugated exterior surfaces or are fitted with a multitude of radial screwed pins on the outside so that for a given diameter of cylinder a very much increased external surface shall be presented to the cooling action of the air, both by radiation and contact. This effect may be increased by a fan action which shall blow air upon the radiating surface of the cylinder and tend by this action to keep it cool. Such cylinders are, of course, heavier than the water-cooled cylinder, but the weight of the water is avoided. Fig. 50 shows a characteristic structure of such air-cooled cylinders (see also par. 87). The limitation in air-cooling seems to be set by the quantity of heat in units which is liberated on the working stroke. As the engine grows more powerful, the difficulty of effective cooling increases, and is set at not far from ten horse-power at present, by the fact that enough air cannot be brought into action in the limited surface to cool a cylinder heated hot by a large weight of fuel. If the cylinder is not adequately cooled, the compression of the charge may result in pre-ignitions or back-firing, and difficulty may be experienced in stopping it quickly without inconvenient use of powerful brakes (see par. 127).

150. The Circulation of the Cooling Water and the Amount Required for Cooling.—In automobile practice it has been observed that it is usual to proportion the surface of the radiator so that the temperature of the cooling water shall rise nearly to the boiling-point. If it were to be allowed to get hotter than this, the water would, of course, generate steam, which would produce a pressure upon surfaces ill adapted to resist pressure and which would result in a dissipation of the water in the tanks, requiring its frequent renewal. This circulation of the water through the jacket and the radiator is accomplished by a pump either of the centrifugal or rotary type in most cases, since a small volume or weight of pump of this design will circulate the greatest weight of water with the small resistance which such circulation offers. The weight of water which is to be circulated to keep a given weight of metal at a certain fixed temperature is given by a simple equation for the transfer of heat involving the temperatures, specific heats, and weights. If W be the weight of the iron to be cooled, and its specific heat C_i, and the range of temperature through which it is to be cooled be denoted by $t_1 - t_2$, while for water the weight be designated by w, specific heat by unity, and its range of temperature $t_2 - t_3$, since the water and the iron will be assumed to have practically the same temperature when the latter has been cooled, the equation will appear

$$W \times C_i \times (t_1 - t_2) = w \times 1 \times (t_2 - t_3).$$

In this equation the temperatures are known or assumed and the unknown quantity to be calculated is w, when the other elements of the equation are known.

It should be carefully observed in designing water-jackets that the different parts of the cylinder casting are not exposed to the same intensity of heating effect. Hence the consequences of distortion from unequal heating and disproportionate cooling are as carefully to be guarded against. The parts should be free further to expand and contract under the changes of heat without distortions of other parts. Valve-seats in particular are points of special importance, and a design of much merit will be one in which both seat and stem are formed in cored castings around which a copious flow of cooling water can be secured. Such cooling also prevents a previous dilation of the mixture by heat on entering the admission-valve, whose consequence will be a loss of weight in a volume of given dimensions. It is best to cool the valve-casings first and with the coolest water; the warmed water may pass thence to the cylinders and so to waste.

CHAPTER XIV.

THE COMBUSTION-CHAMBER AND THE EXHAUST.

151. Introductory. — In the steam-engine and in the air-compressor it is desirable that the space between the head of the piston and the head of the cylinder should be reduced to the lowest possible percentage of the volume swept through by the piston. This is by reason of the fact that this volume in the steam-cylinder is filled at each stroke by steam from the boiler which is not required to do the work of that stroke and which is exhausted with the working steam when the exhaust-valve is opened. In the air-compressor, whatever compressed air remains in this clearance volume expands during that stroke which should be entirely the admission stroke for fresh air, and by its expansion precludes the opening of the inlet-valves until the pressure in the cylinder is less than the atmospheric pressure without. In the gas-engine, on the other hand, the space between the piston and the end of the cylinder, when the engine is on its inner dead-centre, is the space in which the combustible mixture is to be held under compression from the return stroke of the piston, and which must contain a sufficient amount of the combustible mixture to furnish whatever effective pressure is to be exerted when the charge is ignited. It will be apparent, therefore, that the volume of this clearance space is, in effect, the combustion-chamber of the engine, and it must bear such a relation to the volume of the piston displacement as shall give the desired compression pressure at the moment that the charge is to be fired, under normal conditions. What, then, should be the volume of the combustion-chamber?

152. Volume of the Combustion-chamber. — The volume of the combustion-chamber will be very greatly dependent upon the quality of the fuel which is to be burned in the engine. While the greater the pressure caused by the compression, the greater work will be done in the Otto cycle, a definite limit is set with rich mixtures which are readily ignited. With gasoline gas, for instance, it is impossible to carry the compression above 80 pounds per square inch, since the heat generated by the compression within these limits will be sufficient to pre-ignite the charge, causing the piston to make a back stroke. With lean mixtures, such as blast-furnace gas, the compression pressure may be made much greater than this, without danger of pre-ignition. The following table gives some data concerning the limits of pressure with combustibles of different points of ignition:

COMPRESSION PRESSURES IN POUNDS PER SQUARE INCH.

Type of Engine.	Pressures Permissible When Fuel Used is						
	Blast Furnace Gas.	Producer Gas.	Natural Gas.	City or Illum'tg Gas.	Gasoline.	Kerosene.	Alcohol.
1	**2**	**3**	**4**	**5**	**6**	**7**	**8**
Large size							
Rich gas............	120	100	75	60	...	...	...
Lean gas............	190	160	130	100	...	...	...
Average .... {	150	135	...	...	...	...	...
	155	140	115	80	...	...	...
Small size	...	125	60	...	...	...	...
High speed { low........	...	...	...	65	45	45	100
{ high......	...	...	...	90	95	85	200
{ average ...	...	...	...	80	80	65	150
Low speed { low........	...	...	...	...	60	...	...
{ high......	...	...	...	...	85	...	...
{ average.....	...	...	...	...	70	...	...

If the piston be assumed to draw in one pound of the mixture of fuel and air at atmospheric pressure when it has moved to its extreme position (called A in Fig. 205), and such initial volume V_i be assumed to have its weight unaffected by the presence of the gasefied fuel, it may be supposed to occupy the same volume at 62° F. as the same weight of air would. In paragraph 11 the volume at 32° F. was called 12.39 cubic feet; hence at 62° F., by the Mariotte law (par. 43), its volume will be:

$$V_i = \frac{522}{492} \times 12.39 = 13.133 \text{ cubic feet.}$$

If this initial volume v_i be compressed to any other volume V_x, or to a final volume V_f, and the corresponding pressure be assumed, the value will be given when the cylinder walls do not carry off any of the heat due to such compression by the formula of paragraph 51, in which the initial pressure is that of the atmosphere, or 14.7 pounds per square inch, and the exponent $n = 1.41$. Hence

$$V_x = V_i \left(\frac{P_a}{P_x}\right)^{\frac{1}{1.4}} = 13.13 \left(\frac{14.7}{P_x}\right)^{.71},$$

or,

$$V_f = V_i \left(\frac{P_a}{P_f}\right)^{\frac{1}{1.4}} = 13.13 \left(\frac{14.7}{P_f}\right)^{.71}.$$

The table on following page gives the computation of such volumes with assumed ratios of $\frac{P_a}{P_x}$ in atmospheres in order to make the numerator unity for convenience in graphical representation, as well as in pounds per square inch.

A graphical presentation of the results of computation as given in column 4 of the foregoing table appears in Fig. 257 herewith, which is the standard curve for adiabatic air compression, and if used as a template may be applied to actual indicator cards by superposing the template pressure over that realized actually in the card under observation to find the volume ratio, or the volume ratio being known the pressure theoretical for these volumes may be read directly.

But the foregoing discussion concerning the pre-ignition difficulty with the various fuels, and the appropriate compression pressure indicates that the compression pressure ratio cannot be

COMPUTATION TABLE OF VOLUME RATIOS IN COMPRESSION
WHEN PRESSURE RATIOS ARE TO BE ASSUMED.

Compression in Atmospheres.	Pounds per Square Inch.	Value of $\left(\frac{P_a}{P_x}\right)^{.71}$ Computed.	Resulting Volume V_x in Cubic Feet.	Value of $\left(\frac{P_a}{P_b}\right)^{.29}$ Computed.	Value of $5.41\left[1-\left(\frac{P_a}{P_b}\right)^{.29}\right]$ Computed.
1.5	22.05	0.881	11.57	...	...
2.0	29.40	.611	8.02	...	...
2.5	36.75	.5217	6.85	...	...
3.0	44.10	.458	6.01	0.727	1.474
3.5	51.45	.411	5.39	...	...
4.0	58.80	.374	4.91	.669	1.787
4.5	66.15	.343	4.50	...	...
5.0	73.50	.319	4.19	.627	2.014
5.5	80.85	.298	3.91	...	...
6.0	88.20	.281	3.69	.595	2.187
6.5	95.55	.265	3.48	...	...
7.0	102.90	.251	3.29	.569	2.327
7.5	110.25	.239	3.11	...	...
8.0	117.60	.228	2.99	.547	2.446
8.5	124.95	.219	2.875	...	...
9.0	132.30	.209	2.74	.527	2.554
9.5	139.65	.202	2.65	...	...
10.0	147.00	.195	2.56	.513	2.630
10.5	154.35	.188	2.47	...	...
11.0	161.70	.182	2.39	.498	2.711
11.5	169.05	.177	2.33	...	...
12.0	176.40	.171	2.245	.486	2.775
12.5	183.75	.166	2.18	...	...

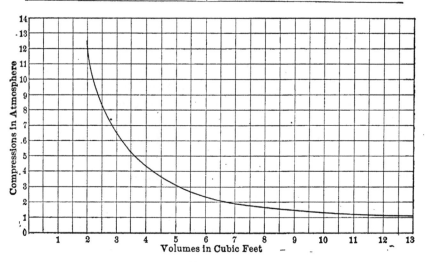

FIG. 257.

arbitrarily assumed, but must be based on the characteristic of the fuel in this respect. .

A second underlying principle for design will be to work back from the mean effective pressure desired. In paragraphs 40 and 47 it was developed that

$$\text{M.E.P.} = 5.4\frac{H}{a+1}\left[1 - \left(\frac{P_a}{P_b}\right)^{.29}\right],$$

an equation which involves the ratio of the final to the initial pressures resulting from the compression process, when H is the calorific power or heating value of the fuel. In column 5 of the preceding table are given the computed values for the factor $\frac{P_a}{P_b}$ and in column 6 the result of multiplying the efficiency factor by 5.41. This column can therefore be used by inserting in the formula

$$\text{M.E.P.} = \frac{H}{a+1}\left[\begin{array}{l}\text{tabular value} \\ \text{from column 6.}\end{array}\right],$$

the value appropriate for the compression which is proper for that fuel, and checking the result of this substitution with the value for M.E.P. which has been used in working out the H.P. To use this effectively, however, and to pass from the theoretical results with standard conditions to those of practice, a knowledge should be available of the values for M.E.P. with various compression values as realized in engines at work and using the various fuels. The following table gives some of these data, but while all are from tests, it does not follow that all are from tests under the best combinations of condition. It will be convenient to have the data of this table greatly extended by series of tests under identical fuel conditions with a given engine, and with different engines. The lack of concurrence in the series indicates the need of further research, and the care and experience which must be the guides in use.

Compressions Pounds per Square Inch Absolute.	Mean Effective Pressure Observed when Fuel is					
	Blast Furnace Gas.	Producer Gas.	Natural Gas.	City or Illuminating Gas.	Gasoline.	Kerosene.
45	...	...	...	45	...	...
46	...	...	...	...	...	40
50	...	...	...	...	...	35
55	...	...	...	...	...	85
60	...	...	...	80	...	...
63	...	...	...	...	...	69
65	...	...	...	...	...	68
66	...	...	...	95	106	...
68	...	...	...	...	...	40
70	...	...	...	60	75	72
75	...	...	...	...	100	...
86	...	...	...	...	70	...
86	...	...	...	...	72	...
90	...	62	...	...	...	...
91	...	...	...	90	...	...
95	...	90	...	...	60	...
95	...	100	...	...	...	...
103	...	63	...	...	...	...
108	...	88	...	...	...	...
115	...	103	...	...	...	...
125	...	51	68	...	...	...
130	...	...	82	...	...	...
135	...	...	90	...	...	...
140	47	83	...	...	...	...
155	81	...	...	...	...	...
170	...	73	...	...	...	...

A third principle of design is that of which an example is given in paragraph 203, in which the actual value of n as discussed in paragraph 56 for the compression stroke is taken as 1.35 to allow for the action of the cylinder walls and other transfers of heat. If as before and in Fig. 205 the initial volume before compression be called V_i = piston displacement + clearance volume, V_f = final volume = clearance volume only; then $V_i - V_f$ = piston displacement, and if R be the ratio between the clearance volume and the piston displacement:

$$\frac{V_f}{V_i - V_f} = \frac{\text{clearance volume}}{\text{piston displacement}} = \frac{1}{\frac{V_i}{V_f} - 1} = R.$$

If then as in paragraph 56 it be taken that:

$$\frac{V_i}{V_f} = \left(\frac{p_a}{p_b}\right)^{\frac{1}{1.35}} = \left(\frac{p_a}{p_b}\right)^{.74},$$

and these pressures be substituted for the volumes in the equation for R, and the appropriate values for the former be inserted the values result;

$$R = \frac{1}{(3.33)^{.74} - 1} = \frac{1}{1.43} = 70\%.$$

For street-gas

$$R = \frac{1}{(4)^{.74} - 1} = \frac{1}{1.79} = 56\%.$$

For producer-gas

$$R = \frac{1}{(8.6)^{.74} - 1} = \frac{1}{3.91} = 25\%.$$

For the Diesel high compression

$$R = \frac{1}{(30)^{.74} - 1} = \frac{1}{11.39} = 8.7\%.$$

From these derivations the origins of the usual values are made to appear. When the builder does not know what fuel is to be used, but is obliged to make a commercial article which will not be too far amiss for any ordinary case, he selects the middle of the series and a mean value. This explains the very general adherence to about 30 per cent.

It will be apparent, from the foregoing equations and discussions, that the less the clearance volume with a given displacement volume, the higher the compression pressure. To guard against pre-ignitions, this will compel a leaner mixture, or one with less fuel in proportion to air, and as the fuel is the costly element, the greater will be the fuel economy for a given displacement volume. But further than this, a diminished clearance volume

gives a less volume of burnt gases to be admixed with the fresh
charge, and to dilute it so as to diminish the effective weight of
fuel. This diminished dilution will lessen the percentage of
neutral gas with its effect on combustion (par. 222). With a small
clearance the cylinder volume is more effectively scavenged when
the exhaust opens.

153. Form of the Combustion - chamber. — It is desirable
for reasons of strength and simplicity in casting that the combustion-chamber be of rounded or spheroidal form, with no corners
or pockets in its volume. The openings leading to the valves
for similar reasons should open small vestibules into the combustion-chamber, so that the gas in these entries may partake of
the facility of ignition belonging to the main volume of mixture.
It has been found that some very curious phenomena result from
such a shape of the combustion volume behind the piston as shall
make the ignition of the mixture resemble a succession of explosions. When these shapes of the combustion-chamber and its
entries are conducive to this action, there are developed pulsations
in the mass of gas which seem to be curiously cumulative in effect
and which, when the right proportions are attained, reach an
intensity close to the limit of the possibility of their being resisted
by the metal. These pulsations reveal themselves, often, to
the indicator-card by a succession of waves in the line of expansion which would naturally be attributed to inertia of the indi-
cator-piston and attached parts, but which can be proved to be
the result of this pulsating action by the very simple experiment
which is described in paragraph 204 of Chapter XIX, to which
the student is referred.

154. Disposal of the Exhaust-gases. — The exhaust from
the cylinder should be mainly carbonic acid gas, steam, and air.
With the gasoline engines there may be present a certain amount
of hydrocarbon gas incompletely burned, which will give a slight
color to the exhaust with a characteristic odor. In damp weather
the exhaust also may contain a little steam from the moist air
drawn into the cylinder and there made into steam to appear
in the cooled exhaust in visible form. The requirement in many

cities for stationary practice is that the exhaust-pipe from the engine must first discharge into a reservoir or pot from which the exhaust-pipe proper shall pass out and shall be carried continuously through the chimney or flue to the open air. The purpose of the pot or chamber is that any explosions due to defective propagation of the flame in the mixture, or due to improperly timed ignitions, may occur in the pot rather than at a place where they might be attended with more danger of fire. In some considerable plants the exhaust from the engines has been taken into a subterranean cistern fitted with a loose cover of planks and weighted with iron. The purpose of this structure was to form a relief-valve for the harmless release of explosion energy which escaped from the cylinders and had to be taken care of in the exhaust circuit.

155. Back-pressure of Exhaust-gases. — In the ordinary Otto cycle the gases at the end of the working stroke will have a pressure of varying intensity, but considerably above that of the atmosphere. This fact explains a considerable noise in the escape of the exhaust-gases, since they are at pressures and temperatures above that of the atmosphere, which will result in a considerable release of energy when they come together. The volume of the exhaust-gases will be increased by the release of the pressure, but diminished by the drop in temperature. If the pressure be nearly that of the atmosphere, the effect of the temperature change is to diminish the volume, while if the pressure is above the atmosphere, the tendency to expand takes precedence over the tendency to contract. The expansion is the occasion of the noisy cough which is more frequent than the other effect.

The Clerk system, which has the exhaust-port uncovered by the traverse of the piston near the end of its stroke, and which is used in most of the two-phase engines, can be used with or without an additional exhaust-valve. When used with a second valve, it will be apparent that the release of pressure by the uncovering of the lateral port in the cylinder takes off much of the pressure against which the lifting poppet-valve in the Otto system has to work. Among the engines using this double-exhaust are the White and Middleton, the New Era, and the Springfield machines. The Clerk port, being without effect on the back-pressure of the return or exhaust stroke, can have its pipe and its noise very effectively muffled (§ 156), while the low pressure prevailing during the cleansing stroke causes little noise in any case. If there are two exhaust-ports, it is convenient to pipe them separately, since the pressures in each pipe will differ widely as will also the phenomena connected with them.

The considerable pressure on the inner or pressure side of the exhaust-valves throws a considerable strain also on the cams by which they are lifted, particularly in large engines where the valve will have a large diameter to give sufficient area. This strain wears the cam and the roller and pins, and tends to buckle the valve-spindle when the cam is in the plane of the valve; or when operated through rock-shaft and arms, the latter are subjected to torsion and flexure. This condition has led to the use of balanced valves of the double-seat or poppet type, and to the use of balancing pistons, so that the pressure of the expanded charge on the area of the valve should be counterbalanced to a great degree, leaving only enough excess on the valve to hold it firmly to its seat, and leaving only to the cams or levers the work of overcoming weight, friction, and this small unbalanced excess of pressure.

156. Muffling of the Exhaust.—In order to diminish the unpleasant noise resulting from the pressure and change of volume of the exhaust as it escapes into the open air, a number of devices have been presented on various designs of machine, which are known as mufflers in America, and as silencers in English and European practice. The theory of the silencers or mufflers is to secure one or all of four results. The first is to reduce the pressure of the gases until they are as nearly as convenient at the same pressure as the atmosphere when they are ready to escape into it. Second, this is secured by allowing the gases to expand in volume and in this process to become cool. Third, the effect of such expansion, which is the manifestation of the reduction of pressure, is to diminish the velocity with which the gases escape into the open air, which velocity is partly the occasion of noise. Fourth, all of these foregoing results are helped if the body of escaping gas is broken into a number of smaller streams, by being baffled and allowed to escape through a large area in divided form. Any construction of muffler which will reach these results will meet the case. It is quite usual to use a piece of pipe or tube of diameter considerably greater than that of the exhaust-pipe (Fig. 90), and to allow the exhaust-gases to enter into that larger pipe or tube through a number of small openings where-

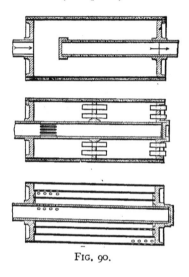

FIG. 90.

by the volume of the gas is gradually increased, its pressure reduced, the volume expanded, and the final discharge takes place through a great number of openings.

Other plans are to fill the enlarged pipe or tube with baffling partitions, so that the velocity of the gas is greatly retarded, while the areas through which it passes are large enough to result in no back-pressure upon the exhaust-pipe, and at the free end the velocity is very low. The muffler can conveniently be at some considerable distance from the engine (Fig. 91), so that in the length of pipe which couples the engine to the muffler, there may be opportunity for the cooling of the gases before they enter the muffler. If this cooling can occur, it is followed by a lowering of the pressure, a diminution of the volume, and a greater effect of the muffler in silencing the noise of the escape. The baffling in the muffler can be done with perforated plates, with coils of wire, with pebbles or balls, through which the gases must pass on their way out.

Unless this baffling device is located in chambers of enlarged

volume, it will be apparent that the tendency of this plan of silencing will be to produce an increased back-pressure on the exhaust

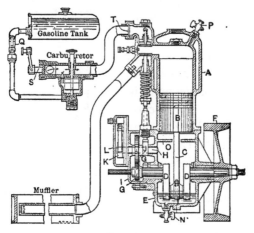

Fig. 91.

stroke of the cylinder. This may be enough to invade the power
of the stroke by acting in the same way as the method of governing by throttling the exhaust. In many forms of the automobile engine a by-pass is arranged so that when the motor is to
propel the carriage slowly, as in city streets, the muffler will be
in action and the exhaust silenced; when the open country is
reached, where noise is of less moment, the muffler is switched
off and the exhaust takes place freely through a direct connection
to the air, with the attendant noise, but diminished back-pressure.
To be effective, a muffler should be of large volume.

In the exhaust-pipes from stationary motors, discharging
into the open air, an effective silencing has been secured by the
simple expedient of cutting slots in the side of the pipe near its
end, so that as the moving column of gas drew near to the opening, whereby it would naturally escape with considerable velocity
into the air, the pressure was allowed to fall by a free but gradual
expansion through these slots sidewise. The principle of this
suggestion is undeniably sound in the light of the foregoing discussion.

CHAPTER XV.

160. Introductory. — The gas-engine appears in so many forms which differ in detail to such an extent that it is not easy to give suggestions as to the manipulation of such machines which shall be applicable in every case, or which shall apply to all forms of the internal-combustion engine. Certain fundamental principles, however, must be observed in all cases, to which attention may properly be directed.

161. Effects of Quality or Richness of the Gas.—If the gas used in the motor comes from a street main, it is usually of a certain standard quality or calorific power and will not be likely to vary at different seasons of the year or on different days. Where the gas comes from a producer, it is likely to vary in richness with variations in the operation of the producer itself and variations in the fuel from which the gas is distilled. But the widest variations and those which are the sources of the greatest trouble occur as the result of varying carburation of the gas in automobile motors, and from such action of the governor as will vary the percentage of hydrocarbon in the volume of the mixture of air and fuel which enters the cylinder in any stroke.

If the engine be adjusted to a normal running, with a certain proportion of air and fuel in the mixture, the mixture may be varied either by impoverishing it below this normal proportion or by enriching it above the normal (§§ 100, 137). If the mixture is impoverished, the effect will be likely to manifest itself in a reluctance to ignite. A failure to ignite will obviously result in no

297

impulse or working stroke at the normal interval of such impulse which will result either in the motor working irregularly, or perhaps in explosions in the exhaust-pipe, or both. Too poor a mixture will be the consequence of defective working of the carburetor or inadequate opening of the fuel inlet-valve. An impoverished mixture will be particularly annoying where hot-tube or hot-surface ignitions are used, since the compression may not produce a temperature sufficient for the ignition to be properly timed with respect to the working stroke. If, on the other hand, the mixture be enriched above the normal, there may result an irregular working due to pre-ignition of the charge, particularly with hot-tube systems of ignition, since the temperature of such rich mixture will be raised to the ignition-point before the full compression stroke is completed. This will make the engine work irregularly and with considerable sacrifice of its full capacity. It is obvious that too rich a mixture is wasteful of fuel, since more than the necessary amount is supplied at each stroke, and while the power of the working stroke is increased, the engine will be noisy and will operate with considerable shock and jar. In view of the principle of the compression cycle it is of advantage, so far as the volume of the cylinder is concerned and the economy of fuel, to compress to a considerable degree before the charge is ignited, and it is desirable not to make the charge so rich as to make it difficult to secure a high compression without danger of pre-ignition. Pre-ignition in starting the engine is particularly annoying, since under these circumstances the motor will start backwards.

162. The Starting of the Engine.—The first step in the process of starting a gas- or gasoline-engine is to see that all the appliances for lubrication are full of oil and in working order. The modern engine is universally supplied with sight-feed oil-cups on the stationary bearings, either as single units or having a common reservoir of oil from which small pipes lead to the various points requiring to be oiled. The advantages of the reservoir system are that it is easy to stop the flow from all cups at once when stopping, and equally to start all cups when the motor is ready to start. After the flow is once adjusted as each bearing may require, it should not need subsequent attention, except to see that the pipes are not clogged by a gumming process nor from impurities in the oil.

After the oil-cups and lubrication have been attended to, the ignition circuit should be next made ready. If the ignition is by the flame system, the open burner is ignited by turning on the gas and lighting it at the outlet. If the hot-tube system or the hot-chamber system is used, the necessary heat in tube or chamber is to be secured by starting the pre-heating lamp or burner. This pre-heating lamp may be of any of the types which will meet the purpose of bringing the surface to the required temperature, and the necessary time must be allowed before the engine is to make its first stroke. If the electric-ignition systems are used, it is desirable to examine the electrical connections and to see whether the spark passes or the arc is formed, so that if ignition should fail, the origin of the difficulty may be known to be in some other element than in the electric circuit. A brief treatment will be given in a later article of this chapter concerning the usual troubles of electric ignitions.

The next step is to make the engine begin its cycle. This has to be done by some mechanical force acting upon the crankshaft to move the piston in the cylinder so as to cause it to draw in the charge of air and fuel on the outgoing stroke and compress that charge on the return stroke and cause the ignition as the piston draws near to and passes the dead-centre. In small engines this rotation of the shaft is done by hand, the engine being turned over by its own fly-wheel, in stationary practice. In automobile practice the motor is released from the driven mechanism by throwing out a clutch, and the motor-shaft is turned by hand by means of a crank. In engines of the middle size it is quite usual to arrange a special cam on the valve-shaft which will release a certain amount of the pressure of the compression stroke which might be sufficient to prevent the hand-starting process from having sufficient power to bring the engine to its inner dead-centre and cause the first ignition.   In large plants some mechanical appliance has to be furnished to give sufficient power to produce these first compressions and first ignitions. This may be (par. 164) a storage of compressed air, or in plants of sufficient magnitude a small independent auxiliary motor may be used to start with.   It is better to make the starting turning of the engine with some speed, since the compression and igni-tion are more certain by this plan than when the turning is more leisurely.   When everything is normal the engine should start within two revolutions of the starting effort on the crank-shaft.

It is an obvious advantage of the multi-cylinder engine that one of its cylinders will reach the phase of compression and igni-tion very shortly after the crank is started.   If the pistons are tight and the electric method of ignition is used, with a button which can make a spark in all four cylinders at once, of a four-cylinder motor, it will be apparent that when the engine was stopped, one of the cylinders had either a partly compressed charge in it or one which was just ready to be ignited.   The partly compressed charge can be used to start the motor-shaft, or by pressing the button and making the electrical connection the unused charge can be ignited and the motor started.   This action is secured by arranging to have the electric ignition dis-connected or retarded just before the motor is stopped.   Otherwise the ignition of the partially compressed charge before the dead-centre is reached will start the motor backwards.   If the opera-tion of the starting revolution of the crank-shaft does not begin the cycle and the ignition is known to be in good order, the diffi-culty is either due to defective carburation, to improper com-pression, to a failure of the fuel-supply, or to improper action of the inlet-valves.   In some forms of motor the moving parts may give difficulty, but as a rule when this is the case the engine is difficult or impossible to start by hand.   After the engine has begun its cycle the cam which releases the compression should be thrown out in engines which are fitted with it, and the machine will at once take up the speed for which the governor adjusts it. In starting motor-carriage engines with the transmission machinery out of gear, the clutch can be thrown in, and usually with the low-speed adjustment in gear the carriage starts more easily for the occupants and with less strain upon the driving mechanism. When the inertia of the carriage and the first friction of starting have been overcome, then the other gears can be successively thrown in. The adjustment for speed in variable-speed engines, such as motor-cars, will be done by the governor methods discussed in Chapter XII. In stationary gas-engine practice, where the supply of gas is regulated by a gas-valve, this can be adjusted to the condition of operation after the engine has reached its speed.

If the water-jacket cooling system is used, operated by a pump driven from the engine itself, it will, of course, go into action with the starting of the motor-shaft. If the water-cooling is done by the circulation from a city supply, the necessary valves are opened and the flow regulated to maintain the desired temperature of the outflowing water. In motor-car practice the circulating water is allowed to go as hot as is consistent with keeping it from vaporizing as steam. In stationary practice more water is used and it is kept cooler.

It is undesirable in any case in starting to have the electrical adjustment of the ignition set for a pre-ignition before the crank reaches its dead-centre. If this precaution is not taken, the motor may back-fire or pre-ignite, starting to revolve in the wrong direction, and the operator at the starting crank is liable to injury and the clutch or pin mechanism may be broken by which the starting crank is released from the motor-shaft.

163. The Stopping of the Engine. — In stationary practice where the design permits, it is desirable to store compressed air or compressed-charge in an auxiliary tank or reservoir before the engine is stopped, so that by connecting this compressed air or mixture with the motor-cylinder it can be used to start the first stroke and save the inconvenience and annoyance of hand starting. This implies that before the engine is stopped the necessary amount of compressed air or mixture shall be stored in the reservoir by throwing in the appliance whereby this is brought about (par. 164). In the smaller engines in motor-cars this detail is disregarded. The motor will stop when the valve is closed by which the supply of fuel for the mixture is brought to the cylinder. Then the ignition apparatus will be disconnected by throwing out the switch in the electric system, or by extinguishing the flame of the hot tube or flame ignition systems. In automobile practice, the ignition apparatus will usually be thrown out first, allowing the aspirated mixture to scavenge the cylinders by the motion of the car before it stops. The oil-cups are then shut off and the cold-water circulation stopped.

164. Restarting after a Stop.—When the mechanism driven by a gas-engine has to be stopped frequently and then started after a short interval, it is by far the most convenient plan to introduce a clutch between the motor-shaft and the driven resistance, so that the latter may be stopped without stopping the motor. This solution, for example, is the universal one in automobile practice, and it is a convenient one in the general case also, since the starting of the motor with the resistance coupled to it in large units might be so difficult as to be almost impossible. But when the motor itself is to be stopped and is to be restarted after a period of rest the condition is very different from that of the steam-engine, where the piston starts by simply opening a valve which connects the piston area with a reservoir of sufficient pressure to overcome the resistance and the engine begins its normal march. The condition to be met is the rotation of the motor-shaft by a proper force through such an angle as shall draw in a charge of fuel and air; shall compress that mixture by a return stroke of the piston, and carry the crank up to and just beyond that point at which ignition of the charge takes place. In single-cylinder engines this will usually require one revolution at least; in multiple-cylinder engines a partial revolution should be enough. By what means shall this starting action be caused, so that the motor may be restarted?

If the normal compression pressure be 60 or 80 pounds per square inch, it will be apparent that only a few inches of area of piston will be required that the resistance to compression may exceed the capacity of the human muscles to meet and overcome it. Hence the meaning and necessity of relieving devices or cams, whereby with even moderate cylinder diameters the exhaust-valves may be opened enough to relieve this compression resistance when starting by hand, and with the resistance thrown out. This may also be done by opening pet-cocks discharging from the compression space, which are closed as soon as the engine will take care of itself. The handle in Fig. 32 which connects the pipe E to the cylinder below A is such a relieving-valve. Obviously, however, this release of the compression makes a weak stroke, and with greatly diminished forward energy. In discussing the methods of starting internal-combustion motors, therefore, the list must begin with

1. Hand-starting with fly-wheel or independent crank. This must be limited to comparatively small cylinder diameters, and demands compression-relieving appliances. Care must be taken lest injury be done with high-speed motors from the cylinder overtaking the human agency and starting the working stroke before the hand or foot can be released from the lever which it is using to start the shaft. In automobiles it is quite usual to make the starting-crank connect to the motor-shaft by a jaw, or clutch, so designed that when the hand-crank drives, surfaces normal to the effort shall receive this action. When the motor overtakes the hand-crank, the contact surfaces are oblique to the effort, and tend to force the crank-hub along the shaft, and disconnect the clutch surfaces. Back-firing or pre-ignitions and a reverse of the motor-shaft are both annoying and dangerous in hand-starting. The limitations of this system for large installations, or for small ones which are to be used by non-muscular

operators, as in automobiles for use by women and children, have turned attention to other systems.

2. In multi-cylinder engines, if the ignition system be electric and switched off before the previous stop, the inertia of the fly-wheel of the motor will have one or two of the cylinders charged with mixture which has been compressed and which has begun to expand unignited in what would have been the working stroke. If, then, with the spark adjustment retarded past the dead-centre, the switch be thrown in, that mixture will be ignited and will turn the engine over. If the spark adjustment were before the dead-centre position, and the charges were fired, the motor would start backward on the charge in process of compression, but not com-pletely compressed. This postulates, of course, that the pistons fit tightly enough to hold the charge of mixture, and that the stop is not so long that even with tight pistons all compression shall have leaked away. This system is available only in the electric systems of ignition, but its possibility is an additional argument for that system.

3. A storage of mechanical energy which shall be potential or available in quantity to start the motor. The simplest system of this group is a storage of compressed air in air-tight tanks. In the Westinghouse system, for example (par. 72), an air-com-pressor or pump is thrown into gear before the engine stops for a period sufficient to fill the necessary tankage with air at 150–200 pounds pressure. When the engine is to be started, one of the cylinders is by-passed as respects the gas-suction, but is operated as a compressed-air engine using air from the storage tanks.

Assuming that the cylinder shown in Fig. 35 is to be used in this way, the action is as follows: By turning a screw on the end of the upper cam-shaft, the cam B is thrown out of action, so that the admission-valve J remains closed. By moving the lever seen on the outside of the crank-case near the cam A, this cam is con-verted into a double-acting one, such that the exhaust-valve E is open on every up stroke of the piston. Another cam on the same shaft A operates a valve in the compressed-air pipe, permitting compressed air to enter on every down stroke of the piston. If now the crank be placed in the proper position, and the air turned on, the cylinder will operate as a single-acting compressed-air engine. In this way momentum enough is secured to compress a charge lightly in one of the remaining cylinders, which, on ignition, augments the speed, so that the air-cylinder may be thrown into its normal working condition. A very simple stop throws the compressed-air valve out of action, and a motion of the lever changes the exhaust-cam to its original condition. By holding a knurled head at the end of the upper shaft, the rotation of the shaft locks the admission-valve cam in its usual position so that the cylinder operates again as a gas-engine.

4. The fourth system of the same general class compresses and stores an explosive mixture of gas and air into a tight reservoir. As carried out in the Clerk system, with an independent cylinder (Fig. 32) for the aspiration and compressing phases, this is done by a by-pass valve between the motor and compressor cylinder, so that an occasional cylinderful from the compressing phase is delivered to the reservoir instead of to the combustion-chamber. This can be done at intervals just before shutting down the motor without seriously affecting it, and until the compression pressure is reached in the reservoir. When the motor is to be restarted after a stop it is barred over into a crank-angle just past the dead-centre; a charge from the pressure-reservoir is admitted to the combustion-chamber behind the piston through a pipe and valve; the mixture is fired by electric spark, and the march of the engine begins.

5. A variation of this plan is to have an auxiliary hand-pump to compress mixture into the space behind the piston, just past its dead-centre. This compressed mixture is then fired either electrically or by working a timing-valve by hand.

6. The sixth system is that providing an auxiliary or external exploding-chamber within which without compression an explosive mixture of gas and air may be gathered and ignited. The pressure from the expansion caused by ignition and explosion

passes through the large passage at the top of Fig. 92 and enters the working cylinder through the inlet-valve, and has force enough to start the engine turning. The explosion is effected by a pilot-light *G*. As long as gas is flowing into *D* through *C* and out at *G* the flame at *G* cannot run back. When *C* is closed after the

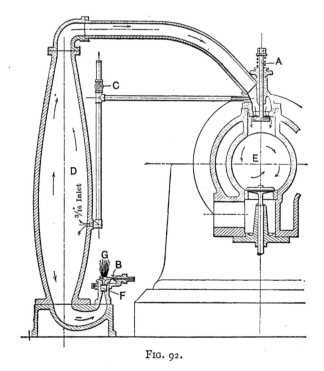

Fig. 92.

mixture in *D* has become explosive the flame runs back past *F* and fires the entire charge.

7. Instead of one explosion being used to start tne working piston, the English designers have used a succession of smaller explosions, coming after each other and acting with cumulative effect.

8. Belonging also to this group is the proposed plan of start-ing by means of a cartridge, introduced in a tube into the cylinder from without and detonated; and the plan of igniting the ârst

mixture by a match, firing by percussion inside the cylinder. Here, of course, the charge was not initially compressed.

9. In large ·plants an auxiliary or " barring " engine independently driven from another source of power can be made a starting feature. A small gas-engine of size to be hand-started will have power enough when started to put the massive machine to turning over. Or the auxiliary may be a small steam-engine or electric motor.

165. The Lubrication of the Engine.—The heat incident to the ignition of the charge in the gas-engine cylinder makes the problem of its lubrication more difficult than that of the steam-engine. If an animal oil is used, or, worse, a vegetable oil, or a lubricant which is adulterated with either, a process of oxidation takes place whereby the oil burns to a hard gum which adheres closely to the surfaces which it is supposed to lubricate. For these reasons the mineral oils are the only proper ones, and these should be of good quality, so that the gumming may not occur. With the mineral oils, on the other hand, a difficulty is sometimes met that the oil in combination with heat and compression will form a gas which will pre-ignite on the compression stroke, giving a back-fire and a tendency for the engine to reverse. The cylinder cannot be lubricated by the ordinary methods used for steam-cylinders, but the oil has to be pumped in either by pumps mechanically operated or by utilizing the varying pressure on the oil acting through suitably arranged check-valves. Horizontal engines are usually lubricated as to the cylinder by an oil cup which supplies the front or cool end of the trunk piston, and the movement of the piston over this lubricated surface of the cool end of the cylinder is intended to secure adequate supply of oil. In many designs, horizontal or vertical, which have a closed crank-pit the lubrication of the piston is effected by a spattering of the oil from the oil-bath in which the crank and connecting-rod-end dip at each· revolution. This method of an oil-bath in the crank-case secures the lubrication of the crank-pin and main shaft-bearing also. The valves of nearly all engines are made to lift, to open, inasmuch as it would be difficult to secure a lubrication of a sliding surface under the conditions of heat to which these valves are exposed. This difficulty has been the occasion of abandoning the sliding valve of the early designs. It is, furthermore, impossible to lubricate with the ordinary oil any surfaces over which a gasoline vapor can have access, since the latter is a solvent for the lubricating oil and destroys its properties for this end. If the cylinder is not properly cooled by its water-jacket, the oil may either gasify or gum, whereupon the piston growing overheated and expanding will offer excessive friction, or will become seized in the bore and stop the engine. An overexpanded piston usually causes a thump or pound in the engine.

For the external bearings any accepted form of lubricator or system of lubrication can be applied which will give a continuous supply of oil as needed.

166. Improper Working of the Engine. The Engine Refuses to Start or Work.—When the engine refuses to start, it is usually by reason of defects either in the ignition, mixture proportion, the carburation, or the compression. The possible ignition difficulties will be different according to the system of ignition used (Chapter XI). If the tube-ignition is used, the most usual causes of failure are due to defects in the platinum or steel tube or in the burner. The hot tube may become cracked, allowing the compressed gas to escape, or the tube may become coated with soot on the inside. The joint between the tube and the cylinder may also leak. The leakage from the joint or from the tube may be detected with a match. The tube may be cleansed by gasoline or by rubbing out the tube with a small piece of emery-cloth on a stick. The burners which heat the tube are usually of the Bunsen class. For gasoline the upper end of the burner is a tube in which is inserted a small plug of asbestos in a sheath of fine brass gauze. The upper end of this tube has a nipple with a minute hole in it, and the gasoline or gas will escape from this small orifice, which is surrounded by a larger tube acting somewhat like a cowl and forming a mixing-tube for the fuel and air which is ignited at a slit in the top of the cowl or mixing-tube. This slit is directly under the ignition-tube. To start such a gasoline burner a little cup under the base of the burner-tube receives alcohol by which the tube is pre-heated and made into a vaporizer for the gasoline. Burners of this class not infrequently jump and put themselves out when first started and improperly heated. The burner should show a blue flame. If it burns yellow, it is usually by reason of being clogged, although excessive pressure of gasoline from the source of supply will blow the flame out, as well as excessive jolts and a high wind in car-motors. These constitute difficulties with this system. If the charge is ignited too early, the flame heating the tube should be moved nearer to its closed end. The tube should be at a good red heat for starting.

For the failure of the electric ignitions the difficulties are likely to originate either in the battery, in the circuit, or in the sparking-plug. From the magneto or dynamo ignitions the battery difficulties are eliminated, but the others remain. Satisfactory conditions of the battery may be assured by any of the ordinary test instruments which will indicate whether the current is flowing between the terminals. In motor-carriage work imperfect insulation is a very usual form of difficulty with the ignition, since it is liable to be burned from contact with hot exhaust-pipes or the metal of the cylinder, to be cut or chafed from the motion of the vehicle, and to have the connection with the binding-posts or other terminals become loose or dirty. The coil (pars. 128, 129) by which the self-induction is increased in the arc system or the secondary current formed in the jump-spark system are likely to break down from failure of insulation whereby the circuit passes across instead of around the coil. The vibrator must be in good order or else the current will fail to form in the secondary circuit. The plug across which the spark passes is liable to fail either by the cracking of the porcelain tube which is used to insulate the two terminals from each other, or the points may become sooty, or a deposit of oil may take place on them. In either case the spark will fail to pass. If they touch, there will be no spark, or if they are too far apart, the spark may not have intensity enough to jump the gap. With dynamo or magneto ignitions they have the same difficulties as the jump-spark systems, with the added difficulty that the contacts may become oxidized.

If the ignition is in good order and properly timed, the carburation may be unsatisfactory for one of several reasons:

(1) Proportions of air and gas badly adjusted.
(2) Carburetor flooded.
(3) Carburetor insufficiently supplied.
(4) Cold weather or damp weather.
(5) Gasoline of poor quality.
(6) Gasoline-valve closed partly or entirely.

The most usual difficulty from improper proportions is the consequence of the mixture being too weak in fuel. This will occur with a governor system which throttles the fuel-supply without throttling the air, or by a leakage of air in excess through a defective joint in the suction circuit, or by the presence of excess of products of combustion in the suction charge which will so impoverish the mixture that with a spark of a given intensity or a hot tube with a given temperature it will fail to light. This difficulty is to be corrected, experimentally, by varying the mixture to see whether by such variation it shall be possible to cause the motor to make its first explosion.

If the carburetor is either supplied with gasoline in excess or not enough flows to it, the adjusting-valves of the carburetor are to be reset to make the mixture right. Too rich a mixture will give trouble by pre-igniting and back-firing on the compression stroke, and it will also give an exhaust with an offensive odor due to the presence of gasoline partially oxidized, but not completely burned. When the mixture has become too rich, the gasoline supply should be cut off and the engine revolved with the air-inlets open until the first explosions succeed. The carburetor may fail to supply gasoline enough by reason of the nipple or spraying-nozzle being stopped up. Of course the valves supplying the carburetor may have been left shut, or the gasoline tank may be empty.

In cold weather the cylinder will be at a low temperature and the carburetor will itself be cold. It has been observed that the spark often fails to ignite the cold mixture, while after the metal of the engine has become thoroughly warm no difficulty is encountered. The vapor is given off in the cold carburetor less readily than when it is warm. This difficulty is met, of course, mainly in automobile engines which are operated in the open air. Artificial heat by a torch or lamp is the most effective cure for this failure. The same is true of the difficulty from damp air, which, carrying a proportion of moisture, will act to cool the charge on the suction or compression stroke, and may keep it low enough in temperature not to ignite with the energy in either spark or hot tube.

Gasoline may be of inferior quality when it has been allowed to stand for some time, particularly under circumstances favorable to its slow vaporization. Such gasoline becomes reluctant to vaporize by the absence of the more volatile constituents. The gasoline may be stale in the carburetor or in the tank. Gasoline sometimes also has water or oil mixed with it which, of course, greatly interferes with its fuel qualities.

The starting of the machine by hand may have been done slowly, so that the passage of the air on the suction stroke was not sufficiently rapid to carry, mechanically, the spray of gasoline into the cylinder. This is, of course, corrected by turning the starting-crank more rapidly.

If it is the compression phase of the cycle which is at fault, it will be the result of leaks either in the fit of the piston in the cylinder, or at joints. The valves, also, which should seat tight under the compression stroke, may be corroded or coated so that they permit an escape of the compressed mixture. Improper working of the compression is revealed to the hand-starting process. A gumming of old oil in the cylinder may also produce the effect of resistance to compression, or a similar gumming or sticking of the packing-rings will allow the compression to escape. This makes it desirable that every motor cylinder have a means of injecting a little gasoline, kerosene, or other solvent to cause this oil to be dissolved as the piston is operated by hand. The phenomena of pre-ignition may be mistaken for those of excessive compression. If the timing arrangements for ignition are set forward, so as to occur before the piston reaches the dead-centre, the motor-shaft will receive a backward impulse.

A motor which has started properly and has been working satisfactorily for some time may fail to work properly and will gradually lose its power and speed and tend to come to rest. This condition may result from one of several causes, or several in combination. The piston may seize in the cylinder, or excessive friction be set up by reason of overheating of the cylinder. The most frequent cause of this difficulty is the failure of the water-cooling system. The failure may be caused by the circulating pump in motors in which the circulation is caused mechanically by this means; the water may have been evaporated off, leaving the quantity in circulation too small to carry away the heat in the cylinder; by a clogging of the pipes by solid matter, or by the formation of an air-lock or steam-lock in the circulation at some point where either air or steam may gather and refuse to be dislodged by the circulation. A thick incrustation or deposit of mineral matter from the circulating water may take place upon the hot surfaces, if the circulating water is used at a point at which such mineral constituents in the water will be deposited upon the hot surfaces. They will form a cake there which will be a non-conducting surface, so far as cooling effect is concerned.

Defective lubrication will cause excessive friction and the same phenomena of overheating will occur. The lubricant may fail to reach the desired point from clogging of the pipes or because the lubricator has been allowed to get empty. Overheating may also result, but less frequently, from the use of too rich a mixture in the cylinder.

The float-feed carburetors not infrequently fail at work by reason of the bending or sticking of the needle-valve or because the float has become punctured and liquid gasoline has leaked inside it so as to destroy its relation of weight to that of the gaso-line on which it is supposed to float. The carburetor also is liable to starvation from dirt stopping up the small orifice through which gasoline passes. Not infrequently, also, the vent-hole in a gravity-fed gasoline-tank becomes stopped up so that the air cannot enter to take the place of the gasoline which the engine would like to withdraw.

167. Usual Causes of Failure to Operate.—In addition to the maladjustments of igniter and carburetor referred to in the preceding section, and treated more fully in Chapters IX, X, and XI, the internal-combustion engine is liable to difficulties which would not come under those heads. The engine is liable to lose its full power, and possibly to slow down gradually till it stops. The complete stoppage is usually due to one or more of the causes referred in the previous section. A loss of power may usually be attributed to leakages or clogging.

Leakages are most troublesome in the valves and piston-rings, and in joints. If the valves either of inlet or exhaust will not close tight, compression is lessened, and the charge escapes through these leaks when fired instead of driving the piston. The valves and seats are liable to warping and cracking from heat, and from erosion. The valves of alcohol-motors are particularly exposed to a corrosive action when the alcohol is decomposed on incomplete combustion so as to become hot acetic acid in part. The tendency to corrode and become leaky makes it impera-tive in the design of the engine that both valves and seats should be easily accessible for removal. Wear of the cam or of the roller operating the valves, and lost motion in joints, or bending of the levers or stems which operate the valves will produce this same loss of normal power. The piston-rings are liable to wear, but if they have become gummed in their grooves from an oxidation of the oil, so that they do not expand easily, they will permit leakage of pressure around them and loss of power. Leaky rings are detectable by a sort of barking noise when the ear is near the open end of the piston-trunk, and by the appearance of smoky air at the same point.

The inlet-passages of the motor are usually protected at the air-intake by a gauze screen of some sort, and in the carburetor passage is also likely some mixing or distributing surface. Both of these are liable to become clogged with dirt or dust or soot, when of course the power of the motor begins to fail. In carburetors of the liquid-surface type, where the carburation process gradually cools the liquid fuel and surrounding metal, it may easily happen in damp, cool weather that the watery part of the incoming air may grow cold enough to freeze into anchor-ice, gradually stopping the flow.

A leak from the water-jacket into the working parts of the motor may lower the temperature of a part which should normally be hot. This of course causes a loss of power, but is, as a rule, of most annoyance in making the engine reluctant to start and reach the first high temperature required. In a hot-tube-igniter system a failure to have the tube hot enough will make the engine miss firing its charges occasionally.

The occasional missing of the proper firing of a charge not only reduces the power of the motor, but is usually the occasion of annoying exhaust explosions. Misfiring is probably due to improper mixing or carburation (Chapters IX and X) or due to improper ignition (Chapter XI). It will be obvious that in gasoline-motors a poor quality of gasoline will cause both kinds of trouble, particularly unreliable ignitions. In hot-tube systems the tube may not be hot enough, or may be heated at a point so near its closed end that with the compression used the gases held in that tube are not compressed enough to let the first mixture reach the hot part and be ignited. Or the tube itself or the pipe heating it may be clogged. In electric ignitions the trouble will be with battery, coil, or circuit, as above treated. The explosive charge not ignited will pass out when the exhaust-valve is opened, and may be fired in the pipe or passages, where it will cause a detonating noise.

Back-firing into the inlet connections will result from a delayed combustion in the cylinder which is not completed by the time the inlet-valves open for the next stroke, so that the explosive mixture in the inlet-passages is fired from the cylinder and through the opened inlet valve back to the source of the fuel-supply. Slow-burning mixtures are due either to too little or too much fuel. An excess of liquid fuel is particularly liable to cause this trouble. In two-cycle engines where the cylinder is supplied from the slightly compressed charge of mixture in a closed crank-case (§§ 73, 90), when back-firing occurs the motor stops until new fresh charges can be introduced into the crank-case or in front of the piston before it will start again.

Finally, of course, the motor may lose power by the over-heating of its mechanical bearings at shaft, crank-pin, or cross-head, causing these to seize and begin to cut. These difficulties should be met naturally by the usual remedies common to any machine and familiar to all skilled operators.

The engine may thump or pound from lost motion at any of its joints in the mechanism, which should of course be investigated and located and the cause removed. But a pre-ignition of the charge produces a deep, heavy pound, differing from the mechanical pound, and similar to that in a steam-engine due to excessive lead. It is to be corrected by properly timing the ignition. (Chapter XI.)

168. Concluding Summary.—It will be apparent from the foregoing that the gas-engine differs from the steam-engine and other forms of motor which make use of stored energy, for each stroke or cycle is an independent event and is, therefore, liable to interference from purely instantaneous causes. In the steam-engine or compressed-air engine, for example, the failure to work will be a gradually manifested phenomenon, while in the gas-engine the failure to work may be effected instantaneously by any one of a number of different causes. This indicates the necessity for a careful attention to all details necessary for successful operation before the engine is started, and that when it is stopped it should also be inspected to see whether any defects have developed during the run. It is the indifference or the ignorance of the operator concerning the action of the elements which affect the running of the motor which has given rise to the impression that the explosive engine is tricky and uncertain. This has been, doubtless, aggravated by the introduction of appliances still in an experimental stage, but as more experience is gathered and these appliances become reduced to standard forms and proportions, the uncertainties of this class will disappear.

CHAPTER XVI.

THE PERFORMANCE OF GAS-ENGINES BY TEST.

170. Introductory.—All motors of the piston class have two standards of performance. The first may be called the indicated horse-power, which is based upon the general formula (par. 40) in which

$$HP = \frac{PALN}{33,000}.$$

In this the factor P in the second member denotes the observed or calculated mean pressure in the cylinder prevailing during the working stroke; A is the area of the cylinder in square inches; L, the length of the stroke in feet; and N, the number of explosions or ignitions which occur in a minute. It will be apparent that for a two-phase or four-phase single-cylinder engine the explosions are not as frequent as the number of revolutions. In engines of the hit-or-miss governor system the number of explosions may be considerably less than the number of revolutions when the engine is running lightly loaded.

The other standard for the capacity of the engine is called the brake horse-power and is the actual work in foot-pounds delivered at the revolving shaft of the engine as measured by an apparatus devised to determine the net output in foot-pounds. This brake horse-power takes no account of the energy delivered to the cylinder on the explosion stroke, but does take account of the energy stored in the fly-wheel in excess on that stroke and given out during the other phases of the piston operation to overcome the resistance. It averages out these inequalities and gives a

317

mean result independent of the variations of piston effort. This brake horse-power is the commercially valuable unit, since the resistance to be overcome is the factor which determines the size of the cylinder required. It is apparent that the brake horse-power will usually be less than the indicated horse-power even in four-cylinder engines by reason of the losses between the head of the piston and the revolving crank-shaft.

The brake horse-power can be determined by fitting on the engine-shaft a drum or pulley which can be surrounded with a flexible band which shall constitute a brake, or by the ordinary brake-blocks which can be pressed against the face of the pulley. If the power is large, a projecting arm from the brake-block or from the band resting upon a scale platform permits the effort in pounds to be measured which the friction surface is exerting. (Fig. 93.) That effort in pounds, if the surfaces did not slip,

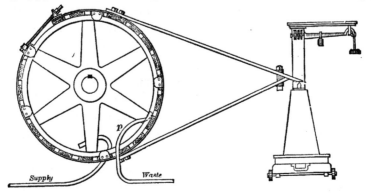

Supply   *Waste*

FIG. 93.

would be exerted through a space per revolution which is the circumference of the circle whose radius is the distance from the centre of the shaft to the point which exerts the pressure on the scale. This number of pounds multiplied by the computed number of feet gives the brake foot-pounds per minute.

It has been found more convenient in small sizes to make use of the device called a rope brake on the fly-wheel of the engine (Fig. 94). The fly-wheel is surrounded by a rope band which may have several plies in it, kept in place upon the wheel by blocks of wood. The friction of the rope tends to lift a weight or to exert a pressure upon a scale, while a weight or spring appliance maintains the necessary tension upon the rope to hold it to the wheel with the necessary friction. The pounds on the scale

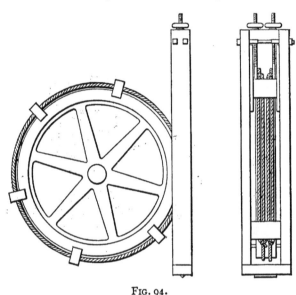

FIG. 94.

(less the tension effort if the weight or spring is attached to a fixed point) gives an indication in pounds as before, and the space per minute through which any point of the circumference of the fly-wheel passes is the feet through which those pounds are exerted. It has been found inconvenient to use the rope brake when the speed of the circumference of the fly-wheel exceeded a rate of 400 linear feet per minute per horse-power to be absorbed. For example, if P be the net effort downward on the scale, and l the length of the lever-arm with which this pressure is exerted on the weighing-scale, and N be the number of revolutions per minute, then

$$\text{B.H.P. per minute} = \frac{P \times 2\pi \times l \times N}{33,000}.$$

171. The Indicator for Gas-engine Testing.—The apparatus used in determining the indicated horse-power to measure the mean pressure is the apparatus known for steam-engine testing as the steam-engine indicator. The gas-engine indicator differs only from the steam form in that the demand upon it is particularly severe, due to the sudden way in which the pressure is applied at the instant of ignition. The high pressure and

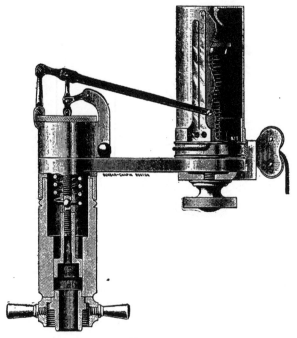

Fig. 95.

temperature of the charge also make great demands upon the indicator. For this reason it is convenient to use a piston of smaller area in the indicator than is usual in steam practice, with a view to eliminating the inaccuracies which would be caused by inertia in the piston and attachments. The high speed of the gas-engine makes it desirable also that the indicator-drum should be of small diameter, that inertia effects in the length of the card may also be reduced to their lowest terms. Fig. 95 shows a successful form of gas-engine indicator.

To impart motion to the indicator-drum when the latter is of small diameter, some form of reducing motion is necessary. The forms which are acceptable in steam-engine practice are not serviceable with the gas-engine, and the cord from the drum to the reducing motion should be as short as possible. The reducing mechanism should be positively driven in both directions. Forms of reducing motion which have been found convenient and satisfactory are shown in Fig. 96. The device in either

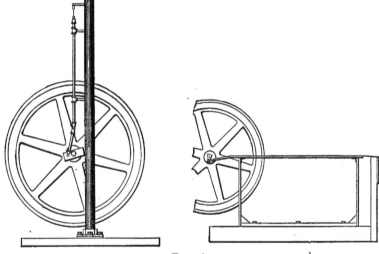

FIG. 96.

form receives its motion from the shaft of the engine by a small crank, and the two forms are adapted to horizontal and vertical engines respectively.

The piping of the indicator to the combustion-chamber should be very short and direct, so that no loss of time or effect may follow in communicating the change of pressure in the cylinder to the piston of the indicator.

172. The Apparatus for a Test.—Besides the indicator and the brake equipment, the test of the gas-engine must also give the consumption of fuel per horse-power per hour of the run. With a gas-engine using ready-made gas, the requirement is simply for a calibrated gas-meter on the suction connection of the engine which shall read closely and accurately enough to give reliable data concerning the cubic feet of gas consumed during the period of the test. In gasoline or liquid-fuel engines the same information is required concerning the weight or volume of the liquid fuel used by the engine per horse-power per hour, and this can be ascertained by drawing the supply of liquid fuel from a vessel mounted upon scales for direct measurement of the weight used, or by drawing the fuel from a calibrated vessel which shall read directly the consumption by volume.

It is further necessary in a complete test that the weight and temperature of the cooling water circulated in the jackets may be observed so that the amount of heat carried away by this cooling water may be subtracted from the heat furnished to the engine in its working charges. This supply of cooling water may be measured by calibrated meters, or it can pass through weighing-tanks upon scales whereby its weight can be observed directly. The temperatures before entry and after leaving are measured by thermometers.

It is also interesting and serviceable to measure the temperature of the exhaust-gases to determine the quantity of heat which escapes by this channel. The following method of making this observation has appeared to be a distinct improvement upon any of its predecessors.

173. Fernald & Lucke's Apparatus to Observe Exhaust Temperatures.— The exhausted products of combustion carry with them an amount of energy which is observably present in the form of temperature, but which is also present in the form of elastic tension which cannot be observed. The problem is, therefore, to reduce the exhaust-gases to atmospheric pressure without losing temperature in the process and then to observe the temperature of the expanded gases. This result was attained after many trials and the rejection of uncertain solutions by the

device illustrated in Fig. 97. The exhaust from the engine passe
into a chamber of fire-brick or common brick, or made of fire·
clay as shown in the left-hand half of the cut. It enters the cham
ber through a throttling device shown in the right-hand half
which consists of a T of the proper size with a plug in the top an
a nipple and cap at the bottom. The plug is drilled to carry
½-inch bolt, with spring and nut at the top; the bolt at the bot
tom carries a flat iron disk which rests against the perforated bot

FIG. 97.

tom of the cap. The gases enter into the branch of the tee, an
any desired resistance to their entry into the chamber can b
secured by tightening the nut and spring, but no wire-drawin
occurs as would occur if the passage were throttled by a fixe
valve. By so adjusting the nut and spring, and by by-passin
part of the exhaust if necessary, the exhaust-gases enter the bric
chamber so as to leave it at atmospheric pressure. This is don
by setting up the bricks dry if a made-brick chamber is used
or by surrounding the notched bottom of a flue-chamber, a.

shown in Fig. 97, with a rubber band acting as a flap-valve to open outward and release any pressure within. Thermometers giving the sensible temperature, without effect of radiation from the walls, may be read for the actual temperature and amount of heat energy escaping with the products of combustion. To use the results of a test employing this apparatus, the first com· putations involve the determination of the combined volumes of air and gas per stroke and their temperatures; after that the temperature of the final mixture in the cylinder is to be found. For the first step, assuming the temperatures of gas and air to be the same, the data and computations will be as follows:

Data Given.

Mins. = No. minutes in time interval.

$$g = \text{gas per min.} = \frac{\text{item 19}}{\text{mins.}}.$$

$$a = \text{air per min.} = \frac{\text{item 13}}{\text{mins.}}.$$

T_2 = absolute temp. gas.

T_1 = absolute temp. air.

= item (21 or 15) + 459°.

Ex. P. M. = explosions p. min. = item 6.

R. P. M. = revolutions p.min. = item 4.

Ms. P. M. = explosions missed per min.

= $\frac{1}{2}$ R. P. M. − Ex. P. M. for single-cylinder four-cycle engine.

To Find—

v' = cu. ft. gas per explosion at T_1.

v'' = cu. ft. air per explosion at T_2.

T_{ag} = absolute temperature in F.° re-sulting from combining air and gas at T_1 and T_2.

v_{ag} = combined vol. in cu. ft. of air and gas per explosion at T_{ag}.

Solution.

$$v' = \frac{g}{Ex. \, P. \, M.}$$

$$v'' = \frac{a - v' \, [Ms. \, P. \, M.]}{[\frac{1}{2} R. \, P. \, M.]}$$

Since $T_{ag} = (T_1 + T_2) \times \frac{1}{2}$, [see Note,]

$$v_{ag} = v' + v''.$$

NOTE.—If a and g have the same spec. heat and same spec. gravity. If not, then work out on the basis of heat lost by one equal to the heat gained by other.

The "item" with its number refers to the scheme of a log record, presented hereafter in paragraph 174. If the gas and air are not at the same temperatures, the second alternative method to find the same data is as follows:

Data Given. ·

$Mins.$ = No. minutes in time interval.

g = gas per min. = $\dfrac{\text{item } 19}{\text{mins.}}$.

a = air per min. = $\dfrac{\text{item } 13}{\text{mins.}}$.

T_1 = absolute temp. air in F.°.
  = item $15 + 459°$.

T_2 = absolute temp. gas in F.°.
  = item $21 + 459°$.

$Ex. P. M.$ = explosions per minute = item 6.

$R. P. M.$ = revolutions per minute = item 4.

$Ms. P. M.$ = explosions missed per min.
  = $\frac{1}{2}R. P. M. - Ex. P. M.$ for single-cylinder four-cycle engine.

w_1 = wgt. per cu. ft. air at T_1 = item 16.

w_0 = wgt. per cu. ft. air at 32° F. = .0807 lb.

T_0 = absolute temp. corresponding to 32° F. = 491°.

v' = cu. ft. gas per explosion at T_2 } as found under Case 1.

v'' = cu. ft. air per explosion at T_1

C_p = specific heat of air at constant pressure.

To Find—

w_2 = wgt. cu. ft. gas at T_2.

w_a = wgt. of air per explosion at T_1.

w_g = wgt. of gas per explosion at T_2.

T_{ag} = absolute temp. in F.° resulting from combining air at T_1 and gas at T_2.

v''' = cu. ft. gas per explosion at T_{ag}.

v'''' = cu. ft. air per explosion at T_{ag}.

v_{ag} = combined vol. in cu. ft. of air and gas per explosion at T_{ag}.

Solution.

The general equation from which T_{ag} can be computed is

$$c_p(T_1 - T_{ag})w_a = c_p(T_{ag} - T_2)w_g.$$

It is now necessary to find w_a and w_g.

If it is not convenient to obtain the weight of gas per cubic foot, the best that can be done is to take the weight of gas the same as that of air at the same temperature. The error involved by so doing is not serious.

Then
$$w_2 = w_0 \frac{T}{T_2};$$

$$w_a = w_1 \times v'';$$

$$w_g = w_2 \times v'.$$

T_{ag} can now be computed from the equation above.

$$v''' = v' \frac{T_{ag}}{T_2};$$

$$v'''' = v'' \frac{T_{ag}}{T_1};$$

$$v_{ag} = v''' + v''''.$$

For the second part of the computation, the problem is from the data of the first part to determine the temperature of the final mixture in the cylinder after the air and gas have united with the exhaust-gases in the clearance space—unless the engine is of the scavenging type.

It is to be noticed that if the governor is of the hit-or-miss type, the exhaust stroke following a miss corresponds to a scavenging stroke.

Data Given.

Assume the weight of the final mixture equal to the weight of air at the same temperature. Assume the specific heats of the different mixtures the same as for air.

w_0 = weight cu. ft.. air at 32° F. = .0807 lb.

T_0 = absolute temp. corresponding to 32° F. = 491°.

C_p = specific heat air at constant pressure.
= specific heat air and gas at constant pressure.
= specific heat final mixture at constant pressure.

T_{ag} = absolute temp. of air and gas entering cylinder as computed in Part I.

v_{ag} = cu. ft. air and gas at T_{ag} as computed in Part I.

T_e = absolute temp. exhaust-gases at atmospheric pressure.
= temp. observed and recorded in item 14 of log + 459°.

v_b = vol. of clearance in cu. ft.

To Find—

w_3 = weight cu. ft. of air and gas at T_{ag}.

w_4 = weight cu. ft. of exhaust-gases at T_e.

w_{ag} = weight of air and gas per explosion at T_{ag}.

w_e = weight of exhaust-gases per explosion at T_e.

T_m = absolute temp. of final mixture in cylinder.

$T_m - 459°$ = F.° as in item 29.

Solution.

$$w_3 = w_0 \frac{T_0}{T_{ag}};$$

$$w_4 = w_0 \frac{T_0}{T_e};$$

$$w_{ag} = w_3 \times v_{ag};$$

$$w_e = w_4 \times v_b.$$

T_m can now be computed from the equation

$$c_p(T_m - T_{ag})w_{ag} = c_p(T_e - T_m)w_e;$$

$$T_m = \frac{T_{ag}w_{ag} + T_e w_e}{w_{ag} + w_e}.$$

Another method would be to use a surface condenser with water-cooling so as to reduce pressure and temperature together. If the cooling was active enough the gases could be nearly always brought down to atmospheric pressure before leaving the condensing apparatus, and the weight and temperature range of the cooling water would give the heat energy which it had absorbed to produce this result.

It may or may not be desirable according to the desired completeness of the test, to make observations as to the composition of the exhaust-gases. A full and complete analysis demands that not only the composition of the inlet gas and its calorific power be made, but also that the supply of inlet air be measured as well as the composition of the outgoing products of combustion.

174. The Observations in a Test.—The extent and number of the observations to be made in a gas-engine test are determined by the results which are sought. If the object is simply to determine the cubic feet of gas or the measure of liquid fuel per horse-power per hour, a preliminary run can be made to determine the conditions or adjustments which give greatest efficiency and to make sure that all details of the engine are performing their functions, as well as possible. The engine having been put in its most favorable condition, the test is then begun with the observation of the quantities desired. For a full and exhaustive investigation to determine not only these fundamental data, but also questions connected with the utilization of the heat and the quantity of heat furnished to the engine, the log will require to cover a very much wider range of observations. The accompanying list gives a full series of headings for the log of such test, together with the columns for the computations. It embodies the practice found serviceable in the gas-engine laboratories of Columbia University, where its form originated.

In comment and explanation as to the accompanying blank for data it may be desirable to add concerning the various items the following computation-methods.

Date, ...

Report of ....

"A" .... Gas-engine test

Test No. ....

By ....

Object .... Efficiency, proportions of air to gas ....

	Number.	1	2	3	4	5	6	7	8	9	10	Totals.	Average.
1.	Number												
2.	Time intervals, minutes												
3.	Total revolutions												
4.	Revolutions per minute (mean)												
5.	Total explosions												
6.	Explosions per minute (mean)												
7.	Ratio of revolutions to explosions												
8.	Weight in lbs												
9.	Weight in lbs. per hour												
10.	Temperature range												
11.	Max. velocity, feet per second												
12.	Heat absorbed, B. T. Units												
13.	Cubic feet												
14.	Cubic feet per hour												
15.	Temperature, Fahr. deg.												
16.	Weight per cubic foot												
17.	Max. velocity, feet per second												
18.	Specific heat, C_p												
19.	Cubic feet or gals												
20.	Cubic feet or gals. per hour												
21.	Temperature, Fahr. deg.												
22.	Weight per cubic foot												
23.	" " gallon												
24.	Max. velocity, 1 feet per second												
25.	Specific heat, C_p												
26.	Cubic feet standard gas per hour (60° Fahr., 14 7 lbs. pressure)												
	Gas												
27.	Air												

Side group labels: Time and Explosions / Revolut's — Jacket water — Air — Fuel — Index Reading / V

28. Values of n in Eq. $pv^n = p_1v_1^n$ for expansion curve.

Pressures from Indicator-cards.

29. Temperature, Fahr. deg.
30. Weight per cubic foot.
31. Max. velocity, feet per second.
32. Pressure at end of Comp.
33. Max. pressure or pressure at beginning of expansion.
34. Pressure at end of expansion.
35. Pressure if expansion be carried to end of stroke.
36. Mean effective pressure.

Exhaust.

37. Temperature, Fahr. deg.
38. Max. velocity, feet per second.
39. Specific heat, C_p.

Mixture.

40. Air to gas to... Neutrals...

Ratios.

41. Stroke to expansion.
42. Vols v_2 to v_1.
43. Max. pressure to M.E.P.
44. Max. pressure to comp. pressure.

Value of R.

45. Value of R...
46. Temperature, Fahr. deg. at comp.
47. Max. temperature, Fahr. deg.

Energies.

48. Brake-work, foot-pounds.
49. Brake-work, foot-lbs. per hour.
50. Brake horse-power.
51. B.T.U. equivalent to B.H.-P.
52. Indicated horse-power.
53. B.T.U. equivalent to I.H.-P.
54. Gas horse-power.
55. B.T.U. equiv. to gas H.-P. $= H_1$.
56. Heat supplied B.T.U. from indicator-card $= H'$.
57. Heat extracted B.T.U. by observation $= H_2$.
58. Heat extracted B.T.U. from indicator-card $= H_2$.
59. I.H.-P. − B.H.-P.

Number.	1	2	3	4	5	6	7	8	9	10	Totals	Average.
60. Throttling, cubic feet.												
61. Throttling, per cent.												
62. Work gained by complete expansion.												
63. Mech. eff. = Item 50												
64. Thermal for B.H.-P. $= \dfrac{\text{Item } 52}{\text{B.H.-P.}} = \dfrac{\text{Item } 51}{H_1} = \text{Item } 55$												
65. Thermal for B.H.-P. $= \dfrac{\text{B.H.-P.}}{H_1'} = \dfrac{\text{Item } 51}{} = \text{Item } 56$												
66. Thermal for I.H.-P. $= \dfrac{\text{I.H.-P.}}{H_1} = \dfrac{\text{Item } 53}{} = \text{Item } 55$												
67. Thermal for I.H.-P. $= \dfrac{\text{I.H.-P.}}{H_1'} = \dfrac{\text{Item } 52}{} = \text{Item } 56$												
68. Thermal $= H_1 - H_2$												
69. Thermal $= H_1' - H_2'$												
70. Thermal $= H_1'$												
71. Fuel per I.H.-P.												
72. Fuel and igniter per I.H.-P.												
73. Fuel per B.H.-P.												
74. Fuel and igniter per B.H.-P.												
75. Cost per I.H.-P. per hour, cents, gas at $1.00 per 1000 cubic feet.												

Efficiencies. {63.–70.}

Stand'd Gas per Hour. {71.–74.}

Heat balance.

	B.T.U.	Per Cent.
Heat turned into work.	.....	.....
Heat rejected into jacket-water.	.....	.....
Heat rejected in exhaust.	.....	.....
Heat rejected in condensation and radiation.	.....	.....

Analysis of fuel by.
Analysis of exhaust by.
Heat equivalent by analysis, B.T.U., to fuel.
"   "   "   to exhaust.

Remarks.—

No. 12. *Heat in the Jackets.*

Since the specific heat of water is taken as unity, the calculation consists only in multiplying the number of pounds of water used during the interval by the range of temperature, this range of temperature being equal to the number of heat-units absorbed per pound of water.

Data Given.	*Solution.*
s = specific heat of water = 1.	
t_r = temperature range.	
W = weight water for time interval.	$h_w = s \times t_r \times W.$
To Find—	
h_w = B.T.U. for the time interval.	

No. 13. *Cubic Feet.* No. 14. *Cubic Feet per Hour.*

In reading the ordinary meter it is not sufficiently accurate to catch the readings by noting the positions of the index hands at the beginning and close of the time interval, but it is necessary to keep an observer at the meter and require the readings to be taken from the hand which indicates the single cubic feet, and whose complete revolution records 10 cubic feet.

The cubic feet per hour are readily calculated from the data for the given time interval.

No. 16. *Weight per Cubic Foot*

This weight is that of a cubic foot of air at the temperature given in item 15, and is found as follows:

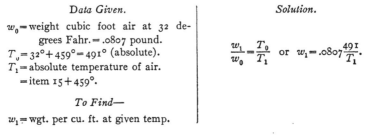

Data Given.	*Solution.*
w_0 = weight cubic foot air at 32 degrees Fahr. = .0807 pound.	
T_v = 32° + 459° = 491° (absolute).	$\dfrac{w_1}{w_0} = \dfrac{T_0}{T_1}$ or $w_1 = .0807\dfrac{491}{T_1}.$
T_1 = absolute temperature of air.	
= item 15 + 459°.	
To Find—	
w_1 = wgt. per cu. ft. at given temp.	

No. 25. *Cubic feet of Standard Gas per Hour, at 60° Fahr. and at 14.7 lbs. pressure.*

Data Given.

$v_g =$ cu. ft. of gas per hour at $t_g°$ F.

$T_g =$ absolute temp. of gas $= t_g° + 459°$.

$p_g =$ pressure under which gas is flowing.

 $= 14.7$ lbs. + pressure shown by manometer.

$T_s = 60° + 459° = 519°$ F., absolute temp. of standard gas.

$p_0 =$ atmospheric pressure.

 $= 14.7$ lbs. per square inch.

To Find—

$v_s =$ cu. ft. standard gas per hour.

Solution.

$$\frac{v_s}{v_g} = \frac{p_g T_s}{p_c T_g};$$

$$v_s = v_g \frac{p_g T_s}{p_0 T_g}.$$

No. 28. The values of the coefficient n are to be computed carefully from the data and formulæ discussed in paragraph 56.

No. 29 has been referred to separately in the previous paragraph (173).

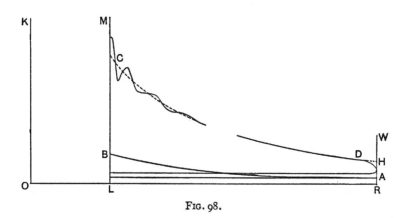

FIG. 98.

No. 32 is the length of the line LB in Fig. 98.

No. 33 only requires care in case these are explosion waves as in Fig. 98. By marking the centre points of these waves, and continuing the curve of the expansion line through some lower point and these centre points, a fairly accurate determination can be made.

No. 34. The pressure at end of expansion is usually the point of inflection at D. The pressure, of course, is measured in this and in No. 33 from the line of zero pressure.

No. 35. *Pressure if Expansion were Carried to End of Stroke.*

This value is readily obtained from the equation of the expansion curve, the value of the exponent n having been computed in 28. If the expansion were thus continued it would give the point H, as shown in Fig. 98. The pressure corresponding to the point H is deduced as follows:

<table>
<tr><td align="center">*Data Given.*</td><td align="center">*Solution.*</td></tr>
<tr><td></td><td align="center">$P_1V_1{}^n = P_2V_2{}^n;$</td></tr>
<tr><td>

$V_1 =$ volume at some point of the card.
$P_1 =$ pressure corresponding to V_1.
$n =$ value deduced in 28.
$V_2 =$ total volume of cylinder.

</td><td align="center">

$P_2 = P_1\left(\dfrac{V_1}{V_2}\right)^n;$

$\log P_2 = \log P_1 + n \log \dfrac{V_1}{V_2}.$

</td></tr>
<tr><td align="center">

To Find—
$P_2 =$ pressure corresponding to V_2; i.e.,
if expansion continued to H.

</td><td>

The volumes being used as a ratio, the piston area may be omitted, the ratio of lengths being the same as the ratio of volumes, as is customary in working with indicator-cards.

</td></tr>
</table>

No. 36. *Mean Effective Pressure.*

The area of the diagram in Fig. 98 should be measured by the planimeter, using the lengths between the perpendiculars LM and RW, or LR. Then the area in square inches divided by the length of the diagram in inches, multiplied by the scale of the spring used, will give the mean effective pressure in pounds per square inch.

<table>
<tr><td align="center">*Data Given.*</td><td align="center">*Solution.*</td></tr>
<tr><td>

$A =$ area diagram, sq. ins.
$L =$ length diagram, ins.
$S =$ scale of spring used.

</td><td></td></tr>
<tr><td align="center">

To Find—
$M.E.P. =$ mean effective pressure.

</td><td align="center">

$M.E.P. = \dfrac{A}{L}S.$

</td></tr>
</table>

No. 37. The difficulties of exhaust-gas measurement have been referred to in paragraph 173 and the apparatus convenient for the test. The computations refer to Fig. 98 and involve:

<table>
<tr><td>

Data Given.

(See Fig. 96.)

p_a = pressure at A = atmosph. pressure = 14.7 lbs.

p_h = pressure at H = value of item 35.

T_m = absolute temp. of mixture at atmosph. pres. = item 29 + 459°.

To Find—

T_h = absolute temp. of exhaust.

</td><td>

Solution.

The volumes at A and H being equal,

$$\frac{T_h}{T_m} = \frac{p_h}{p_a} \quad \text{or} \quad T_h = T_m \frac{p_h}{p_a};$$

$$T_h - 459° = \text{F.°}.$$

</td></tr>
</table>

No. 39. Specific heat C_v for exhaust-gases may be taken the same as for air $C_v = .1691$ unless it is convenient to measure it directly.

No. 40. *Air to Gas to Neutrals.*

By "neutrals" is meant the products of combustion left in the cylinder of a non-scavenging engine after exhaust—an amount equal in volume to that of the clearance space.

In determining the proportions called for, the number of cubic feet of gas is taken as unity, and the temperature of gas is taken as the basis for the computation. The quantities of air and neutrals must be reduced to corresponding amounts at this temperature.

The cubic feet of air used in ten minutes or an hour cannot be taken as a basis of comparison, without modification, owing to the misses of explosions, in which case air is taken into the cylinder without gas.

Data Given.

T_1 = absolute temperature of air = item 15 + 459°.

T_2 = absolute temperature of gas = item 21 + 459°.

T_e = absolute temp. of exhaust-gases = item 14 of log + 459°.
at atmospheric pressure.

v' = cu. ft. gas per explosion at T_2 as computed in 29.

v'' = cu. ft. air per explosion at T_1 as computed in 29.

v_e = cu. ft. neutrals per explosion.
= volume of clearance = v_b.

To Find—

v_x = cu. ft. air per explosion at T_2.

v_z = cu. ft. neutrals per explosion at T_2.

$v_x : v' : v_z = ?$

Solution.

$$\frac{v_x}{v''} = \frac{T_2}{T_1}, \quad v_x = v'' \frac{T_2}{T_1};$$

$$\frac{v_z}{v_e} = \frac{T_2}{T_1}, \quad v_z = v_e \frac{T_2}{T_e}.$$

Taking v' as the basis, i.e., calling v' unity, then

$$v_x : v' : v_z = v'' \frac{T_2}{T_1} : 1 : v_e \frac{T_2}{T_e}.$$

No. 41. Is the ratio of the full stroke LR in Fig. 98 to the length from the perpendicular LM to the point D where the exhaust opens.

No. 42. In Fig. 98 v_1 is proportional to OL and v_2 is proportional to OR; or

$$\frac{v_2}{v_1} = \frac{OR}{OL}.$$

No. 45. *Value of R.*

R is the constant which enters into the mathematical statement of the law; $PV = RT$.

Data Given.

P_0 = atmospheric pressure per sq. ft.
= 2117 lbs. per sq. ft

v_2 = total vol. of cylinder in cu. ft.

T_m = absolute temperature of mixture filling cylinder before compression begins.
= item 29 + 459°.

To Find—

R = a constant.

Solution.

By the above law:

$$P_0 v_2 = RT_m;$$

$$R = \frac{P_0 v_2}{T_m}$$

No. 46. *Temperature, Degrees Fahr., at Compression.*

Having determined R, as in 45, the temperatures corresponding to any point in the diagram are readily determined by the general formula used in obtaining R after solving for T.

<table>
<tr><td>

Data Given.

In general,
P = pressure in lbs. per sq. ft.
V = corresponding volume in cu. ft.
R = constant determined in 45.

To Find—

T = absolute temperature corresponding to the point of the diagram selected for the temperatures of compression.

</td><td>

Solution.

$$T = \frac{PV}{R};$$

$$T - 459° = \text{F.°}.$$

</td></tr>
</table>

No. 47. *Maximum Temperature, Degrees Fahr.*

The formula in No. 46 will be used in general, and if the ignition line rises vertically from the point of maximum compression, and the expansion curve drops at once from the maximum pressure, the computation can take the following form:

<table>
<tr><td>

Data Given.

p_c = maximum pressure, lbs. per sq. in. = item 36.
p_b = compression pressure, lbs. per sq. in. = item 35.
T_b = absolute temp. at compression = item 46 + 459°.

To Find—

T_c = absolute maximum temperature.

</td><td>

Solution.

$$T_c = T_b \frac{p_c}{p_b}.$$

</td></tr>
</table>

If it is not apparent just where the maximum product of pressure into corresponding volume did actually occur, the explosion end of the diagram may be divided by vertical lines at several points on the line of that part of the stroke, and the volume being measured and the pressure scaled off, the maximum product may be those found experimentally, and these values used for P and V as in No. 46, when R is known from 45.

No. 51. *B.T.U. Equivalent to Brake H.-P.*

Data Given.	*Solution.*
1 H.-P. = 33,000 ft.-lbs. per min. 1 B.T.U. = 778 ft.-lbs. B.H.-P. = brake horse-power of 50. int. = time interval of item 2. *To Find—* B.T.U. per int. equivalent to B.H.-P.	B.T.U. per min. for 1 H.-P. $= \dfrac{33,000}{778} = 42.4.$ B.T.U. per int. $= 42.4 \times$ B.H.-P. $\times$ int

No. 54. *Gas H.-P.* No. 55. *B.T.U. Equivalent to Gas H.-P. $= H_1$.*

Data Given.	*Solution.*
$Hf =$ heat of combustion of fuel determined by analysis or calorimeter. $F =$ lbs. of coal or oil, or cu. ft. of standard gas per interval. *To Find—* $H_1 =$ heat equivalent to G.H.-P. $Gas\ H.-P. = \dfrac{H_1 + 778}{33,000 \times \text{int.}}.$	$H_1 = Hf \times F.$

No. 56. *Heat Supplied B.T.U.* from *Indicator-card $= H_1'$.*

If the specific heat of the gases be assumed to be the same before and after explosion, and assumed to be the same as for air, the computation is much simplified and no serious error introduced.

Data Given.

T_m = absolute temperature of mixture in cylinder before compression = item 49 + 459°.

T_{ag} = absolute temp. of entering air and gas as computed in paragraph 29.

v_{ag} = vol. of entering air and gas per explosion at T_{ag} as computed in Part I of 29.

v_b = clearance vol. of cylinder.

w_s = wgt. per cu. ft. of mixture at T_m as computed in 30.

C_v = specific heat at constant vol. = .1691.

T_c = absolute temp. of point C of diagram.

T_b = absolute temp. of compression = item 46 + 459°.

Exps. = total explosions per time interval.

To Find—

v_t = vol. entering air and gas at T_m.

v = total vol. per explosion of mixture before compressing.

w_m = total wgt. of mixture in cylinder.

H_1' = heat supplied in B.T.U. per time interval.

Solution.

$$v_t = v_{ag} \frac{T_m}{T_{ag}};$$
$$v = v_t + v_b ;$$
$$w_m = v \times w_s ;$$
$$H_1' = C_v(T_c - T_b)w_m \times Exps.$$

No. 57. *Heat Extracted, B.T.U., by Observation* = H_2.

This value is determined by an analysis of the exhaust-gases from which the heat equivalent of a cubic foot of these gases is determined.

Data Given.

T_e = absolute temp. exhaust at atmospheric pressure.
= item 14 of log + 459°.
T_m = absolute temp. of entering mixture = item 29 + 459°.
C_p = specific heat at constant pressure.
T_{ag} = absolute temp. of combined air and gas as found in 29.
v_{ag} = combined vol. of air and gas per explosion at T_{ag}.
T_k = absolute temp. of exhaust as analyzed.
h = B.T.U. per cu. ft. exhaust gases at T_k as found by analysis.
$Exps.$ = explosions per time interval.
w_0 = weight per cu. ft. by analysis.

To Find—

v_k = vol. in cu. ft. at T_k per explosion.
w_k = total weight per explosion.
H_2 = total heat exhausted, B.T.U., per time interval.

Solution.

$$v_k = v_{ag} \frac{T_k}{T_{ag}};$$
$$w_k = w_0 \times v_k;$$
$$H_2 = [C_p(T_e - T_m)w_k \times Exps.] + [h \times v_k \times Exps.]$$

NOTE.—If the exhaust-gases are incombustible, the second quantity in brackets becomes zero.

58. *Heat Extracted, B.T.U., from Indicator-card* = H_2'.

This is the heat thrown off in the exhaust as derived from the pressures shown by the indicator-card.

Data Given.

T_h = absolute temp. of exhaust = item 37 + 459°.
T_m = absolute temp. of mixture at atmospheric pressure = item 29 + 459°.
$K = C_v \times w_m \times Exps.$ as found in 57.

To Find—

H_2' = heat rejected, B.T.U., per time interval.

Solution.

$$H_2' = K(T_h - T_m).$$

The elements concerning which there may be a difference of opinion in the computations will be principally those which involve the specific heat of the mixture and of the products of combustion. The specific heat is obviously not that of either the gas by itself or the air by itself, but is the specific heat of a mechanical mixture. This actual or effective specific heat (par. 55) is a quantity which will affect the computations of the diagram of expansion since the ratio which it bears to the specific heat at constant volume will determine the exponent to be used in treating the expansion according to the adiabatic law.

175. The Precautions against Error in a Test.—The precautions which must be observed in conducting a gas-engine test are the same as those which should be taken in conducting a test with any high-speed engine in addition to certain others which are the consequence of the peculiarities of the engine itself. In the first place the spring of the indicator is particularly liable to error from the heat and from friction, and the inertia effects from weight of the parts, and particularly from the paper drum, introduce notable errors into the indicator diagram. The effects upon the lines of the diagram, due to defective operation of the igniter, the carburetor and the valves, are specially liable to be masked by defective methods of actuating and connecting the indicator. Peculiarities also in the behavior of the phenomenon of propagation of the flame in the mixture are liable to be confounded with inertia effects and a wrong interpretation is very easily made. Fig. 98, which bears all the appearance of a diagram suffering from inertia of the indicator piston, has really no relation to such inertia but solely to the presence of pockets in the ignition chamber from which the propagation of energy was undulating. (Paragraph 214.)

In tests which are made to ascertain the satisfactory working of the engine, great care must be taken to eliminate the effects of other causes upon those which are being particularly studied. The timing of the ignition, for instance, may be so masked by a variation in the rate of ignition through the mass as to make it very difficult to separate accurately the effect due to each separate cause. As discussed in the previous chapter, the engine is particularly liable to defective working as the result of improper lubrication, and as each stroke or each cycle stands by itself,

and independent of every other, the forming of an average condition or standard is more difficult than in the case of the steam-
engine.

It must be observed, furthermore, that if the governing operations are in action during the test, that these will introduce wide
variations from the conditions which are found best for one particular resistance and speed. It will be obvious that if the mixture is varied in composition a number of attendant and coinci-
dent changes should be made if the engine is to be equally efficient under these changed conditions of resistance. These, of
course, it is difficult to meet so that a test is only fair to the motor
when it is made under practically constant conditions, so far as
resistance and speed are concerned. This is particularly true
so far as liquid fuel engines are dependent upon the effective
action of their carburetors for the mixture supplied to the cylinder.
Again, variations in the quality of the fuel, either gaseous or liquid,
supplied to the engine during a test affect that part of the test
which immediately follows such change. In the steam-engine,
on the other hand, the effect of any such changes are averaged
up into the run, by reason of the storage action which takes place
in the boiler when heat from the fuel is stored in the water.

176. The Conclusions from a Test.—The list of columns
in a complete log given under the foregoing paragraph indicates
the conclusions which are usually required and deduced from the
engine test. If the test is made to determine the economy cr
consumption of an engine, only those conclusions are drawn from
the observations which are required for the purpose in hand. It
is plainly from the results of actual tests, completely and accurately made, that the development of the gas-engine along sound
lines is to be looked for.

177. Records of Performance and Economy.—It is difficult to present records of tests of gas- or gasoline-engines which
shall not be misleading by reason of a lack of definite statements
concerning all the elements which entered into the test. For
instance, the quality of the gas as to calorific power and source

| 1 | 2 | 3 | 4 | 5 | 6 | 7 | 8 | 9 | 10 | 11 | 12 | 13 | 14 | 15 | 16 |
Fuel.	H.-P.	Cylinder — Diam.	Cylinder — Stroke.	Builder or Design.	Fuel per H.-P. per Hour, Cu. Ft.	Fuel per H.-P. per Hour, Lbs.	Calorific Power in B.T.U.	Date.	Revolutions per Minute.	Location.	Test by	Heat Distribution — Jacket, Per Cent.	Heat Distribution — Exhaust, Per Cent.	Heat Distribution — Rest, Per Cent.	Heat Distribution — Work, Per Cent.
City gas	5.04	6.7	13.4	Otto	28.3			1881	156.7	Deutz	Slaby	51	31	2	16
"	4.4	8.5	14	"	32.4				158	Hoboken	Thurston	52	15.5	15.5	17
"	9.6				24.5			1885	146	Glasgow	Garrett		15.5		16
"	8.1	7	12	Clerk	29.1			1885	132						
"	7.2				24.3										
"	23.2	9	20		30.4										
"	5.5	7.5	9.25	Atkinson	24.1			1887		London	Unwin	19.4			20.6
"	4.8				19.8										
"	11.1	9.5	11.13	"	22.5			1888	131	"	Kennedy Tower	27			22.8
"	9.5				22.61										
"	19.25	9.5	18	Crossley	22.61			1892	160	Manchester					
"	15.75				21.2										
"	73.9	12	18	Wells	25.9				164	Nottingham					
"	64.0	11¼	18	Otto	16.1		770	1900	228	Boston		20–25			15
"	36	9	18	"	18.6		640	1900	160			36	44		19
"	15				21										
"	8.18	6	12¼	Amer. Otto	25.0			1902	281	New York	Fernald	43.0	28.5	7.9	20.6
"	7.03				19.01										
"	44	13	22	Erie gas-eng.	21.83			1899	200	Buffalo	Carlton				
"	45	13	22		12.8			1899	200	Jamestown					
"					10.8										
Producer-gas	118	17	24	Crossley 2-cyl. / Simplex 2-cyl.		0.873 lbs.		1890	156	Paris	Witz				
Dowson gas	100			Westinghouse 3-cyl.		1.34 lbs.									
Natural gas, Ind.	86.9	13	14		14.18	.9511*	1000 p. cu. ft.	1899	270	Lafayette	Robertson	30.2	48.8		20.7
"	73.2				17.96										
Taylor producer-gas	100	14⅜	25	Schleicher-Schumn		.315†	125	1893	160	Philadelphia	Spangler	30.6	28.0	13.7	27.7
"	92														

* Combustible per H.-P. hour .83.　　† Combustible per H.-P. hour 1.15.

Fuel	H.P.	Cylinder Diam.	Cylinder Stroke	Builder or Design.	Fuel per H.-P. per Hour, Cu. Ft.	Fuel per H.P. per Hour, Lbs.	Calorific Power in B.T.U.	Date.	Revolutions per Minute.	Location.	Test by	Heat Distribution Jacket, Per Cent.	Exhaust, Per Cent.	Rest, Per Cent.	Work, Per Cent.
1	2	3	4	5	6	7	8	9	10	11	12	13	14	15	16
Blast-furnace gas	50	51.18	55.12	John Cockerill	109.88		112	1900	80	Séraing, Belgium		50	20		30
Producer-gas	27.5			Crossley		1.4		1900		England					
Kerosene	31	14.5	17	Horsnby-Akroyd		0.63	19,100	1899	202	"	Robinson	50.4	28.7		20.9
"	26			Priestman		0.74	19,957	1890	207	"	Unwin				
"	7.40	8.5	12			0.95									
"	6.76														
"	21.5	9.5	18	Campbell		0.656	18,847	1900	210	"	Stanfield	39.9	39.8	3.9	16.4
"	17.3					0.815									
"	21	10.23	16.16	Amer. Diesel		0.546	18,604	1000	187	U.S.A.	Denton	35.4	27.4		37.2
"	17	9.86	15.7	Germ'n Diesel		0.540	18,371	1000	154	Augsburg	Schroeder	40.3	25		34.7
"	15	10.23	16.16	Amer. Diesel		0.594	18,604	1000	191	U.S.A.	"	37.4	21.4		41.2
"	9	10.23	16.16	Diesel		0.721	18,604	1000	154						
"	5.2	8.5	12	Priestman		1.063	19,700	1890	180	Plymouth	Unwin	53.39			12.6
"	4.5					1.243					"				
"	7.7					0.694									
"	8.3	8.5	12	"		0.842	21,490	1892	207	Hull	Capper				18.8
"	8.5					0.919									
"	7.9	10	15	Hornsby-Akroyd		0.73	21,180	1892	240	Cambridge	Capper	26.8			15.3
"	7.0	7	10	Crossley		0.82	21,180	1892	210	"	"				
Gasoline	20	10.23	16.16			0.86 pts.									
Alcohol	50	10	15			0.99 pts.	19,980	1900	184	U.S.A.	Delahaye				
Fuel oil	70	16	24	Diesel		0.613		1903		"	Denton				
"	80	16	24	"		0.44		"		"	"				
"	85	16	24			0.482									
"						0.498									
"						0.487									

is often omitted from the record and only the cubic feet of gas per horse-power given, and no measure taken of the quality. The character of kerosene oil used in various tests is indicated often merely by the trade name and with kerosene and gasoline the quality is often a variable within considerable limits. Stale gasoline is, of course, less favorable to the engine than fresh, but a difference in its specific gravity will make a considerable difference in the quantity required to do a given work.

The most reliable tests of gas-engines on a large scale have been made in England under competitive conditions at exhibitions and in the table which precedes several of these competitive tests are reported. The American records of tests are much less complete than the English. In the case of the gasoline-engine using carbureted air it is particularly important to compare only engines operated under somewhat similar conditions as to speed and fuel-supply. The effect of speed in varying the fuel supply at high numbers of revolutions introduces an important variable in such records. The table on pages 284 and 285 presents a series of the data which have been taken from various sources.

178. Sources of Loss in Actual Engines as Compared with the Ideal.—By an examination of the columns headed Heat Distribution, it will be apparent that there are four channels through which the expenditure of the heat energy occurs. There is, first, the mechanical work done upon the piston of the engine and which is the net output which should be made as large as possible. It will be noted that it ranges between 20 and 22 in the higher figures. The limitations which prevent this figure from reaching higher values are set by the necessity for keeping the metal of the cylinder and the seats of the valves at a low temperature, so that they shall not undergo too rapid deterioration or deformation from the high heats inside the cylinder, and so that it shall be possible to lubricate the surfaces which are in contact. Recent experiments have shown that the total heat distribution is little affected by changes in the temperature of the jacket-water between 40° F. and the boiling-point. The combined withdrawal of heat by the two sources of loss, jacket-water and exhaust temperature, ranges between 70 and 80 per cent. The balance, which is friction, radiation, leakage and the like, is usually a small percentage and ought to be less than ten. These deductions make it evident that the directions open for the most manifest improvement are those which have to do with the transformation of a greater proportion of heat into mechanical work and the reduction of the heat which in the present forms of motor has to be disposed of by the cylinder either through the jackets or at the exhaust. It means an increase in the temperature range in the cylinder without securing this by means which have no direct relation to the mechanical energy utilized.

The most obvious of the methods to accomplish this purpose are those which have been presented by Mr. Atkinson. His constructions in 1885 and 1887 were known as the differential and cycle engines and were designed so that by the mechanism driving of the crank of the engine, the expansion stroke of the piston should be longer in travel than any other stroke of the cycle. By this means the volumes appropriate to compression were expanded after ignition to a volume greater before the exhaust opened than they had before compression, and as a consequence the terminal pressure—and therefore also the mean pressure—was lowered while the external work was being done. The difficulty connected with both designs is the complication of mechanism which has to be introduced in order to bring about the variable length of piston traverse. The same object has been further sought by injecting water or steam into the mass of mixture in the expanding cylinder, with the idea that the result of this action would be to compel the mixture of steam and air to partake more nearly of the expanding action of steam which, in expanding, parts rapidly with its heat, while the permanent gases, like air, are reluctant to lower their temperature by expansion. The difficulty here has been that the injection of the steam cooling the mixture lowers the mean effective pressure and diminishes the net driving effort of the piston.

CHAPTER XVII.

THEORETICAL ANALYSIS OF THE GAS-ENGINE.

180. Introductory. — In the foregoing chapters the treatment of the internal combustion engine has been mainly from the practical or experimental point of view. It has been the purpose to point out the operation of the internal combustion motor working under the usual forms in which the theoretical cycles have been reduced to practice. In Chapter IV and those which preceded it, a certain amount of attention was given to the cycle independent of the motors which utilized it and to certain theoretical considerations having their origin and deductive treatment from the science of thermodynamics. It will be the purpose of the present chapter to treat in a mathematical way, on a basis of pure theory, the cycles which are available for use with the internal combustion motor, and to deduce from the theoretical equations which appear in such analysis some serviceable statements as to the limits which theory imposes upon the development of this class of motor. Certain suggestive equations bearing upon the design and proportioning of cylinders will also result from this theoretical treatment.

It will be necessary, however, to supplement the fundamental treatment and definitions in the first three chapters by a brief reference to the use made in this chapter of the diagram whose coordinates are the absolute temperature for ordinates and the value of the entropy factor as abscissas.

181. The Temperature Entropy Diagram.—It has been shown in paragraph 40 that with a piston motor the work done in foot-

346

pounds could be conveniently represented, graphically, by an area of a diagram whose coordinates are the pressure in pounds per square foot, as ordinates, and with the volumes in cubic feet of the cylinder or piston displacement as abscissas. Such a diagram is conveniently called the *PV* diagram. It shows at a glance how the work of the motor varies with pressure and volume as the piston reciprocates, but it shows nothing at all concerning the variation of work done as heat is added or withdrawn as temperature varies. If, on such a diagram, the line representing the increase of volume and decrease of pressure be drawn which shall closely resemble the variation of pressure and volume in adiabatic expansion, it is impossible to say whether the gas undergoing that expansion is gaining or losing heat. If the line drawn is above the computed adiabatic line, the gas must be receiving heat. If it is below such computed adiabatic line, it is losing heat. But in the absence of such computation, the diagram is silent concerning the gain or loss of heat with temperature.

It is very necessary to know what the action is of a metal wall on a mass of expanding gas in the matter of gain or loss of heat energy. Such a method has been proposed and is now in quite general use. It has for its object the presentation of a diagram with two coordinates, of which one shall be the absolute temperature (conveniently the vertical ordinate) and the abscissa or horizontal measurement such that the area will show the quantity of heat energy in British thermal units gained or lost by the gas during any change. If the horizontal coordinates be designated by the Greek letter phi (ϕ) then for any small change in the total quantity of heat at a temperature T that small gain or loss in heat designated by dH will become equal to $T(d\phi)$. It has been quite usual when the absolute temperature is associated with ϕ as a coordinate, that it should be written θ, and the coordinates θ and ϕ give what has been called the theta-phi diagram. It is capable of demonstration by the method of the calculus that the coordinate ϕ is the factor which was designated by Clausius by the name *entropy* as the value for a convenient

factor made necessary by the process of integration. This fact
gives to the theta-phi diagram its other name of temperature-
entropy diagram. If, in Fig. 100 the curved line ab represents
an addition of heat to a mass of gas at a constant pressure, it will
be apparent that the temperature will vary with such addition
of heat. For a very small change in the temperature $d\theta$ it will
be true to say that

$$dH = c_p d\theta.$$

Hence it will be true to write

$$C_p(d\theta) = \theta(d\phi)$$

and

$$d\phi = C_p \frac{d\theta}{\theta}.$$

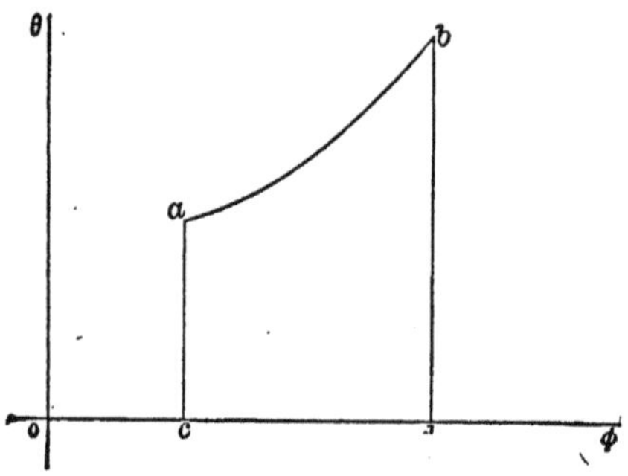

FIG. 100. 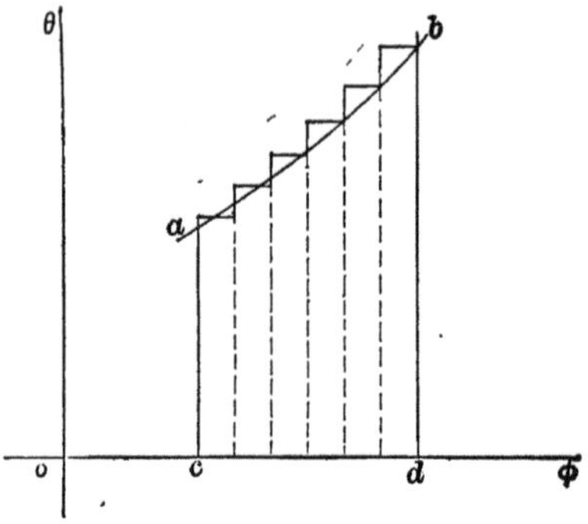   FIG. 101.

If the value of θ be assumed constant, as it may be during
any one of the infinitesimal theta-phi diagrams, the successive
elements of which that change is made up would appear as in
Fig. 101 so that the whole change in the quantity of heat due to
the successive additions will be the sum of all the small elements or

$$\phi_1 - \phi_2 = \text{the sum of } \Sigma\left[C_p \frac{d\theta}{\theta}\right].$$

If, obviously, $d\theta$ is taken very small, the steps forming the broken line become very short and the change in the value of the area under the curve becomes

$$C_p \int \frac{d\theta}{\theta}.$$

The appearance of that quantity suggests that by the processes of the calculus the sum of a number of such infinitesimal increases will take the form

$$\phi_1 - \phi_2 = C_p \text{ hyp } \log \frac{\theta_2}{\theta_1}.$$

The hyperbolic logarithm bears to the ordinary logarithm whose base is ten, the ratio of 2.3026, so that in terms of the common logarithms, this equation appears

$$\phi_1 - \phi_2 = 2.3026 \, C_p \log \frac{\theta_2}{\theta_1}.$$

182. Changes in Value of Phi when Heat is Added to Air.— It will be recalled (par. 54) that the quantity of heat necessary to raise a unit weight of air will differ according to the condition of that air with respect to the constant value of pressure or volume. In the foregoing paragraph the pressure was assumed constant under the addition of heat. Under this circumstance

$$C_p = 0.2375 \text{ B.T.U.}$$

Substituting in the foregoing equation,

$$\phi_1 - \phi_2 = 0.547 \log \frac{\theta_1}{\theta_2}.$$

If the change be at constant volume, while the pressure is allowed to vary, the specific heat C_v is 0.169, so that the equation should be

$$\phi_3 - \phi_1 = 2.3026 \times .169 \log \frac{\theta_3}{\theta_1} = .389 \log \frac{\theta_3}{\theta_1}.$$

If the temperature remains constant, the addition of heat is called isothermal (par. 40), in which case

$$\phi_4 - \phi_1 = R \text{ hyp log } \frac{V_4}{V_1}.$$

Inserting the appropriate figures,

$$\phi_4 - \phi_1 = 53.35 \times 2.3026 \log\frac{V_4}{V_1} = 122.84 \log\frac{V_4}{V_1}.$$

Since this last change takes place at constant temperature of the gas, it will be apparent that the line which is a curve in Fig. 100 becomes a straight line, parallel to the horizontal axes of coordinates. If a vertical line be drawn at any point on that horizontal coordinate axis, which shall represent the value of ϕ_1 measured from the point O and having a vertical height equal to the value of the temperature T at which the addition of heat

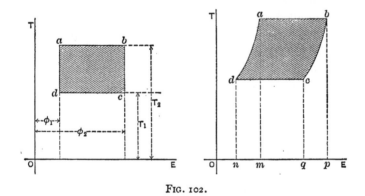

FIG. 102.

was made at constant temperature, the area $T \times (\phi_2 - \phi_1)$ (Fig. 102) will denote the addition of heat which is the quantity H.

Finally, if the change in pressure and volume be that which is designated as an adiabatic change (par. 50) there will be, by definition, no heat added or subtracted during that expansion or change of relation between pressure and volume. Under this condition the $\theta\phi$ diagram must be such that the area between the line representing this change and the line of zero temperature be zero. The only way that this can be realized with a finite value for the temperature will be to have the value for the abscissa ϕ zero. Hence an adiabatic line on the plane of the $\theta\phi$ coordinates is a vertical line.

183. Analysis of the Possible Cycles of the Internal Combustion Engine.—By a reference to paragraph 61, it will be noted that the medium used in the gas-engine is subjected usually, though not always, first, to a compression, then to a heating process, and that after the heating the gas is expanded, doing work against the piston and is then cooled. This paragraph also presented a table indicating the possible forms which these processes might take, which is here reproduced. In the following diagrams an attempt has been made to place side by side a typical work diagram with the PV coordinates and a temperature-entropy diagram on the $\theta\phi$ coordinates. In each case the odd number at the left hand is the PV diagram and the even number at the right the corresponding $\theta\phi$, for each cycle. For clearness of presentation, Figs. 111–151 are not drawn to the same or to any definite scale. For purposes of comparison of cycles a second series of PV areas on the same scale for the various cycles is presented in Figs. 153 to 161.

The cycles in the first group, without compression, and the cycles of the sixth to tenth groups, where the cyclic operations take place at or below atmospheric pressure, are of insignificant importance in any practical way.

The early gas-engines previous to Otto (Lenoir, Barnet, Hugon, Langen, and Bischof) belong to this class, but the introduction of the compression so greatly increased the efficiency and economy of the gas-engine that they do not deserve detailed consideration at this date. The cycle IB, for example, is that of the free piston engines, such as Barsanti and Matteucci in 1854 and the Otto and Langen of 1866, in which the piston was not connected positively to the shaft for the expansion stroke, but was thrown freely upward to a point beyond that at which atmospheric pressure would have resulted from the increase of volume.

CLASSIFICATION OF CYCLES.

1.	2.	3.	4.	5.	6.
Cycle No.	Compression.	Heating.	Expansion.	Cooling.	Cooling.
I	....	Isometric	Adiabatic	Isopiestic	
I A	....	"	"	Isometric	Isopiestic
I B	....	"	"	Isothermal	"
I C	....	"	"	"	
II	Adiabatic	Isometric	Adiabatic	Isopiestic	
II A₁	"	"	"	Isometric	Isopiestic
II A₂	"	"	"	"	
II B	"	"	"	Isothermal	Isopiestic
II C	"	"	"	"	
III	Adiabatic	Isopiestic	Adiabatic	Isopiestic	
III A	"	"	"	Isometric	Isopiestic
III B	"	"	"	Isothermal	"
III C	"	"	"	"	
IV	Adiabatic	Isothermal	Adiabatic	Isopiestic	
IV A	"	"	"	Isometric	Isopiestic
IV B	"	"	"	Isothermal	"
IV C	"	"	"	"	
V	Adiabatic	Any law	Adiabatic	Isopiestic	
V A	"	"	"	Isometric	Isopiestic
V B	"	"	"	Isothermal	"
V C	"	"	"	"	
VI	....	Atmospheric	....	Isometric	Isothermal
VII	Adiabatic	"	Adiabatic	Isopiestic	
VIII	....	"	"	Isothermal	
IX	Adiabatic	"	"	Isometric	
X	"	"	"	Any law	

Similarly, the cycles numbered six and upwards are of comparatively slight importance. It would be, furthermore, entirely possible to make combinations or differentiations of the typical cycles other than those selected if it were worth while.

Let Fig. 111 be a *PV* diagram and Fig. 112 be a $\theta\phi$ diagram for the cycle of the Lenoir engine.

Then in this case:

From *B* to *C*. Addition of heat isometrically from atmospheric pressure.

From *C* to *D*. Adiabatic expansion to atmospheric pressure.

From *D* to *B*. Cooling at atmospheric pressure.

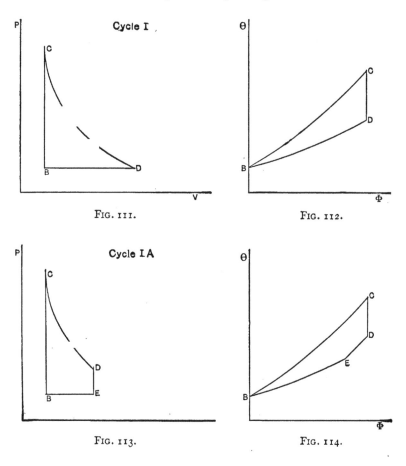

Cycle I

FIG. 111.

FIG. 112.

Cycle I A

FIG. 113.

FIG. 114.

The first modification would be that in which the expansion was incomplete, as in an ideal Lenoir engine where the cut-off was too late to secure complete expansion. Calling this cycle *IA*, we have Figs. 113 and 114 as follows:

In this case:

From *B* to *C*. Addition of heat isometrically from atmospheric pressure.

From *C* and *D*. Adiabatic expansion to point above atmospheric pressure.

From *D* and *E*. Cooling isometrically to atmospheric pressure.
From *E* to *B*. Cooling at atmospheric pressure.

The second modification is that in which the expansion goes below atmosphere before the end of the strok as in the designs of Otto and Langen (1866) and Barsanti and Matteucci (1854), which were called free-piston engines. The pair of diagrams will be as follows:

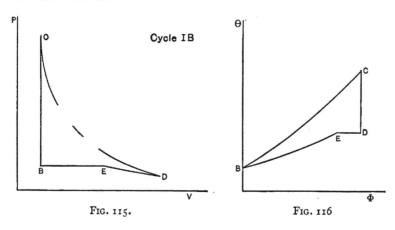

FIG. 115.   FIG. 116

In these:

From *B* to *C*. Addition of heat isometrically from atmospheric pressure.

From *C* to *D*. Adiabatic expansion to below atmospheric pressure.

From *D* to *E*. Cooling isothermally to atmospheric pressure.
From *E* to B. Cooling at atmospheric pressure.

No engine has ever been built to operate on the third modification of this cycle in which work is received from the gas during its cooling phase. Calling this cycle *IC*, Figs. 117 and 118 result:

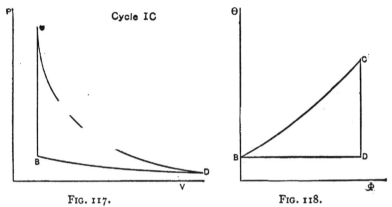

FIG. 117.   FIG. 118.

So that:

From *B* to *C*. Addition of heat isometrically from atmospheric pressure.

From *C* to *D*. Adiabatic expansion to pressure below atmosphere such that

From *D* to *B*. Cooling isothermally to original volume and atmospheric pressure.

The analysis for cycle *I* is as follows:

Let H_1 be the heat added from *B* to *C*.

Let C_v be the specific heat of gas at constant volume, and here assumed constant for simplification. It is probably a variable, but to make this assumption gives unmanageable formulæ. A correction may afterward be applied, if desired. C_v = heat to raise one pound gas 1° F. at constant volume.

Let v_b be the volume of the gases at point *B* of the diagram, i.e., before heating and expressed in cubic feet.

Let p_b be the corresponding pressure in pounds per square foot.

Let T_b be the corresponding temperature in absolute degrees Fahrenheit.

Then will the increase in temperature be given by

$$T_c - T_b = \frac{H_1}{C_v},$$

or

$$T_c = T_b + \frac{H_1}{C_v}. \quad \cdot \quad \cdot \quad \cdot \quad \cdot \quad \cdot \quad \cdot \quad \text{(1)}$$

Since volume is constant from B to C,

$$\frac{p_c}{p_b} = \frac{T_c}{T_b};$$

whence

$$p_c = p_b \frac{T_c}{T_b}.$$

From (1),

$$\frac{T_c}{T_b} = 1 + \frac{H_1}{C_v T_b}.$$

Since this quantity

$$1 + \frac{H_1}{C_v T_b}$$

will enter into many of the equations, let it be denoted by

$$1 + \frac{H_1}{C_v T_b} = X;$$

whence

$$p_t = p_b X. \quad \cdot \quad \cdot \quad \cdot \quad \cdot \quad \cdot \quad \cdot \quad \cdot \quad \text{(2)}$$

The adiabatic relation

$$p_d v_d{}^\gamma = p_c v_c{}^\gamma$$

gives

$$v_d = v_c \left(\frac{p_c}{p_d} \right)^{\frac{1}{\gamma}}.$$

But $p_d = p_b$ by hypothesis, hence

$$v_d = v_c \left(\frac{p_c}{p_b} \right)^{\frac{1}{\gamma}} = v_b (X)^{\frac{1}{\gamma}}. \quad \cdot \quad \cdot \quad \cdot \quad \cdot \quad \cdot \quad \cdot \quad \text{(3)}$$

Another adiabatic relation gives

$$\frac{T_d}{T_c} = \left(\frac{p_d}{p_c}\right)^{\frac{\gamma-1}{\gamma}},$$

whence

$$T_d = T_c \left(\frac{p_d}{p_c}\right)^{\frac{\gamma-1}{\gamma}};$$

remembering $p_d = p_b$ and substituting the value of T_c,

$$T_d = T_b(X)\left(\frac{1}{X}\right)^{\frac{\gamma-1}{\gamma}} = T_b X^{\frac{1-\gamma+\gamma}{\gamma}} = T_b X^{\frac{1}{\gamma}}. \quad \cdot \quad \cdot \quad (4)$$

Let H_2 be the heat discharged. Then

$$H_2 = C_p(T_d - T_b),$$

where $C_p =$ specific heat at constant pressure and assumed constant. Hence substituting

$$H_2 = C_p\left(T_b\left[1 + \frac{H_1}{C_v T_b}\right]^{\frac{1}{\gamma}} - T_b\right) = C_p T_b(X^{\frac{1}{\gamma}} - 1). \quad \cdot \quad \cdot \quad (5)$$

The work done in heat-units will be

$$W = H_1 - H_2 \quad \cdot \quad \cdot \quad \cdot \quad \cdot \quad \cdot \quad \cdot \quad \cdot \quad \cdot \quad \cdot \quad \cdot \quad (6)$$

$$= H_1 - C_p T_b(X^{\frac{1}{\gamma}} - 1). \quad \cdot \quad \cdot \quad \cdot \quad \cdot \quad \cdot \quad \cdot \quad (7)$$

And in foot-pounds

$$W = J[H_1 - C_p T_b(X^{\frac{1}{\gamma}} - 1)].$$

This work of expansion could have been obtained by temperatures and by integration as well.

The work W will be:

$$W = C_v(T_c - T_b) - C_p(T_d - T_b).$$

But

$$C_p = C_v + \frac{p_0 v_0}{T_0} \times \frac{1}{J};$$

$$\therefore W = C_v T_c - C_v T_b - C_v T_d + C_v T_b - \frac{1}{J}\left[\left(\frac{p_0 v_0}{T_0}\right)T_d + \left(\frac{p_0 v_0}{T_0}\right)\right]T_b.$$

It is true also that

$$\frac{p_0 v_0}{T_0}T_d = \frac{p_d v_d}{T_d}T_d = p_d v_d,$$

$$\frac{p_0 v_0}{T_0}T_b = p_b v_b,$$

and

$$p_b = p_d.$$

$$\therefore W = C_v(T_c - T_d) - \frac{1}{J}p_b(v_d - v_b)$$

in heat-units. This second term is the area of the rectangle between $\begin{cases} p = 0 \\ p = \text{atmosphere} \end{cases}$ and $\begin{cases} v = v_b \\ v = v_d \end{cases}$ and lying below atmosphere is not available for work.

By integration $W' = \int_{v_d}^{v_c} p\,dv = $ area between expansion **curve** and axis of volumes. The expansion is adiabatic.

$$\therefore W' = p_1 v_1^{\gamma}\int_{v_d}^{v_c}\frac{dv}{v^{\gamma}} = -\frac{p_1 v_1^{\gamma}}{\gamma - 1}\left[\frac{1}{v^{\gamma-1}}\right]_{v_d}^{v_c};$$

$$\therefore W' = \frac{p_d v_d^{\gamma}v_d^{1-\gamma}}{1-\gamma} - \frac{p_c v_c^{\gamma}v_c^{1-\gamma}}{1-\gamma}.$$

$$\therefore W' = \frac{p_d v_d - p_c v_c}{1-\gamma}.$$

Since

$$1 - \gamma = \frac{C_v - C_p}{C_v};$$

$$W' = \frac{T_d}{\dfrac{C_v - C_p}{C_v}}\left(\frac{p_d v_d}{T_d}\right) - \frac{T_c}{\dfrac{C_v - C_p}{C_v}}\left(\frac{p_c v_c}{T_c}\right)$$

$$= J C_v(T_c - T_d) \text{ in foot-pounds.}$$

Subtracting the rectangle $p_b(v_d - v_b)$, we get

$$W_1 = JC_v(T_c - T_d) - p_b(v_d - v_b)$$

in foot-pounds, or in heat-units

$$W = H_1 - C_p T_b(X^{\frac{1}{r}} - 1)$$

as before.

Applying at this stage a test to each of the states B, C, D from the law of perfect gases:

$$\frac{p_b v_b}{T_b} = R;$$

$$\frac{p_c v_c}{T_c} = \frac{p_b X v_b}{T_b X} = \frac{p_b v_b}{T_b} = R;$$

$$\frac{p_d v_d}{T_d} = \frac{p_b v_b (X)^{\frac{1}{r}}}{T_b (X)^{\frac{1}{r}}} = \frac{p_b v_b}{T_b} = R;$$

hence these are identities, as they should be.

Denote the volume swept through or volume range by R_v. Then will

$$R_v = v_d - v_b = v_d - v_c = v_b[X^{\frac{1}{r}} - 1]. \quad \ldots \ldots (8)$$

Whence mean effective pressure

$$\text{M.E.P.} = \frac{W}{R_v} = J\frac{\{H_1 - C_p T_b(X^{\frac{1}{r}} - 1)\}}{v_b[X^{\frac{1}{r}} - 1]}. \quad \ldots \ldots (9)$$

Efficiency

$$E = \frac{W}{H_1} = 1 - \frac{H_2}{H_1} = 1 - \frac{C_p T_b(X^{\frac{1}{r}} - 1)}{H_1}. \quad \ldots (10)$$

The entropy range is given by

$$\phi_c - \phi_b = C_v \log_e \frac{T_c}{T_b} = C_v \log_e X. \quad \ldots \ldots (11)$$

Mean effective temperature

$$\text{M.E.T.} = \frac{1}{2}\left(\frac{H_1 + C_p T_b(X^{\frac{1}{r}} - 1)}{C_v \log_e X}\right). \quad \ldots \quad (12)$$

The temperature range

$$R_T = T_c - T_b = \frac{H_1}{C_v}. \quad \ldots \ldots \quad (13)$$

The pressure range

$$R_p = p_c - p_b = p_b(X - 1). \quad \ldots \ldots \quad (14)$$

Whence an expression for a mean effective volume may be written

$$\text{M.E.V.} = \frac{W}{R_p} = J\frac{\{H_1 - C_p T_b(X^{\frac{1}{r}} - 1)\}}{p_b(X - 1)}. \quad \ldots \quad (15)$$

These results are here tabulated for reference and comparison with what follows:

It would be possible to take a set of formulæ derived elsewhere for mean effective temperature, but as these were the results of a comparison of cycles, none of which ran below atmospheric pressure, it would be better to take another standard here. Taking arbitrarily as the mean effective temperature one-half the sum of the mean temperature of heat addition and the mean temperature of heat abstraction, there results:

$$\text{CYCLE I.} \quad X = 1 + \frac{H_1}{C_v T_b}.$$

Symbol.	Formula as Derived.	Formula Reduced to Initial Conditions.
p_b	Arbitrary	p_b
B $\begin{cases} v_b \end{cases}$	"	v_b
T_b	$\dfrac{p_b v_b}{R}$	$\dfrac{p_b v_b}{R}$

Symbol.	Formula as Derived.	Formula Reduced to Initial Conditions.

$$C \begin{cases} p_c & \dots\dots\dots\dots p_b\dfrac{T_c}{T_b}\dots\dots\dots\dots\dots\dots p_b X \\[2ex] v_c & \dots\dots\dots\dots v_b \dots\dots\dots\dots\dots\dots v_b \\[2ex] T_c & \dots\dots\dots\dots T_b + \dfrac{H_1}{C_v}\dots\dots\dots\dots\dots T_b X \end{cases}$$

$$D \begin{cases} p_d & \dots\dots\dots\dots p_b\dots\dots\dots\dots\dots\dots\dots p_b \\[2ex] v_d & \dots\dots\dots\dots v_c\left(\dfrac{p_c}{p_d}\right)^{\frac{1}{r}}\dots\dots\dots\dots v_c X^{\frac{1}{r}} \\[2ex] T_d & \dots\dots\dots\dots T_c\left(\dfrac{p_d}{p_c}\right)^{\frac{1}{r}}\dots\dots\dots\dots T_b X^{\frac{1}{r}} \end{cases}$$

$$H_2 \dots\dots\dots\dots C_p(T_d - T_b)\dots\dots\dots\dots C_p T_b(X^{\frac{1}{r}} - 1)$$

$$W \dots\dots\dots\dots J(H_1 - H_2)\dots\dots\dots J\{H_1 - C_p T_b(X^{\frac{1}{r}} - 1)\}$$

$$E \dots\dots\dots\dots \frac{W}{H_1}\dots\dots\dots\dots 1 - \frac{C_p T_b(X^{\frac{1}{r}} - 1)}{H_1}$$

$$R_v \dots\dots\dots\dots v_d - v_c \dots\dots\dots\dots v_b(X^{\frac{1}{r}} - 1)$$

$$\text{M.E.P.}\dots\dots\dots\dots J\frac{W}{R_v}\dots\dots\dots\dots J\left(\frac{H_1 - C_p T_b(X^{\frac{1}{r}} - 1)}{v_b(X^{\frac{1}{r}} - 1)}\right)$$

$$R_p \dots\dots\dots\dots p_c - p_b \dots\dots\dots\dots p_b(X - 1)$$

$$\text{M.E.V.}\dots\dots\dots\dots \frac{W}{R_p}\dots\dots\dots\dots J\left(\frac{H_1 - C_p T_b(X^{\frac{1}{r}} - 1)}{p_b(X - 1)}\right)$$

$$R_\phi \dots\dots\dots\dots C_v \log_e\frac{T_c}{T_b}\dots\dots\dots\dots C_v \log_e X$$

$$\text{M.E.T.}\dots\dots\dots\dots \frac{1}{2}\left(\frac{H_1 + H_2}{R_\phi}\right)\dots\dots\dots \frac{1}{2}\left(\frac{H_1 + C_p T_b(X^{\frac{1}{r}} - 1)}{C_v \log_e X}\right)$$

$$R_T \dots\dots\dots\dots T_c - T_b \dots\dots\dots\dots T_b(X - 1)$$

Cycle I. A.

As in *Cycle* I for point C:

$$v_c = v_b; \qquad \cdots \cdots \cdots \quad (1)$$
$$p_c = p_b X; \qquad \cdots \cdots \cdots \quad (2)$$
$$T_c = T_b X. \qquad \cdots \cdots \cdots \quad (3)$$

Assume

$$p_c > p_d > p_b.$$

Then from the adiabatic relation

$$p_d = p_c \left(\frac{v_c}{v_d} \right)^r,$$

or

$$v_d = v_c \left(\frac{p_c}{p_d} \right)^{\frac{1}{r}}. \qquad \cdots \cdots \quad (4)$$

Also

$$T_d = T_c \left(\frac{p_d}{p_c} \right)^{\frac{r-1}{r}}. \qquad \cdots \cdots \quad (5)$$

Substituting values of p_c and T_c in (4) and (5),

$$v_d = v_b \left(\frac{p_b X}{p_d} \right)^{\frac{1}{r}} = v_b X^{\frac{1}{r}} \left(\frac{p_b}{p_d} \right)^{\frac{1}{r}}; \qquad \cdots \cdots \quad (6)$$

$$T_d = T_b X \left(\frac{p_d}{p_b X} \right)^{\frac{r-1}{r}} = T_b X^{\frac{1}{r}} \left(\frac{p_d}{p_b} \right)^{\frac{r-1}{r}}. \qquad \cdots \cdots \quad (7)$$

If it be granted that

$$\frac{p_b}{p_d} = n,$$

then

$$v_d = v_b (Xn)^{\frac{1}{r}}; \qquad \cdots \cdots \cdots \quad (8)$$

$$T_d = T_b X^{\frac{1}{r}} n^{\frac{1}{r} - 1} = \frac{T_b (Xn)^{\frac{1}{r}}}{n}; \qquad \cdots \cdots \quad (9)$$

$$v_e = v_d = v_b (Xn)^{\frac{1}{r}}; \qquad \cdots \cdots \cdots \quad (10)$$
$$p_e = p_b;$$

$$T_e = T_b \frac{v_e}{v_b} = T_b (Xn)^{\frac{1}{r}}. \qquad \cdots \cdots \cdots \quad (11)$$

Applying the perfect gas law to the points B, C, D, and E,

$$\frac{p_b v_b}{T_b} = R,$$

$$\frac{p_c v_c}{T_c} = \frac{p_b X v_b}{T_b X} = R,$$

$$\frac{p_d v_d}{T_d} = \frac{p_d v_b (Xn)^{\frac{1}{r}}}{T_b^{\frac{1}{n}} (Xn)^{\frac{1}{r}}} = R,$$

$$\frac{p_e v_e}{T_e} = \frac{p_b v_b (Xn)^{\frac{1}{r}}}{T_b (Xn)^{\frac{1}{r}}} = R.$$

Heat is abstracted in two parts, the first at constant volume from D to E and the second at constant atmospheric pressure from E to B.

Hence

$$H_2 = C_v(T_d - T_e) + C_p(T_e - T_b)$$

$$= C_v\left[T_b\frac{1}{n}(Xn)^{\frac{1}{r}} - T_b(Xn)^{\frac{1}{r}} \right] + C_p\left[T_b(Xn)^{\frac{1}{r}} - T_b \right]$$

$$= C_v T_b(Xn)^{\frac{1}{r}}\left[\frac{1}{n} - 1 \right] + C_p T_b\left[(Xn)^{\frac{1}{r}} - 1 \right]. \quad \dots \quad (12)$$

The work done in foot-pounds is

$$W = J(H_1 - H_2). \quad \dots \dots \dots \quad (13)$$

$$\therefore W = J\left(H_1 - C_v T_b(Xn)^{\frac{1}{r}}\left(\frac{1}{n} - 1\right) + C_p T_b[(Xn)^{\frac{1}{r}} - 1] \right).$$

$$\therefore E = 1 - \frac{H_2}{H_1}. \quad \dots \dots \dots \quad (14)$$

$$\therefore E = 1 - \frac{C_v T_b(Xn)^{\frac{1}{r}}\left(\frac{1}{n} - 1\right) + C_p T_b[(Xn)^{\frac{1}{r}} - 1]}{H_1}. \quad (15)$$

The mean effective pressure

$$\text{M.E.P.} = \frac{W}{R_v}. \quad \cdots \cdots \cdots \quad (16)$$

But

$$R_v = v_d - v_b = v_b[(Xn)^{\frac{1}{r}} - 1]. \quad \cdots \cdots \quad (17)$$

$$\therefore \text{M.E.P.} = J \left\{ \frac{H_1 - C_v T_b (Xn)^{\frac{1}{r}} \left(\frac{1}{n} - 1\right) + C_p T_b[(Xn)^{\frac{1}{r}} - 1]}{v_b[(Xn)^{\frac{1}{r}} - 1]} \right\} \quad (18)$$

$$R_p = p_c - p_b = p_b(X - 1); \quad \cdots \cdots \quad (19)$$

$$\text{M.E.V.} = \frac{W}{R_p}$$

$$= J \left\{ \frac{H_1 - C_v T_b (Xn)^{\frac{1}{r}} \left(\frac{1}{n} - 1\right) + C_p T_b[(Xn)^{\frac{1}{r}} - 1]}{p_b(X - 1)} \right\}. \quad (20)$$

As before, the entropy range is

$$R_\phi = C_v \log_e X. \quad \cdots \cdots \cdots \quad (21)$$

Taking the mean effective temperature as the mean of the average heating temperature and the average cooling temperature,

$$\text{M.E.T.} = \frac{T' + T''}{2} = \frac{1}{2} \left(\frac{H_1 + H_2}{R_\phi} \right) \quad \cdots \cdots \quad (22)$$

$$= \frac{1}{2} \left\{ \frac{H_1}{C_v \log_e X} + \frac{C_v T_b (Xn)^{\frac{1}{r}} \left(\frac{1}{n} - 1\right) + C_p T_b[(Xn)^{\frac{1}{r}} - 1]}{C_v \log_e X} \right\}. \quad (23)$$

The temperature range is

$$R_T = T_c - T_b = T_b(X - 1) \quad \cdots \cdots \quad (24)$$

The pressure range is

$$R_p = p_c - p_b = p_b(X - 1) \quad \cdots \cdots \quad (25)$$

Whence

$$\text{M.E.V.} = \frac{W}{R_p}$$

$$= J \frac{H_1 - C_v T_b (Xn)^{\frac{1}{r}} \left[\left(\frac{1}{n'} \right)^{\frac{r-1}{r}} - (1)^{\frac{1}{r}} \right] - C_p T_b \left[(Xn)^{\frac{1}{r}} - 1 \right]}{p_b (X - 1)}$$

(26)

$$= J \left\{ \frac{H_1 - C_v T_b (Xn)^{\frac{1}{r}} \left(\frac{1}{n} - 1 \right) - C_p T_b [(Xn)^{\frac{1}{r}} - 1]}{p_b (X - 1)} \right\}.$$

Tabulating these results:

CYCLE I.   A.

Symbol.	Formula as First Derived.	Formula Reduced to Initial Conditions.
B $\begin{cases} p_b \\ v_b \\ T_b \end{cases}$	Arbitrary.	p_b
	"	v_b
	$\dfrac{p_b v_b}{R}$	$\dfrac{p_b v_b}{R}$
C $\begin{cases} p_c \\ v_c \\ T_c \end{cases}$	$p_b \dfrac{T_c}{T_b}$	$p_b X$
	v_b	v_b
	$T_b \left(1 + \dfrac{H_1}{C_v T_b} \right)$	$T_b X$
D $\begin{cases} p_d \\ v_d \\ T_d \end{cases}$	$p_c > p_d > p_b.$	$p_c > p_d > p_b$
	$v_c \left(\dfrac{p_c}{p_d} \right)^{\frac{1}{r}}$	$v_b (Xn)^{\frac{1}{r}}$
	$T_c \left(\dfrac{p_d}{p_c} \right)^{\frac{r}{r-1}}$	$\dfrac{T_b (Xn)^{\frac{1}{r}}}{n}$
E $\begin{cases} p_e \\ v_e \\ T_e \end{cases}$	p_d	p_b
	v_d	$v_b (Xn)^{\frac{1}{r}}$
	$T_b \dfrac{v_e}{v_b}$	$T_b (Xn)^{\frac{1}{r}}$

| Symbol. | Formula as First Derived. | Formula Reduced to Initial Conditions. |

$H_2 \ldotp\ldotp C_v(T_d - T_e) + C_p(T_e - T_b)$

$$= C_v T_b(Xn)^{\frac{1}{r}}\left(\frac{1}{n}-1\right) + C_p T_b[(Xn)^{\frac{1}{r}}-1]$$

$$W \ldotp\ldotp J(H_1 - H_2) = J\left\{H_1 - C_v T_b(Xn)^{\frac{1}{r}}\left(\frac{1}{n}-1\right) - C_p T_b[(Xn)^{\frac{1}{r}}-1]\right\}$$

$$E \ldots\ldots 1 - \frac{H_2}{H_1} \ldots 1 - \frac{C_v T_b(Xn)^{\frac{1}{r}}\left(\frac{1}{n}-1\right) + C_p T_b[(Xn)^{\frac{1}{r}}-1]}{H}$$

$$R_v \ldots\ldots v_d - v_b \ldots\ldots\ldots\ldots v_b[(Xn)^{\frac{1}{r}}-1]$$

$$\text{M.E.P.}\ldots \frac{W}{R_v} = J\left\{\frac{H_1 - C_v T_b(Xn)^{\frac{1}{r}}\left(\frac{1}{n}-1\right) - C_p T_b[(Xn)^{\frac{1}{r}}-1]}{v_b[(Xn)^{\frac{1}{r}}-1]}\right\}$$

$$R_\phi \ldots\ldots C_v \log_e \frac{T_c}{T_b} \ldots\ldots\ldots\ldots C_v \log_e X$$

$$\text{M.E.T.} = \frac{1}{2}\left(\frac{H_1 + H_2}{R_\phi}\right)$$

$$= \frac{1}{2}\left\{\frac{H_1 + C_v T_b(Xn)^{\frac{1}{r}}\left(\frac{1}{n}-1\right) + C_p T_b[(Xn)^{\frac{1}{r}}-1]}{C_v \log_e X}\right\}$$

$$R_p \ldots\ldots p_c - p_b \ldots\ldots\ldots\ldots p_b(X-1)$$

$$\text{M.E.V.}\ldots \frac{W}{R_p} \ldotp\ldotp J\left\{\frac{H_1 - C_v T_b(Xn)^{\frac{1}{r}}\left(\frac{1}{n}-1\right) - C_p T_b[(Xn)^{\frac{1}{r}}-1]}{p_b(X-1)}\right\}$$

$$R_T \ldots\ldots T_c - T_b \ldots\ldots\ldots\ldots T_b(X-1).$$

CYCLE I.　B.

As the operations up to the point C, i.e., after addition of heat, are the same as in Cycle I, these results may be assumed:

$$v_c = v_b, \quad \ldots\ldots\ldots\ldots\ldots (1)$$
$$p_c = p_b X, \quad \ldots\ldots\ldots\ldots (2)$$
$$T_e = T_b X. \quad \ldots\ldots\ldots\ldots (3)$$

Choose p_d so that

$$p_b > p_d > 0. \quad \cdots \cdots \cdots \quad (4)$$

Expansion CD gives

$$v_d = v_c \left(\frac{p_c}{p_d}\right)^{\frac{1}{r}}, \quad = v_b X^{\frac{1}{r}} \left(\frac{p_b}{p_d}\right)^{\frac{1}{r}}, = v_b (Xn)^{\frac{1}{r}} \quad \text{if} \quad n = \frac{p_b}{p_d}. \quad (5)$$

Also

$$T_d = T_c \left(\frac{p_d}{p_c}\right)^{\frac{r-1}{r}} = T_b X^{\frac{1}{r}} \left(\frac{p_d}{p_b}\right)^{\frac{r-1}{r}} = \frac{T_b (Xn)^{\frac{1}{r}}}{n}. \quad \cdots \quad (6)$$

From the isothermal relation along DE,

$$T_e = T_d = \frac{T_c (Xn)^{\frac{1}{r}}}{n}; \quad \cdots \cdots \cdots \quad (7)$$

$$p_e = p_b \text{ by hypothesis}; \quad \cdots \cdots \cdots \quad (8)$$

$$\therefore \ v_e = \frac{p_d v_d}{p_b},$$

or

$$v_e = \frac{p_d v_b (Xn)^{\frac{1}{r}}}{p_b} = \frac{v_b (Xn)^{\frac{1}{r}}}{n}. \quad \cdots \cdots \quad (9)$$

Applying the perfect gas law to the various points,

$$\frac{p_b v_b}{T_b} = R;$$

$$\frac{p_c v_c}{T_c} = \frac{p_b v_b X}{T_b X} = \frac{p_b v_b}{T_b} = R;$$

$$\frac{p_d v_d}{T_d} = \frac{p_d v_b (Xn)^{\frac{1}{r}}}{T_b \frac{p_d}{p_b} (Xn)^{\frac{1}{r}}} = \frac{p_b v_b}{T_b} = R;$$

$$\frac{p_e v_e}{T_e} = \frac{p_b v_b \dfrac{(Xn)^{\frac{1}{r}}}{n}}{T_b \dfrac{(Xn)^{\frac{1}{r}}}{n}} = \frac{p_b v_b}{T_b} = R.$$

Heat is abstracted in two parts, first, a part isothermally, and second, a part at atmospheric pressure. The part abstracted isothermally is most easily calculated with the aid of the $\theta\phi$ diagram and its relations.

The entropy range along BC has been found to be

$$R_\phi = \phi_c - \phi_b = C_v \log_e \frac{T_b}{T_e} = C_v \log_e X. \quad \ldots \quad (10)$$

Now it is evidently the same so far as entropy range is concerned whether the cooling is at constant pressure from E to B or heating is done isopiestically from B to E, thus

$$\phi_e - \phi_b = C_p \log_e \frac{T_e}{T_b}. \quad \ldots \ldots \quad (11)$$

Hence the entropy range for the isothermal operation will be given by

$$\phi_d - \phi_e = C_v \log_e X - C_p \log_e \frac{T_e}{T_b} \quad \ldots \ldots \ldots \quad (12)$$

$$= C_v \log_e X - C_p \log_e \left[X^{\frac{1}{r}} \left(\frac{p_d}{p_b} \right)^{\frac{r-1}{r}} \right]. \quad \ldots \ldots \quad (13)$$

This latter isothermal change taking place at temperature $T_e = T_d$, the heat of cooling will be given by

$$T_d \left(C_v \log_e X - C_p \log_e \frac{T_e}{T_b} \right). \quad \ldots \ldots \quad (14)$$

Hence the total heat abstracted is

$$H_2 = C_p (T_e - T_b) + T_d \left[C_v \log_e X - C_p \log_e \frac{T_e}{T_b} \right] \quad \ldots \quad (15)$$

$$= C_v \left[\frac{T_b (Xn)^{\frac{1}{r}}}{n} - T_b \right] + \frac{T_b (Xn)^{\frac{1}{r}}}{n} \left[C_v \log_e X - C_p \log_e \frac{(Xn)^{\frac{1}{r}}}{n} \right].$$

But

$$C_p \log_e \frac{(Xn)^{\frac{1}{r}}}{n} = C_p \log_e X^{\frac{1}{r}} + C_p \log_e n^{\frac{1}{r} - 1}$$

$$= C_v \log_e X + (C_v - C_p) \log_e n.$$

Since

$$\frac{1}{\gamma} = \frac{C_v}{C_p},$$

·and

$$\frac{1}{\gamma} - 1 = \frac{C_v - C_p}{C_p};$$

hence

$$H_2 = C_p T_b \left[\frac{(Xn)^{\frac{1}{\gamma}}}{n} - 1 \right] + \frac{T_b (Xn)^{\frac{1}{\gamma}}}{n} \log_e n^{C_v - C_p}. \quad . \quad (16)$$

The work in foot-pounds is

$$W = J(H_1 - H_2);$$

$$\therefore W = J \left\{ H_1 - C_p T_b \left[\frac{(Xn)^{\frac{1}{\gamma}}}{n} - 1 \right] - \frac{T_b X(n)^{\frac{1}{\gamma}}}{n} \log_e {}^{C_v - C_p} \right\} \quad (17)$$

$$E = 1 - \frac{H_2}{H_1};$$

$$\therefore E = 1 - \frac{C_p T_b \left[\dfrac{(Xn)^{\frac{1}{\gamma}}}{n} - 1 \right] + \dfrac{T_b (Xn)^{\frac{1}{\gamma}}}{n} \log_e n^{C_v - C_p}}{H_1} \quad (18)$$

$$R_v = v_e - v_b = v_b [(Xn)^{\frac{1}{\gamma}} - 1]. \quad . \quad . \quad . \quad . \quad (19)$$

$$\therefore M.E.P. = J \left\{ \frac{H_1 - C_p T_b \left[\dfrac{(Xn)^{\frac{1}{\gamma}}}{n} - 1 \right] + \dfrac{T_b X(n)^{\frac{1}{\gamma}}}{n} \log_e n^{C_v - C_p}}{v_b [X(n)^{\frac{1}{\gamma}} - 1]} \right\} \quad (20)$$

$$R_p = p_c - p_d = p_b X - p_d. \quad . \quad . \quad . \quad . \quad (21)$$

$$\therefore M.E.V. = J \left\{ \frac{H_1 - C_p T_b \left[\dfrac{(Xn)^{\frac{1}{\gamma}}}{n} - 1 \right] + \dfrac{T_b (Xn)^{\frac{1}{\gamma}}}{n} \log_e n^{C_v - C_p}}{p_b X - p_d} \right\} \quad (22)$$

The mean effective temperature being the mean of the heating and cooling means is given by

$$M.E.T. = \frac{1}{2} \left(\frac{H_1 + H_2}{R_\phi} \right),$$

where R_ϕ is the same as in previous cycle.

$$\therefore \text{M.E.T.} = \frac{1}{2\log_e X}\left\{H_1 + C_p T_b\left[\frac{(Xn)^{\frac{1}{r}}}{n} - 1\right]\right.$$

$$\left. + \frac{T_b(Xn)^{\frac{1}{r}}}{n}\log_e n^{C_v - C_p}\right\} \quad (23)$$

$$R_T = T_b - T_d = T_b\left[1 - \frac{(X)n^{\frac{1}{r}}}{n}\right] \quad . \quad . \quad . \quad . \quad (24)$$

Tabulating:

Cycle I B.

Symbol.	Formula as First Derived.	Formula Reduced to Initial Conditions.
B $\begin{cases} p_b \\ v_b \\ T_b \end{cases}$	Atmosphere.	Atmospheric p_b
	Arbitrary	v_b
	$\dfrac{p_b v_b}{R}$	$\dfrac{p_b v_b}{R}$
C $\begin{cases} p_c \\ v_c \\ T_c \end{cases}$	$p_b\dfrac{T_c}{T_b}$	$p_b X$
	v_b	v_b
	$T_b\left(1 + \dfrac{H_1}{C_v T_b}\right)$	$T_b X$
D $\begin{cases} p_d \\ v_d \\ T_d \end{cases}$	$p_b > p_d > 0$	$p_b > p_d > 0$
	$v_c\left(\dfrac{p_c}{p_d}\right)^{\frac{1}{r}}$	$v_b(Xn)^{\frac{1}{r}}$
	$T_c\left(\dfrac{p_d}{p_c}\right)^{\frac{r-1}{r}}$	$\dfrac{T_b(Xn)^{\frac{1}{r}}}{n}$
E $\begin{cases} p_e \\ v_e \\ T_e \end{cases}$	p_b	p_b
	$\dfrac{p_d v_d}{p_d}$	$\dfrac{v_b(Xn)^{\frac{1}{r}}}{n}$
	T_d	$\dfrac{T_b(Xn)^{\frac{1}{r}}}{n}$

$$H_2\ldots\ldots C_p(T_e - T_b) + T_d\left[C_v\log_e X - C_p\log_e\frac{T_e}{T_b}\right]$$

$$= C_p T_b\left[\frac{(Xn)^{\frac{1}{r}}}{n} - 1\right] + \frac{T_b(Xn)^{\frac{1}{r}}}{n}\log_e n^{C_v - C_p}$$

$$W\ldots\ldots\ldots J(H_1 - H_2)$$

$$= J\left\{H_1 - C_p T_b\left[\frac{(Xn)^{\frac{1}{r}}}{n} - 1\right] - \frac{T_b(Xn)^{\frac{1}{r}}}{n}\log_e n^{C_v - C_p}\right\}$$

$$E\ldots\ldots\ldots 1 - \frac{H_2}{H_1}$$

$$= 1 - \frac{C_p T_b\left[\frac{(Xn)^{\frac{1}{r}}}{n} - 1\right] + T_b\frac{(Xn)^{\frac{1}{r}}}{n}\log_e n^{C_v - C_p}}{H_1}$$

$$R_v\ldots\ldots\ldots(v_e - v_b)\ldots\ldots\ldots\ldots v_b[(Xn)^{\frac{1}{r}} - 1]$$

$$\text{M.E.P.}\ldots\ldots\ldots\frac{W}{R_v}$$

$$= J\left\{\frac{H_1 - C_p T_b\left[\frac{(Xn)^{\frac{1}{r}}}{n} - 1\right] - T_b\frac{(Xn)^{\frac{1}{r}}}{n}\log_e n^{C_v - C_p}}{v_b[(Xn)^{\frac{1}{r}} - 1]}\right\}$$

$$R_p\ldots\ldots\ldots p_c - p_d\ldots\ldots\ldots\ldots p_b X - p_d$$

$$\text{M.E.V.}\ldots\ldots\ldots\frac{W}{R_p}$$

$$= J\left\{\frac{H_1 - C_p T_b\left[\frac{(Xn)^{\frac{1}{r}}}{n} - 1\right] + \frac{T_b(Xn)^{\frac{1}{r}}}{n}\log_e n^{C_v - C_p}}{p_b X - p_d}\right\}$$

$$R_\phi\ldots\ldots\ldots\log_e\frac{T_e}{T_b}\ldots\ldots\ldots\ldots\log_e X$$

$$\text{M.E.T.}\ldots\ldots\frac{1}{2}\left(\frac{H_1 + H_2}{R_\phi}\right)$$

$$= \frac{1}{2\log_e X}\left\{H_1 + C_p T_b\left[\frac{(Xn)^{\frac{1}{r}}}{n} - 1\right] + \frac{T_b(Xn)^{\frac{1}{r}}}{n}\log_e n^{C_v - C_p}\right\}$$

$$R_T \dots\dots\dots T_b - T_d \dots\dots\dots\dots\dots T_b\left[1 - \frac{(Xn)^{\frac{1}{r}}}{n}\right]$$

CYCLE I C.

FIG. 117.   FIG. 118.

Assume all results to point C from Cycle I:

$$p_c = p_b X, \quad \dots\dots\dots \quad (1)$$

$$v_c = v_b, \quad \dots\dots\dots \quad (2)$$

$$T_c = T_b X. \quad \dots\dots\dots \quad (3)$$

From the adiabatic CD

$$v_d = v_c \left(\frac{p_c}{p_d}\right)^{\frac{1}{r}} = v_b \left(\frac{p_c}{p_d}\right)^{\frac{1}{r}}. \quad \dots\dots \quad (4)$$

This adiabatic must meet the isothermal from B in point D, hence

$$v_d = v_b \frac{p_b}{p_d}. \quad \dots\dots\dots \quad (5)$$

Equate (4) and (5),

$$v_b \left(\frac{p_c}{p_d}\right)^{\frac{1}{r}} = v_b \frac{p_b}{p_d};$$

$$\therefore \; p_b{}^y = p_1{}^v \frac{p_b X}{p_d} p_d{}^{r-1} p_b X;$$

$$\therefore \; p_d = p_b X^{\frac{1}{1-r}} = \frac{p_b}{X^{\frac{1}{r-1}}}. \quad \dots\dots \quad (6)$$

This is the pressure at which the isothermal through B will meet the adiabatic through C. Its corresponding volume is

$$v_d = v_b \frac{p_b}{p_b X^{\frac{1}{1-r}}} = v_b X^{\frac{1}{r-1}}; \quad \dots\dots \quad (7)$$

$$T_d = T_b. \quad \dots\dots\dots \quad (8)$$

The heat abstracted by the isothermal cooling is found as before from $\Theta\phi$ relation,

$$\phi_c - \phi_b = C_v \log_e \frac{T_c}{T_b} = C_v \log_e X, \quad \cdots \quad (9)$$

Hence

$$\phi_d - \phi_b = C_v \log_e X; \quad \cdots \cdots \quad (10)$$

$$\therefore H_2 = T_b C_v \log_e X. \quad \cdots \cdots \quad (11)$$

The work done in foot-pounds is

$$W = J(H_1 = H_2) = J(H_1 - T_b C_v \log_e X). \quad \cdots \quad (12)$$

The efficiency is

$$E = 1 - \frac{H_2}{H_1} = 1 - \frac{T_b C_v \log_e X}{H_1} \quad \cdots \cdots \quad (13)$$

The volume range is

$$R_v = v_d - v_b = v_b(X^{\frac{1}{\gamma-1}} - 1). \quad \cdots \cdots \quad (14)$$

Hence

$$\text{M.E.P.} = \frac{J(H_1 - T_b C_v \log_e X)}{v_b(X^{\frac{1}{\gamma-1}} -)1.} \quad \cdots \cdots \quad (15)$$

The pressure range is

$$R_p = p_c - p_d = p_b X - \frac{p_b}{X^{\frac{1}{\gamma-1}}} = p_b\left(X - \frac{1}{X^{\frac{1}{\gamma-1}}}\right). \quad \cdots \quad (16)$$

$$\text{M.E.V.} = \frac{J(H_1 - T_b C_v \log_e X)}{p_b\left(X - \frac{1}{X^{\frac{1}{\gamma-1}}}\right)} \quad \cdots \cdots \quad (17)$$

The entropy range was found $R_\phi = C_v \log_e X$, hence

$$\text{M.E.T.} = \frac{1}{2}\left(\frac{H_1 + H_2}{R_\phi}\right) = \frac{1}{2C_v \log_e X}(H_1 + T_b C_v \log_e X)$$

$$= \frac{1}{2}\left(\frac{H_1}{C_v \log_e X} + T_b\right).$$

And

$$R_T = T_c - T_b = T_b(X - 1).$$

Tabulating:

CYCLE I C.

Symbol.	Formula as First Derived.	Formula Reduced to Initial Conditions.

B $\begin{cases} p_b \dots \text{Atmospheric} \dots \text{Atmospheric } p_b \\ v_b \dots \text{Arbitrary} \dots v_b \\ T_b \dots \dfrac{p_b v_b}{R} \dots \dfrac{p_b v_b}{R} \end{cases}$

C $\begin{cases} p_c \dots p_b \dfrac{T_b}{T_c} \dots p_b X \\ v_c \dots v_b \dots v_b \\ T_c \dots T_b\left(1 + \dfrac{H_1}{C_v T_b}\right) \dots T_b X \end{cases}$

D $\begin{cases} p_d \dots \left(\dfrac{p_b}{X^{\frac{1}{r-1}}}\right) \dots \dfrac{p_b}{X^{\frac{1}{r-1}}} \\ v_d \dots v_b X^{\frac{1}{r-1}} \dots v_b X^{\frac{1}{r-1}} \\ T_d \dots T_b \dots T_b \end{cases}$

$H_2 \dots T_b C_v \log_e \dfrac{T_c}{T_b} \dots T_b C_v \log_e X$

$W \dots J(H_1 - H_2) \dots J(H_1 - T_b C_v \log_e X)$

$E \dots 1 - \dfrac{H_2}{H_1} \dots 1 - \dfrac{T_b C_v \log_e X}{H_1}$

$R_v \dots v_d - v_b \dots v_b(X^{\frac{1}{r-1}} - 1)$

$\text{M.E.P.} \dots \dfrac{W}{R_v} \dots J\left\{\dfrac{H_1 - T_b C_v \log_e X}{v_b(X^{\frac{1}{r-1}} - 1)}\right\}$

$R_p \dots p_c - p_d \dots p_b\left(X - \dfrac{1}{X^{\frac{1}{r-1}}}\right)$

$\text{M.E.V.} \dots \dfrac{W}{R_p} \dots J\left\{\dfrac{H_1 - T_b C_v \log_e X}{p_b\left(X - \dfrac{1}{X^{\frac{1}{r-1}}}\right)}\right\}$

Symbol.	Formula as First Derived.	Formula Reduced to Initial Conditions.

$$R_\phi \ldots\ldots\ldots\ldots\ldots C_v \log_e \frac{T_c}{T_b} \ldots\ldots\ldots\ldots C_v \log_e X$$

$$\text{M.E.T.} \ldots\ldots\ldots \frac{1}{2 \log_e X}(H_1 + H_2) \ldots\ldots \frac{1}{2}\left(\frac{H_1}{C_v \log_e X} + T_b\right)$$

$$R_T \ldots\ldots\ldots\ldots\ldots T_c - T_b \ldots\ldots\ldots\ldots T_b(X - 1)$$

The relatively poor showing of Cycle I and its modifications with respect to efficiency and mean effective pressure as compared with the compression cycles are the reasons for its minor importance.

184. Compression Cycle with Isometric Heating.—This group includes as No. II A the ideal Otto cycle, where the gas is heated and cooled at constant volume. The PV and $\theta\phi$ diagrams and the mathematical analysis are as follows:

The first or typical case is that of the ideal Atkinson engine, or some compound engines, with an expansion complete down to atmospheric pressure.

Let Fig. 119 be the PV and Fig. 120 the $\theta\phi$ diagram of this cycle.

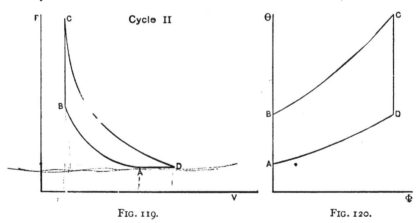

FIG. 119. FIG. 120.

From *A* to *B.* Adiabatic compression from atmospheric pressure.

From B to C.   Addition of heat isometrically.

From C to D.   Adiabatic expansion to atmospheric pressure.

From D to A.   Cooling at atmospheric pressure.

In the compression cycles the volume ratio $\dfrac{v_a}{v_b}$ will enter into many of the formulæ so that it will be found convenient to write

$$\frac{v_a}{v_b} = \gamma.$$

The compression is adiabatic, hence

$$p_b = p_a\left(\frac{v_a}{v_b}\right)^{\gamma} = p_a\gamma^{\gamma}; \quad \bullet \quad \bullet \quad \bullet \quad \bullet \quad \bullet \quad \bullet \quad \text{(1)}$$

$$T_b = T_a\left(\frac{v_a}{v_b}\right)^{\gamma-1} = T_a\gamma^{\gamma-1}. \quad \bullet \quad \bullet \quad \bullet \quad \bullet \quad \text{(2)}$$

During addition of heat $v_c = v_b$, and therefore

$$p_c = p_b\frac{T_c}{T_b}.$$

If $\dfrac{T_c}{T_b} = 1 + \dfrac{H_1}{C_v T_b} = X$ as in the previous cycles,

$$p_c = p_b X = p_a\gamma^{\gamma}X; \quad \bullet \quad \bullet \quad \bullet \quad \bullet \quad \bullet \quad \text{(3)}$$

$$T_c = T_a\gamma^{\gamma-1}X. \quad \bullet \quad \bullet \quad \bullet \quad \bullet \quad \bullet \quad \text{(4)}$$

Adiabatic expansion gives

$$v_d = v_c\left(\frac{p_c}{p_d}\right)^{\frac{1}{\gamma}} \text{ or if } p_d = p_a,$$

$$v_d = v_b\left(\frac{p_a\gamma^{\gamma}X}{p_a}\right)^{\frac{1}{\gamma}} = v_b\gamma X^{\frac{1}{\gamma}} = \frac{v_a}{\gamma}\gamma X^{\frac{1}{\gamma}} = v_a X^{\frac{1}{\gamma}}. \quad \bullet \quad \bullet \quad \text{(5)}$$

$$T_d = T_a\frac{v_d}{v_a} = T_a X^{\frac{1}{\gamma}}. \quad \bullet \quad \bullet \quad \bullet \quad \bullet \quad \bullet \quad \text{(6)}$$

Applying the perfect gas law

$$\frac{p_a v_a}{T_a} = R,$$

$$\frac{p_b v_b}{T_b} = \frac{p_a \gamma^r v_a}{T_a \gamma^{r-1} \gamma} = R,$$

$$\frac{p_c v_c}{T_c} = \frac{p_a \gamma^r X v_a}{T_a \gamma^{r-1} X \gamma} = R,$$

$$\frac{p_d v_d}{T_d} = \frac{p_a v_a X^{\frac{1}{r}}}{T_a X^{\frac{1}{r}}} = R.$$

The heat discharged

$$H_2 = C_p(T_d - T_a) = C_p T_a(X^{\frac{1}{r}} - 1). \quad \ldots \quad (7)$$

The work done is

$$W = H_1 - H_2 = H_1 - C_p T_a(X^{\frac{1}{r}} - 1), \quad \ldots \quad (8)$$

$$E = \frac{W}{H_1} = 1 - \frac{C_p T_a(X^{\frac{1}{r}} - 1)}{H_1}, \quad \ldots \quad (9)$$

$$R_v = v_d - v_b = v_a\left(X^{\frac{1}{r}} - \frac{1}{\gamma}\right). \quad \ldots \quad (10)$$

$$\text{M.E.P.} = \frac{W}{R_v} = J\frac{H_1 - C_p T_a(X^{\frac{1}{r}} - 1)}{v_a\left(X^{\frac{1}{r}} - \frac{1}{\gamma}\right)}. \quad \ldots \quad (11)$$

$$R_\phi = C_v \log_e X; \quad \ldots \quad (12)$$

$$\therefore \text{M.E.T.} = \frac{1}{2}\left(\frac{H_1 + H_2}{R_\phi}\right) = \frac{1}{2}\left(\frac{H_1 + C_p T_a(X^{\frac{1}{r}} - 1)}{C_v \log_e x}\right) \cdot (13)$$

$$R_p = p_c - p_a = p_a(\gamma^r X - 1). \quad \ldots \quad (14)$$

$$\text{M.E.V.} = J\frac{H_1 - C_p T_a(X^{\frac{1}{r}} - 1)}{p_a(\gamma^r \cdot X - 1)}. \quad \ldots \quad (15)$$

CYCLE II.

Symbol.	Formula as First Derived.	Formula Reduced.
p_b	$p_a\left(\dfrac{v_a}{v_b}\right)^{\gamma}$	$p_a\gamma^{\gamma}$
v_b	$\dfrac{v_a}{\gamma}$; (γ arbitrary)	$\dfrac{v_a}{\gamma}$
T_b	$T_a\left(\dfrac{v_a}{v_b}\right)^{\gamma-1}$	$T_a\gamma^{\gamma-1}$
p_c	$p_b\dfrac{T_c}{T_b}$	$p_a\gamma^{\gamma}X$
v_c	v_b	$\dfrac{v_a}{\gamma}$
T_c	$T_b\left(1+\dfrac{H_1}{C_vT_b}\right)$	T_bX
p_d	p_a	p_a
v_d	$v_c\left(\dfrac{p_c}{p_d}\right)^{\frac{1}{\gamma}}$	$v_aX^{\frac{1}{\gamma}}$
T_d	$T_a\dfrac{v_d}{v_a}$	$T_aX^{\frac{1}{\gamma}}$
H_2	$C_p(T_d-T_a)$	$C_pT_a(X^{\frac{1}{\gamma}}-1)$
W	H_1-H_2	$H_1-C_pT_a(X^{\frac{1}{\gamma}}-1)$
E	$1-\dfrac{H_2}{H_1}$	$1-\dfrac{C_pT_a(X^{\frac{1}{\gamma}}-1)}{H_1}$
R_v	v_d-v_b	$v_a\left(X^{\frac{1}{\gamma}}-\dfrac{1}{\gamma}\right)$
M.E.P.	$J\dfrac{W}{R_v}$	$J\dfrac{H_1-C_pT_a(X^{\frac{1}{\gamma}}-1)}{v_a(X^{\frac{1}{\gamma}}-1)}$
R_ϕ	$C_v\log_e\dfrac{T_c}{T_b}$	$C_v\log_e X$
M.E.T.	$\dfrac{1}{2}\left(\dfrac{H_1+H_2}{R_\phi}\right)$	$\dfrac{1}{2}\left(\dfrac{H_1+C_pT_a(X^{\frac{1}{\gamma}}-1)}{C_v\log_e X}\right)$

Symbol.	Formula as First Derived.	Formula Reduced.
R_p	$p_c - p_a$	$p_a(\gamma^\tau X - 1)$
M.E.V.	$J\dfrac{W}{R_p}$	$J\dfrac{H_1 - C_p T_a(X^{\frac{1}{\tau}} - 1)}{p_a(\gamma^\tau X - 1)}$
R_T	$T_c - T_a$	$T_a(\gamma^{\tau-1} X - 1)$

In Cycle II A, one case of which is the normal Otto or Beau de Rochas cycle:

From A to B. Adiabatic compression from atmospheric pressure.

From B to C. Addition of heat isometrically.

From C to D. Adiabatic expansion to pressure above atmosphere.

From D to E. Cooling isometrically to atmosphere.

From E to A. Cooling at atmospheric pressure.

Let Fig. 121 be the PV and Fig. 122 the $\theta\phi$ diagram for the cycle.

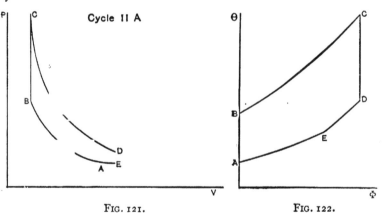

FIG. 121.   FIG. 122.

Then, since the compression is as in Cycle II,

$$\frac{v_a}{v_b} = \gamma, \quad \cdots \cdots \cdots \quad (1)$$

$$p_b = p_a \gamma^\tau, \quad \cdots \cdots \cdots \quad (2)$$

$$T_b = T_a \gamma^{\tau-1}. \quad \cdots \cdots \cdots \quad (3)$$

Also **for** *C,* the heat addition being as before,

$$v_c = v_b, \quad \cdots \quad \cdots \quad \cdots \quad (4)$$

$$p_c = p_b X = p_a \gamma^r X, \quad \cdots \quad \cdots \quad (5)$$

$$T_c = T_b X = T_a \gamma^{r-1} X. \quad \cdots \quad \cdots \quad (6)$$

The point **D** lies arbitrarily between *C* and the atmospheric line on **the adia**batic

$$p_d = p_c \left(\frac{v_c}{v_d}\right)^r = p_c \left(\frac{v_b}{v_d}\right)^r = p_a \gamma^r X \left(\frac{v_a}{\gamma v_d}\right)^r = p_a X \left(\frac{v_a}{v_d}\right)^r. \quad \cdot \quad (7)$$

From **this** point two cases **may** be considered: $1°$, the general case where v_d is greater than v_a, and $2°$, a particular case where $v_d = v_a$. **This** latter results when by reason of a throttling-governor **action** the gases at the end of expansion have the same volume as **before** compression.

$1°$.	$2°$.
$v_d > v_a$ and $p_d > p_a$.	$v_d = v_a$.

Then we have

$p_d = p_a X \left(\dfrac{v_a}{v_d}\right)^r; \quad (8)$	$p_d' = p_a X. \quad (8')$
$\therefore T_d = T_a X \left(\dfrac{v_a}{v_d}\right)^{r-1}, \quad (9)$	$T_d' = T_a X. \quad (9')$
$v_e = v_d, \quad (10)$	$v_e' = v_a \quad (10')$
$p_e = p_a, \quad (11)$	$p_e' = p_a \quad (11')$
$T_e = T_d \dfrac{p_e}{p_d} = T_d \dfrac{p_a}{p_d};$	
$\therefore T_e = T_d \dfrac{v_d}{v_a}. \quad (12)$	$T_e' = T_a. \quad (12')$

Apply the perfect gas law:

$$\frac{p_a v_a}{T_a} = R,$$

$$\frac{p_b v_b}{T_b} = \frac{p_a \gamma^\gamma v_a}{T_a \gamma^{\gamma-1} \gamma} = R,$$

$$\frac{p_c v_c}{T_e} = \frac{p_a \gamma^\gamma X v_a}{T_a \gamma^{\gamma-1} X \gamma} = R.$$

1°.

$$\frac{p_d v_d}{T_d} = \frac{p_a X \left(\dfrac{v_a}{v_d}\right)^\gamma v_d}{T_a X \left(\dfrac{v_a}{v_d}\right)^{\gamma-1}} = R,$$

$$\frac{p_e v_e}{T_e} = \frac{p_a v_d}{T_a \dfrac{v_d}{v_a}} = R.$$

2°.

$$\frac{p_d' v_d'}{T_d'} = \frac{p_a X v_a}{T_a X} = R,$$

$$\frac{p_e' v_e'}{T_e'} = \frac{v_a p_a}{T_a} = R.$$

Heat is abstracted as follows:

$$H_2 = C_v(T_d - T_e)$$
$$+ C_p(T_e - T_a) \quad \cdot \quad (13)$$

$$= C_v T_a \left[X \left(\frac{v_a}{v_d} \right)^{\gamma-1} - \frac{v_d}{v_a} \right]$$

$$+ C_p T_a \left[\frac{v_d}{v_a} - 1 \right].$$

$$\therefore H_2 = C_v T_a \frac{v_d}{v_a} \left[\left(\frac{v_a}{v_d} \right)^\gamma X - 1 \right]$$

$$+ C_p T_a \left[\frac{v_d}{v_a} - 1 \right].$$

or

$$H_2' = C_v(T_d - T_a) \quad \cdot \quad (13')$$
$$= C_v T_a [X - 1],$$

$$H_2' = \frac{C_v T_a H_1}{C_v T_a \gamma^{\gamma-1}}$$
$$= \frac{H_1}{\gamma^{\gamma-1}}.$$

The work is given by

$$W = H_1 - H_2$$

$$= H_1 - C_v T_a \frac{v_d}{v_a} \left[X \left(\frac{v_a}{v_d} \right)^\gamma - 1 \right]$$

$$- C_p T_a \left[\frac{v_d}{v_a} - 1 \right]. \quad (14)$$

$$W' = H_1 - \frac{H_1}{\gamma^{\gamma-1}}. \quad (14')$$

Volume range is

$$R_v = v_d - v_b = v_d - \frac{v_a}{\gamma} \quad . \quad (15)$$

$$R_v' = v_a - v_b = v_a\left(1 - \frac{1}{\gamma}\right) \quad (15')$$

$$\text{M.E.P.} = J\frac{W}{R_v} = J\frac{W}{v_d - \dfrac{v_a}{\gamma}} \quad (16)$$

$$\text{M.E.P.} = J\frac{W'}{v_a\left(1 - \dfrac{1}{\gamma}\right)} (16')$$

Entropy range is the same for both cases:

$$R\phi = C_v \log_e \frac{T_c}{T_b} = C_v \log_e X. \quad . \quad . \quad . \quad (17)$$

Mean of mean temperatures of heat addition and abstraction:

$$\text{M.E.T.} = \frac{1}{2}\left(\frac{H_1 + H_2}{R\phi}\right)$$

$$= \frac{1}{2}\frac{H_1 + H_2}{C_v \log_e X}$$

$$\text{M.E.T.}'$$

$$= \frac{1}{2}\left\{\frac{H_1 + C_v T_a(X - 1)}{C_v \log_e X}\right\} \quad (18')$$

$$= \frac{1}{2}\left\{\frac{H_1 + C_v T_a\dfrac{v_d}{v_a}\left[X\left(\dfrac{v_a}{v_d}\right)^\gamma - 1\right] - C_p T_a\left(\dfrac{v_d}{v_a} - 1\right)}{C_v \log_e X}\right\} (18)$$

Pressure range is same for both cases,

$$\therefore R_p = p_c - p_a = p_a(\gamma^\gamma X - 1). \quad . \quad . \quad . \quad (19)$$

Mean effective volume:

$$\text{M.E.V.} = J\frac{W}{p_a(\gamma^\gamma X - 1)} \quad (20)$$

$$\text{M.E.V.} = J\frac{W}{p_a(\gamma^\gamma X - 1)} \quad (20')$$

Temperature range is also the same for both

$$R_T = T_c - T_a = T_a(\gamma^{\gamma-1} X - 1). \quad . \quad . \quad . \quad (21)$$

CYCLE II B.

The third type of cycle in the second group is one which has never been applied to an actual engine. It gives

From *A* to *B*. Adiabatic compression from atmospheric pressure.

From *B* to *C*. Addition of heat isometrically.

From *C* to *D*. Adiabatic expansion to pressure below atmosphere.

From *D* to *E*. Cooling isothermally to atmospheric pressure.

From *E* to *A*. Cooling at atmospheric pressure.

Let Fig. 123 and Fig. 124 be the *PV* and $\theta\phi$ diagram respectively of the cycle.

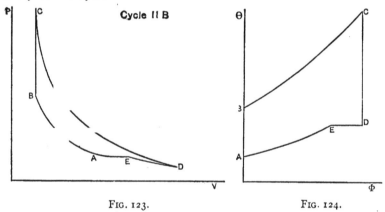

FIG. 123.   FIG. 124.

Assume same results as before up to the point *c*. Take p_d something less than atmosphere, i.e.,

$$p_a > p_d > 0; \quad \cdots \cdots \cdots \quad (1)$$

then

$$v_d = v_c\left(\frac{p_c}{p_d}\right)^{\frac{1}{\gamma}} = \frac{v_a}{\gamma}\left(\frac{p_a\gamma^\gamma X}{p_d}\right)^{\frac{1}{\gamma}} = v_a\left(\frac{p_a X}{p_d}\right)^{\frac{1}{\gamma}}, \quad \cdots \quad (2)$$

and

$$T_d = T_c\left(\frac{p_d}{p_c}\right)^{\frac{\gamma-1}{\gamma}} = T_a\gamma^{\gamma-1}X\left(\frac{p_d}{p_a\gamma^\gamma X}\right)^{\frac{\gamma-1}{\gamma}}$$

$$T_d = T_a X^{1-\frac{\gamma-1}{\gamma}}\left(\frac{p_d}{p_a}\right)^{\frac{\gamma-1}{\gamma}} = T_a X^{\frac{1}{\gamma}}\left(\frac{p_d}{p_a}\right)^{\frac{\gamma-1}{\gamma}}. \quad \cdots \quad (3)$$

Through D and a point E whose volume is greater than the original an isothermal is drawn,

$$T_c = T_d,$$

$$v_e = v_d \frac{p_d}{p_e} = v_a \left(\frac{p_a X}{p_a} \right)^{\frac{1}{\gamma}} \frac{p_d}{p_a}. \quad \dots \dots \quad (4)$$

Hence

$$v_e = v_a X^{\frac{1}{\gamma}} \left(\frac{p_d}{p_a} \right)^{\frac{\gamma-1}{\gamma}}, \quad \dots \dots \dots \quad (5)$$

$$p_e = p_a. \quad \dots \dots \dots \dots \quad (6)$$

Apply the perfect gas law to the points

$$\frac{p_a v_a}{T_a} = R,$$

$$\frac{p_b v_b}{T_b} = \frac{p_a v_a \gamma^\gamma}{\gamma T_a \gamma^{\gamma-1}} = R,$$

$$\frac{p_c v_c}{T_c} = \frac{p_a \gamma^\gamma X v_a}{T_a \gamma^{\gamma-1} X \gamma} = R,$$

$$\frac{p_d v_d}{T_d} = \frac{p_d v_a p_a^{\frac{1}{\gamma}} X^{\frac{1}{\gamma}}}{T_a X^{\frac{1}{\gamma}} \left(\frac{p_d}{p_a} \right)^{\frac{\gamma-1}{\gamma}} p_d^{\frac{1}{\gamma}}} = R,$$

$$\frac{p_e p_e}{T_e} = \frac{p_a v_a X^{\frac{1}{\gamma}} (p_d)^{\frac{\gamma-1}{\gamma}}}{T_a X^{\frac{1}{\gamma}} \left(\frac{p_d}{p_a} \right)^{\frac{\gamma-1}{\gamma}} (p_a)^{\frac{\gamma-1}{\gamma}}} = R.$$

During the isothermal compression heat must be abstracted; the amount can best be calculated by $\theta \phi$ coordinates. Call this amount m, then

$$m = T_d(\phi_d - \phi_e)$$
$$= T_e(\phi_d - \phi_e).$$

But

$$\phi_d - \phi_e = (\phi_c - \phi_b) - (\phi_e - \phi_a)$$

and

$$\phi_e - \phi_b = C_v \log_e \frac{T_c}{T_b} = C_v \log_e X,$$

$$\phi_e - \phi_a = C_v \log_e \frac{T_e}{T_a}$$

$$= C_p \log_e \left[X^{\frac{1}{r}} \left(\frac{p_d}{p_a} \right)^{\frac{r-1}{r}} \right].$$

Besides this amount m a quantity $C_p(T_e - T_a)$ must be abstracted isopiestically, whence

$$H_2 = C_p(T_e - T_a) + T_e \left\{ C_v \log_e X + C_p \log_e \left(X^{\frac{1}{r}} \left(\frac{p_d}{p_a} \right)^{\frac{r-1}{r}} \right) \right.$$

$$= C_p T_a \left[X^{\frac{1}{r}} \left(\frac{p_d}{p_a} \right)^{\frac{r-1}{r}} \right] + T_a X^{\frac{1}{r}} \left(\frac{p_d}{p_a} \right)^{\frac{r-1}{r}} \right\} C_v \log_e X$$

$$+ C_p \log_e \left[X^{\frac{1}{r}} \left(\frac{p_d}{p_a} \right)^{\frac{r-1}{r}} \right], \quad \cdots \quad (7)$$

$$W = H_1 - H_2, \quad \cdots \quad \cdots \quad (8)$$

$$E = 1 - \frac{H_1}{H_2}. \quad \cdots \quad \cdots \quad (9)$$

The volume range is

$$R_v = v_d - v_b = v_a X^{\frac{1}{r}} \left(\frac{p_a}{p_d} \right)^{\frac{1}{r}} - \frac{v_a}{r}$$

$$= v_a \left[X^{\frac{1}{r}} \left(\frac{p_a}{p_d} \right)^{\frac{1}{r}} - \frac{1}{r} \right], \quad \cdots \quad \cdots \quad (10)$$

$$\text{M.E.P.} = J \frac{W}{R_v} = \frac{JW}{\left[v_a X^{\frac{1}{r}} \left(\frac{p_a}{p_p} \right)^{\frac{1}{r}} = \frac{1}{r} \right]}, \quad \cdots \quad (11)$$

$$\text{M.E.T.} = \frac{1}{2} \left(\frac{H_1 + H_2}{C_v \log_e X} \right), \quad \cdots \quad \cdots \quad (12)$$

$$R_p = p_c - p_d = p_a(r)^r X - p_d, \quad \cdots \quad \cdots \quad (13)$$

$$\text{M.E.V.} = J \frac{W}{p_a(r)^r X - p_d}, \quad \cdots \quad \cdots \quad (14)$$

$$R_T = T_a((r)^{r-1} X - 1) \text{ as for II.} \quad \cdots \quad \cdots \quad (15)$$

Cycle II C.

In the fourth type of the second group, the final temperature becomes equal to the initial temperature and the cycle is closed by an isothermal corresponding to the change of volume by compression to get back to the state of pressure and volume at A. Hence there is

From A to B. Adiabatic compression from atmospheric pressure.

From B to C. Addition of heat isometrically.

From C to D. Adiabatic expansion to pressure below atmosphere such that we get

From D to A. Cooling isothermally to original volume and atmospheric pressure.

No engine has been built to utilize this cycle.

Let Fig. 125 be the PV and Fig. 126 the $\theta\phi$ diagrams of the cycle.

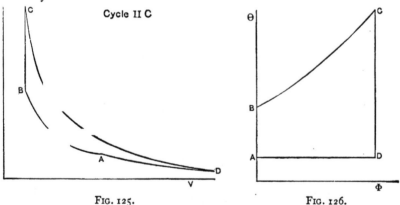

Cycle II C

FIG. 125.   FIG. 126.

All values for the compression and heat addition found in Cycle II may here be assumed. The point D lies at the intersection of two curves, one an adiabatic through C, the other an isothermal through A, and the relations can be written. From the adiabatic relation

$$v_d = v_c\left(\frac{p_c}{p_d}\right)^{\frac{1}{r}}.$$

From the isothermal relation

$$v_d = v_a \frac{p_a}{p_d}.$$

Equating,

$$v_c \left(\frac{p_c}{p_d}\right)^{\frac{1}{r}} = v_a \frac{p_a}{p_d},$$

$$\frac{v_a}{r}\left(\frac{p_a r^r X}{p_d}\right)^{\frac{1}{r}} = v_a \frac{p_a}{p_d};$$

$$\therefore p_d = \frac{p_a}{X^{\frac{1}{r-1}}} = p_a X^{\frac{1}{1-r}}. \quad \ldots \ldots \quad (1)$$

This is the pressure at which the intersection will take place. By substitution,

$$v_d = v_a X^{\frac{1}{r-1}}, \quad \ldots \ldots \ldots \quad (2)$$

$$T_d = T_a. \quad \ldots \ldots \ldots \quad (3)$$

Applying the perfect gas law to D,

$$\frac{p_d v_d}{T_d} = \frac{p_a v_a (X)^{\frac{1}{r-1}}}{X^{\frac{1}{r-1}} T_a} = \frac{p_a v_a}{T_a} = R.$$

All the heat is abstracted at constant temperature during the compression D to A. The entropy range is evidently the same as for heat addition, and this is

$$R\phi = C_v \log_e X; \quad \ldots \ldots \ldots \ldots \quad (4)$$

whence

$$H_2 = T_a(\phi_d - \phi_a) = T_a C_v \log_e X, \quad \ldots \ldots \quad (5)$$

work

$$W = H_1 - H_2 = H_1 - T_a C_v \log_e X, \quad \ldots \ldots \quad (6)$$

$$E = 1 - \frac{H_2}{H_1} = 1 - \frac{T_a C_v \log_e X}{H_1}, \quad \ldots \ldots \quad (7)$$

$$\text{M.E.T.} = \frac{1}{2}\left(\frac{H_1 + H_2}{R\phi}\right) = \frac{1}{2}\left(\frac{H_1 + T_a C_v \log_e X}{C_v \log_e x}\right);$$

whence

$$\text{M.E.T.} = \frac{1}{2}\left(\frac{H_1}{C_v \log_e X} - T_a\right), \quad \cdots \quad \textbf{(8)}$$

$$R_v = v_d - v_b = v_a\left[X^{\frac{1}{\gamma-1}} - \frac{1}{\gamma}\right], \quad \cdots \quad \textbf{(9)}$$

$$\text{M.E.P.} = \frac{JW}{R_v} = J\frac{H_1 - T_a C_v \log_e X}{v_o\left[X^{\frac{1}{\gamma-1}} - \frac{1}{\gamma}\right]}, \quad \cdots \quad \textbf{(10)}$$

$$R_p = p_c - p_d = p_a[\gamma^\gamma X - X^{\frac{1}{1-\gamma}}], \quad \cdots \quad \textbf{(11)}$$

$$\text{M.E.V.} = \frac{JW}{R_p} = J\frac{H_1 - T_a C_v \log_v X}{p_a[\gamma^\gamma X - X^{\frac{1}{1-\gamma}}]}, \quad \cdots \quad \textbf{(12)}$$

$$R_T = T_a(\gamma^{\gamma-1}X - 1) \quad \text{as before.} \quad \cdots \quad \textbf{(13)}$$

Tabulate.

CYCLE II C.

Symbol.	Formula as First Derived.	Formula Reduced.
p_b	$p_a\left(\dfrac{v_a}{v_b}\right)^\gamma$	$p_a\gamma^\gamma$
v_b	$\dfrac{v_a}{\gamma}$	$\dfrac{v_a}{\gamma}$
T_b	$T_a\left(\dfrac{v_a}{v_b}\right)^{\gamma-1}$	$T_a\gamma^{\gamma-1}$
p_c	$p_b\dfrac{T_c}{T_b}$	$p_a\gamma^\gamma X$
v_c	v_b	$\dfrac{v_a}{\gamma}$
T_c	$T_b\left(1 + \dfrac{H_1}{C_v T_b}\right)$	$T_a\gamma^{\gamma-1}X$
p_d		$p_a X^{\frac{1}{1-\gamma}}$
v_d		$v_a X^{\frac{1}{\gamma-1}}$
T_d		T_a
H_2	$T_a(\phi_d - \phi_a)$	$T_a C_v \log_e X$

Symbol.	Formula as First Derived.	Formula Reduced.
W	$H_1 - H_2$	$H_1 - T_a C_v \log_e X$
E	$1 - \dfrac{H_2}{H_1}$	$\dfrac{1 - T_a C_v \log_e X}{H_1}$
R_ϕ	$C_v \log_e \dfrac{T_c}{T_b}$	$C_v \log_e X$
M.E.T.	$\dfrac{1}{2}\left(\dfrac{H_1 + H_2}{R_\phi}\right)$	$\dfrac{1}{2}\left(\dfrac{H_1}{C_v \log_e X} - T_a\right)$
R_v	$v_d - v_b$	$v_b\left(X^{\frac{1}{\gamma-1}} - \dfrac{1}{\gamma}\right)$
M.E.P.	$J\dfrac{W}{R_v}$	$J\left\{ \dfrac{H_1 - T_a C_v \log_e X}{v_a\left(X^{\frac{1}{\gamma-1}} - \dfrac{1}{\gamma}\right)} \right\}$
R_p	$p_c - p_d$	$p_a(\gamma^r X - X^{\frac{1}{1-r}})$
M.E.V.	$J\dfrac{W}{R_p}$	$J\left\{ \dfrac{H_1 - T_a C_v \log_e X}{p_a(\gamma^r X - X^{\frac{1}{1-r}})} \right\}$
R_T	$T_c - T_a$	$T_a(\gamma^{r-1}X - 1)$

185. Compression Cycle with Isopiestic Heating.—In this third group of cycles are included those in which the heating is effected at constant pressure. The most notable example of its application to internal combustion heating was the Brayton engine of America and the Simon engine of England. The succession of events is :

From *A* to *B*. Adiabatic compression from atmospheric pressure.

From *B* to *C*. Addition of heat isopiestically.

From *C* to *D*. Adiabatic expansion to atmospheric pressure.

From *D* to *A*. Cooling at atmospheric pressure.

In hot-air engines it is the cycle identified with the work in England of Sir Geo. Cayley, Dr. Joule, and Sir William Thomson (1851). If the compression from *A* to *B* were isothermal, and also the expansion from *CD*, taking place at constant temperature, the cycle would be that of the Ericsson hot-air engine.

Let Fig. **127** be its PV and Fig. **128** its $\theta\phi$ diagram.

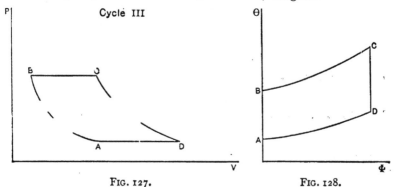

FIG. 127.   FIG. 128.

The compression results of Cycle II may be assumed, hence

$$v_b = \frac{v_a}{\gamma}, \quad\quad\quad\quad\quad\quad\quad (1)$$

$$p_b = p_a \gamma^\gamma, \quad\quad\quad\quad\quad\quad (2)$$

$$T_b = T_a \gamma^{\gamma-1}. \quad\quad\quad\quad\quad (3)$$

Heat is added isopiestically; hence calling C_p the specific heat at constant pressure,

$$T_c = T_b + \frac{H_1}{C_p} = T_b\left(1 + \frac{H_1}{C_p T_b}\right).$$

Write

$$1 + \frac{H_1}{C_p T_b} = Y.$$

$$\therefore T_c = T_b Y = T_a \gamma^{\gamma-1} Y, \quad\quad\quad (4)$$

$$p_c = p_b = p_a \gamma^\gamma, \quad\quad\quad\quad (5)$$

$$v_c = v_b \frac{T_c}{T_b} = v_b Y = \frac{v_a}{\gamma} Y. \quad\quad\quad (6)$$

Adiabatic expansion gives for final pressure of one atmosphere

$$v_d = v_c\left(\frac{p_c}{p_d}\right)^{\frac{1}{\gamma}} = \frac{v_a}{\gamma} Y\left(\frac{p_a \gamma^\gamma}{p_a}\right)^{\frac{1}{\gamma}} = v_a Y = \gamma v_c, \quad\quad (7)$$

$$p_d = p_a, \quad\quad\quad\quad\quad\quad\quad\quad\quad (8)$$

$$T_d = T_c\left(\frac{v_c}{v_d}\right)^{\gamma-1} = T_a \gamma^{\gamma-1} Y\left(\frac{v_a Y}{\gamma v_a Y}\right)^{\gamma-1};$$

$$\therefore T_d = T_a Y = \gamma^{\gamma-1} T_c. \quad\quad\quad\quad\quad (9)$$

Apply the perfect gas law

$$\frac{p_a v_a}{T_a} = R,$$

$$\frac{p_b v_b}{T_b} = R \text{ as in II,}$$

$$\frac{p_c v_c}{T_c} = \frac{p_a \gamma^\gamma v_a Y}{T_a \gamma^{\gamma-1} Y \gamma} = R,$$

$$\frac{p_d v_d}{T_d} = \frac{p_a v_a Y}{T_a Y} = R.$$

Hence the formulæ are verified.

Heat is abstracted isopiestically.

$$\therefore H_2 = C_p(T_d - T_a)$$

$$= C_p T_a(Y - 1) = \frac{H_1}{\gamma^{\gamma-1}}; \quad \cdots \quad (10)$$

$$W = H_1 - H_2 = H_1 - C_p T_a(Y-1) = H_1\left(1 - \frac{1}{\gamma^{\gamma-1}}\right), \cdot \quad (11)$$

$$E = \frac{W}{H_1} = 1 - \frac{H_2}{H_1} = 1 - \frac{C_p T_a(Y-1)}{H_1} = \frac{1}{\gamma^{\gamma-1}}. \quad \cdots \quad (12)$$

Volume range is

$$R_v = v_d - v_b = v_a\left(Y - \frac{1}{\gamma}\right). \quad \cdots \quad (13)$$

Whence for mean effective pressure

$$\text{M.E.P.} = J\frac{W}{R_v} = J\frac{H_1\left(1 - \frac{1}{\gamma^{\gamma-1}}\right)}{v_a\left(Y - \frac{1}{\gamma}\right)}, \quad \cdots \quad (14)$$

$$R\phi = \phi_c - \phi_b = C_p \log_e \frac{T_c}{T_b} = C_p \log_e Y, \quad \cdots \quad (15)$$

$$\text{M.E.T.} = \frac{1}{2}\left(\frac{H_1 + H_2}{R\phi}\right) = \frac{1}{2}\left(\frac{H_1 + C_p T_a(Y-1)}{C_p \log_e Y}\right), \quad \cdot \quad (16)$$

$$R_p = p_b - p_a = p_a(\gamma^\gamma - 1), \quad \cdots \quad (17)$$

$$\text{M.E.V.} = J\frac{W}{R_p} = J\frac{H_1\left(1 - \frac{1}{\gamma^{\gamma-1}}\right)}{p_v(\gamma^{\gamma-1}}, \quad \cdots \quad (18)$$

$$R_T = T_c - T_a = T_a(\gamma^{\gamma-1}Y - 1). \quad \cdots \quad (19)$$

.Tabulate.

Cycle III.

Symbol.	Formula as First Derived.	Formula Reduced.
p_b	$p_a\left(\dfrac{v_a}{v_b}\right)^r$	$p_a r^r$
v_b	$\dfrac{v_a}{r}$	$\dfrac{v_a}{r}$
T_b	$T_a\left(\dfrac{v_a}{v_b}\right)^{r-1}$	$T_a r^{r-1}$
p_c	p_b	$p_a r^r$
v_c	$v_b\dfrac{T_c}{T_b}$	$\dfrac{v_a}{r}Y$
T_c	$T_b\left(1+\dfrac{H_1}{C_p T_b}\right)$	$T_a r^{r-1}Y$
p_d	p_a	p_a
v_d	$v_c\left(\dfrac{p_c}{p_a}\right)^{\frac{1}{r}}$	$v_a Y$
T_d	$T_c\left(\dfrac{v_c}{v_d}\right)^{r-1}$	$T_a Y$
H_2	$C_p(T_d-T_a)$	$\dfrac{H}{r^{r-1}}$
W	H_1+H_2	$H_1\left(1-\dfrac{1}{r^{r-1}}\right)$
E	$1-\dfrac{H_2}{H_1}$	$1-\dfrac{1}{r^{r-1}}$
$R\phi$	$C_p\log_e\dfrac{T_c}{T_b}$	$C_p\log_e Y$
M.E.T.	$\dfrac{1}{2}\left(\dfrac{H_1+H_2}{R\phi}\right)$	$\dfrac{1}{2}\left\{\dfrac{H_1+\dfrac{H_1}{r^{r-1}}}{C_p\log_e Y}\right\}$
R_v	v_d-v_b	$v_a\left(Y-\dfrac{1}{r}\right)$
M.E.P.	$J\dfrac{W}{R_v}$	$J\left\{\dfrac{H_1\left(1-\dfrac{1}{r^{r-1}}\right)}{v_a\left(Y-\dfrac{1}{r}\right)}\right\}$

Symbol.	Formula as First Derived.	Formula Reduced.

R_p .... $p_b - p_a$ .... $p_b(\gamma^\gamma - 1)$

M.E.V. ... . $J\dfrac{W}{R_p}$ ....$J\left\{\dfrac{H_1\left(1 - \dfrac{1}{\gamma^{\gamma-1}}\right)}{p_a(\gamma^\gamma - 1)}\right\}$

R_T.

As in the previous group, if the cut-off is late in a Brayton cycle, and expansion incomplete, the following modification follows which will be called

CYCLE III A.

The first modification of type in Group III presents the following succession, the expansion being incomplete:

From *A* to B. Adiabatic compression from atmospheric pressure.

From *B* to *C*. Addition of heat isopiestically.

From *C* to *D*. Adiabatic expansion to pressure above atmosphere.

From *D* to *E*. Cooling isometrically to atmospheric pressure.

From *E* to *A*. Cooling at atmospheric pressure.

Fig. 129 is its *PV* and Fig. 130 its $\theta\phi$ diagram.

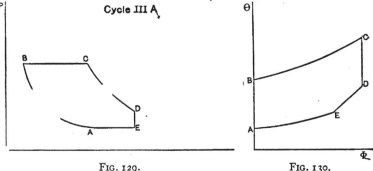

FIG. 129.   FIG. 130.

Assume the results of III up to point *C*. The point *D* is situated anywhere on the adiabatic through *C* between *C* and atmosphere.

Write

$$p_c > p_d > p_a \quad \cdot \quad \cdot \quad \cdot \quad \cdot \quad \cdot \quad \cdot \quad (1)$$

and

$$v_d > v_a. \quad . \quad . \quad . \quad . \quad . \quad . \quad . \quad . \quad (2)$$

This latter (2) will not necessarily follow from (1), but where it does not hold the cycle is decidedly imperfect and this case is here neglected, i.e., the case where the isometric DE cuts the adiabatic AB.

The relations will be:

$$v_d = v_c \left(\frac{p_c}{p_d}\right)^{\frac{1}{r}} = \frac{v_a}{r} Y \left(\frac{p_a r^r}{p_d}\right)^{\frac{1}{r}} = v_a Y \left(\frac{p_a}{p_d}\right)^{\frac{1}{r}}, \quad . \quad . \quad . \quad (3)$$

$$T_d = T_c \left(\frac{p_d}{p_c}\right)^{\frac{r-1}{r}} = T_a r^{r-1} Y \left(\frac{p_d}{p_a r^r}\right)^{\frac{r-1}{r}} = T_a Y \left(\frac{p_d}{p_a}\right)^{\frac{r-1}{r}}, \quad (4)$$

$$p_e = p_a, \quad . \quad . \quad . \quad . \quad . \quad . \quad . \quad . \quad (5)$$

$$v_e = v_d = v_a Y \left(\frac{p_a}{p_d}\right)^{\frac{1}{r}}, \quad . \quad . \quad . \quad . \quad . \quad (6)$$

$$T_e = T_a \frac{v_e}{v_a} = T_a Y \left(\frac{p_a}{p_d}\right)^{\frac{1}{r}}. \quad . \quad . \quad . \quad . \quad (7)$$

Apply the perfect gas law to D and E.

$$\frac{p_d v_d}{T_d} = \frac{p_d v_a Y \left(\frac{p_a}{p_d}\right)^{\frac{1}{r}}}{T_a Y \left(\frac{p_d}{p_a}\right)^{\frac{r-1}{r}}} = R,$$

$$\frac{p_e p_e}{T_e} = \frac{p_a v_a Y \left(\frac{p_a}{p_d}\right)^{\frac{1}{r}}}{T_a Y \left(\frac{p_a}{p_d}\right)^{\frac{1}{r}}} = R.$$

This verifies the formulæ.

Heat is abstracted in two parts and the amount is

$$H_2 = C_v(T_d - T_e) + C_p(T_e - T_a) \quad . \quad . \quad . \quad (8)$$

$$= C_v \left[T_a Y \left(\frac{p_d}{p_a}\right)^{\frac{r-1}{r}} - T_a Y \left(\frac{p_a}{p_d}\right)^{\frac{1}{r}} \right]$$

$$+ C_p \left[T_a Y \left(\frac{p_a}{p_d}\right)^{\frac{1}{r}} - T_a \right] . \quad . \quad . \quad . \quad . \quad (9)$$

$$= C_v T_a Y \left(\frac{p_a}{p_d}\right)^{\frac{1}{r}} \left[\frac{p_d}{p_a} - 1 \right] + C_p T_a \left[Y \left(\frac{p_a}{p_d}\right)^{\frac{1}{r}} - 1 \right].$$

The work done is

$$W = H_1 - H_2, \quad \cdots \cdots \quad (10)$$

and efficiency,

$$E = 1 - \frac{H_2}{H_1}, \quad \cdots \cdots \quad (11)$$

$$R_v = v_d - v_b = v_a \left[Y \left(\frac{p_a}{p_d} \right)^{\frac{1}{\gamma}} - \frac{1}{\gamma} \right], \quad \cdots \quad (12)$$

$$\text{M.E.P.} = J \frac{W}{R_v} = J \frac{W}{v_a \left[Y \left(\frac{p_a}{p_d} \right)^{\frac{1}{\gamma}} - \frac{1}{\gamma} \right]}. \quad \cdots \quad (13)$$

$$R\phi = C_p \log_e Y \text{ as before for III.} \quad \cdots \quad (14)$$

$$\text{M.E.T.} = \frac{1}{2} \left(\frac{H_1 + H_2}{R_v} \right) = \frac{1}{2} \left(\frac{H_1 + H_2}{C_p \log_e Y} \right), \quad \cdots \quad (15)$$

$$R_p = p_c - p_a = p_a(\gamma^\gamma - 1) \quad \text{as in III.} \quad \cdots \quad (16)$$

$$\therefore \text{M.E.V.} = J \frac{W}{R_p} = J \frac{W}{p_a(\gamma^\gamma - 1)}. \quad \cdots \quad (17)$$

As before III the temperature range

$$R_T = T_e - T_a = T_a(\gamma^{\gamma-1} Y - 1). \quad \cdots \cdots \quad (18)$$

Cycle III. B.

In the second modification of Group III the expansion is carried below atmosphere, so that:

From *A* to *B*. Adiabatic compression from atmospheric pressure.

From *B* to *C*. Addition of heat isopiestically.

From *C* to *D*. Adiabatic expansion to pressure below atmosphere.

From *D* to *E*. Cooling isothermally to atmospheric pressure.

From *E* to *A*. Cooling at atmospheric pressure.

No engine has as yet applied this cycle.

Figs. **131** and **132** are its diagrams.

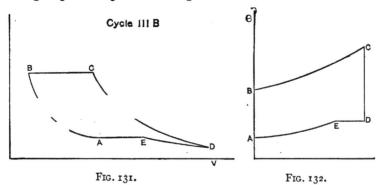

FIG. 131.   FIG. 132.

All results of III A up to period D may be assumed except that p_d which was there arbitrary and was assumed greater than p_a is here less than p_a, i.e.,

$$p_c > p_d > 0. \quad \ldots \ldots \ldots \quad (1)$$

It was found that:

$$v_d = v_a Y \left(\frac{p_a}{p_d}\right)^{\frac{1}{r}} \quad \ldots \ldots \ldots \quad (2)$$

and

$$T_d = T_a Y \left(\frac{p_d}{p_a}\right)^{\frac{r-1}{r}}. \quad \ldots \ldots \ldots \quad (3)$$

Through E and D there must pass an isothermal and

$$v_e > v_a, \quad \ldots \ldots \ldots \quad (4)$$

$$p_e = p_a; \quad \ldots \ldots \ldots \quad (5)$$

$$v_e = v_d \frac{p_d}{p_e} = v_d \frac{p_d}{p_a} = v_a Y \left(\frac{p_a}{p_d}\right)^{\frac{1}{r}} \frac{p_d}{p_a}.$$

$$\therefore \; v_e = v_a Y \left(\frac{p_d}{p_a}\right)^{\frac{r-1}{r}}, \quad \ldots \ldots \ldots \quad (6)$$

$$T_e = T_d = T_a Y \left(\frac{p_d}{p_a}\right)^{\frac{r-1}{r}}. \quad \ldots \ldots \ldots \quad (7)$$

Applying the perfect gas law to E,

$$\frac{p_e v_e}{T_e} = \frac{p_a v_a Y \left(\frac{p_d}{p_a}\right)^{\frac{r-1}{r}}}{T_a Y \left(\frac{p_d}{p_a}\right)^{\frac{r-1}{r}}} \quad \cdots \cdots \cdots \quad (8)$$

Heat abstracted 1° isothermally a quantity m.

2° isopiestically " " n.

$$n = C_p(T_e - T_a)$$

$$= C_p T_a \left[Y \left(\frac{p_d}{p_a}\right)^{\frac{r-1}{r}} - 1 \right],$$

$$m = T_e(\phi_d - \phi_e).$$

But

$$\phi_d - \phi_e = (\phi_b - \phi_c) - (\phi_e - \phi_a),$$

$$= C_p \log_e \frac{T_c}{T_b} = C_p \log_e \frac{T_e}{T_a},$$

$$= C_p \log_e Y - C_p \log_e Y \left(\frac{p_d}{p_a}\right)^{\frac{r-1}{r}},$$

$$= C_p \log Y \left[1 - \left(\frac{p_d}{p_a}\right)^{\frac{r-1}{r}} \right].$$

$$\therefore H_2 = C_p T_a \left\{ Y \left(\frac{p_d}{p_a}\right)^{\frac{r-1}{r}} - 1 + Y \log_e Y \left[1 - \left(\frac{p_d}{p_a}\right)^{\frac{r-1}{r}} \right] \right\}, \quad (9)$$

$$W = H_1 - H_2, \quad \cdots \cdots \cdots \quad (10)$$

$$E = 1 - \frac{H_2}{H_1},$$

$$R_v = v_d - v_b = v_a \left[Y \left(\frac{p_a}{p_d}\right)^{\frac{1}{r}} - \frac{1}{r} \right], \quad \cdots \cdots \quad (13)$$

$$\text{M.E.P.} = J \frac{W}{R_v} = J \frac{W}{v_d \left[Y \left(\frac{p_a}{p_d}\right)^{\frac{1}{r}} - \frac{1}{r} \right]}, \quad \cdots \cdots \quad (14)$$

$$R_p = p_c - p_d = p_a r^r - p_d, \quad \cdots \cdots \cdots \quad (15)$$

$$\text{M.E.V.} = J\frac{W}{p_a\gamma_r - p_d}, \quad \cdots \quad \text{(16)}$$

$$R_T = T_c - T_a = T_a(\gamma^{r-1}Y - 1) \quad \text{as before III.} \quad \text{(17)}$$

CYCLE III C.

In the third and last modification the cooling is isothermal with varying pressure all below the atmosphere, so that an isothermal line is called for to bring the gas back to the state at A with respect to both pressure and volume. Hence

From A to B. Adiabatic compression from atmospheric pressure.

From B to C. Addition of heat isopiestically.

From C to D. Adiabatic expansion to pressure below atmosphere such that we get

From D to A. Cooling isothermally to original volume and atmospheric pressure.

No engine has as yet applied this cycle.

Let Figs. 133 and 134 be its diagrams.

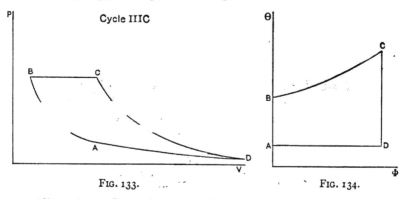

FIG. 133.   FIG. 134.

All results to C may be assumed as already derived. The point D is determined by the intersection of the adiabatic through C with the isothermal through A. From the adiabatic relation

$$v_d = v_c \left(\frac{p_c}{p_d}\right)^{\frac{1}{r}}.$$

From the isothermal relation

$$v_d = v_a \frac{p_a}{p_d};$$

$$\therefore \ v_c \left(\frac{p_c}{p_d}\right)^{\frac{1}{\gamma}} = v_a \frac{p_a}{p_d} = \frac{v_a Y}{\gamma}\left(\frac{p_a \gamma^\gamma}{p_d}\right)^{\frac{1}{\gamma}};$$

$$\therefore \ p_d = \frac{p_a}{Y^{\frac{\gamma}{\gamma-1}}}. \quad \cdot \quad \cdot \quad \bullet \quad \bullet \quad \bullet \quad \bullet \quad \bullet \quad (1)$$

By substitution

$$v_d = v_a Y^{\frac{\gamma}{\gamma-1}}, \quad \cdot \quad \cdot \quad \bullet \quad \bullet \quad \bullet \quad \bullet \quad \bullet \quad (2)$$

$$T_d = T_a, \quad \cdot \quad \cdot \quad \cdot \quad \bullet \quad \bullet \quad \bullet \quad \bullet \quad (3)$$

$$H_2 = T_d(\phi_c - \phi_b) = T_a C_p \log_e Y, \quad \bullet \quad \bullet \quad \bullet \quad \bullet \quad (4)$$

$$W = H_1 - H_2 = H_1 - T_a C_p \log_e Y, \bullet \quad \bullet \quad \bullet \quad \bullet \quad (5)$$

$$E = 1 - \frac{T_a C_p \log_e Y}{H_1},$$

$$R_v = v_d - v_b = v_a \left[Y^{\frac{\gamma}{\gamma-1}} - \frac{1}{\gamma} \right], \quad \bullet \quad \bullet \quad \bullet \quad \bullet \quad (6)$$

$$\text{M.E.P.} = J\frac{W}{R_v} = J\frac{H_1 - T_a C_p \log_e Y}{v_a \left(Y^{\frac{\gamma}{\gamma-1}} = \frac{1}{\gamma} \right)}, \quad \bullet \quad \bullet \quad \bullet \quad (7)$$

$$R_v = p_b - p_d = p_a \left(\gamma^\gamma - \frac{1}{Y^{\frac{\gamma}{\gamma-1}}} \right), \quad \bullet \quad \bullet \quad \bullet \quad \bullet \quad (8)$$

$$\text{M.E.V.} = J\frac{W}{R_p} = J\frac{H_1 - T_a C_p \log_e Y}{p_a \left(\gamma^\gamma - \frac{1}{Y^{\frac{\gamma}{\gamma-1}}} \right)}, \quad \bullet \quad \bullet \quad \bullet \quad (9)$$

$$\text{M.E.T.} = \frac{1}{2}\left(\frac{H_1 + H_2}{R\phi} \right) = \frac{1}{2}\left(\frac{H_1 - T_a C_p \log_e Y}{C_p \log_e Y} \right)$$

$$= \frac{1}{2}\left(\frac{H_1}{C_p \log_e \gamma} + T_a \right), \quad \bullet \quad \bullet \quad \bullet \quad \bullet \quad \bullet \quad \bullet \quad (10)$$

$$R_T = T_c - T_a = T_a[\gamma^{\gamma-1} Y - 1] \text{ as in III.} \quad \bullet \quad \bullet \quad (11)$$

CYCLE III *C.*

Symbol.	Formula as First Derived.	Formula Reduced.

p_b .... $p_a\left(\dfrac{v_a}{v_b}\right)^r$ .... $p_a\gamma^r$

v_b .... $\dfrac{v_a}{\gamma}$ .... $\dfrac{v_a}{\gamma}$

T_b .... $T_a\left(\dfrac{v_a}{v_b}\right)^{r-1}$ .... $T_a\gamma^{r-1}$

p_c .... p_b .... $p_a\gamma^r$

v_c .... $v_b\dfrac{T_c}{T_b}$ .... $\dfrac{v_a}{\gamma}Y$

T_c .... $T_b\left(1+\dfrac{H_1}{C_pT_b}\right)$ .... $T_a\gamma^{r-1}Y$

p_d .... $p_aY^{\frac{r}{r-1}}$

v_d .... $v_dY^{\frac{r}{r-1}}$

T_d .... T_a

H_2 .... $T_d(\phi_c-\phi_b)$ .... $T_aC_p\log_e Y$

W .... H_1-H_2 .... $H_1-T_aC_p\log_e Y$

E .... $1-\dfrac{H_2}{H_1}$ .... $1-\dfrac{T_aC_p\log_e Y}{H_1}$

$R\phi$ .... $C_p\log_e\dfrac{T_c}{T_b}$ .... $C_p\log_e Y$

M.E.T. .... $\dfrac{1}{2}\left(\dfrac{H_1+H_2}{R\phi}\right)$ .... $\dfrac{1}{2}\left(\dfrac{H_1}{C_p\log_e Y}+T_a\right)$

R_v .... v_d-v_b .... $v_a\left(Y^{\frac{r}{r-1}}-\dfrac{1}{\gamma}\right)$

M.E.P. .... $J\dfrac{W}{R_v}$ .... $J\left\{\dfrac{H_1-T_aC_p\log_e Y}{v_a\left(Y^{\frac{r}{r-1}}-\dfrac{1}{\gamma}\right)}\right\}$

R_p .... p_b-p_a .... $p_a\left(\gamma^r-\dfrac{1}{Y^{\frac{r}{r-1}}}\right)$

M.E.V. .... $J\dfrac{W}{R_p}$ .... $J\left\{\dfrac{H_1-T_aC_p\log_e Y}{p_a(\gamma^r-Y^{\frac{1}{1-r}})}\right\}$

R_T .... T_c-T_a .... $T_a(\gamma^{r-1}Y-1)$

18). Compression Cycle with Isothermal Heating. — The fourth group of cycles includes that to which Carnot's name is usually attached by reason of the special study which he gave to it. The special characteristic of the group is the isothermal heating. The modern engine which aims to operate on one of this group most nearly is the Diesel motor.

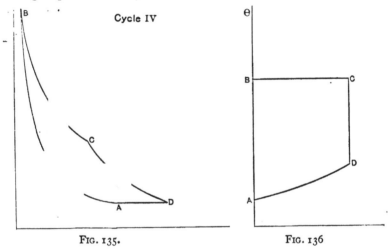

FIG. 135.   FIG. 136

Figs. 135 and 136 are its diagrams, in which

From *A* to B. Adiabatic compression from atmospheric pressure.

From *B* to *C*. Addition of heat isothermally.

From *C* to *D*. Adiabatic expansion to atmospheric pressure.

From *D* to *A*. Cooling at atmospheric pressure.

The results already obtained may be assumed for the compression, but beyond that new conditions arise. By isothermal heating the curve approaches the atmospheric line and there will be a certain quantity of heat which will bring the isothermal down to the atmospheric line, leaving a subsequent adiabatic expansion an impossibility. This quantity of course depends on the location of B, i.e., the amount of previous compression. The higher the previous compression the more heat may be added isothermally before reaching atmospheric pressure.

The quantity of heat which will make adiabatic expansion impossible and stop the isothermal on the atmospheric line can best be determined from $\theta\phi$ relations. Denote this quantity by Q.

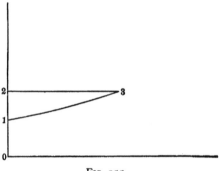

FIG. 155

On the $\theta\phi$ diagram, Fig. 155, the point 3 lies at the intersection of the isothermal 2 3 drawn at temperature o 2, the compression temperature and the isopiestic 1 3 drawn from atmospheric temperature o 1 to the intersection 3. In each case the entropy range is

$$\phi_3 - \phi_1 = C_p \log_e \frac{T_3}{T_1} = \phi_3 - \phi_2;$$

$$\therefore Q = T_2 C_p \log_e \frac{T_2}{T_1}.$$

Apply now to the Cycle IV,

$$\therefore Q = T_b C_p \log_e \frac{T_b}{T_a}$$

$$= T_a \gamma^{\gamma-1} C_p \log_e \gamma^{\gamma-1}. \quad \cdots \quad (1)$$

This is the amount of heat which will bring C down to atmosphere with no adiabatic expansion. In order that the cycle may exist according to the hypothetical definition less heat must be added than this quantity Q. Hence the equation of condition for the existence of the cycle will be

$$H_1 < T_a \gamma^{\gamma-1} C_p \log_e \gamma^{\gamma-1}, \quad \cdots \quad (2)$$

or

$$H_1 < T_b C_p \log_e \frac{T_b}{T_a}.$$

A similar method can be used to find the amount of expansion or resulting pressure and volume after addition of H_1 B.T.U. of heat.

Draw on both diagrams the isopiestic through the termination C of the isothermal and cutting the adiabatic AB at point C'. Then

$$T_c' = T_a \left(\frac{p_c'}{p_a}\right)^{\frac{r-1}{r}}$$

$$= T_a \left(\frac{p_c}{p_a}\right)^{\frac{r-1}{r}},$$

$$\phi_c - \phi_c' = C_p \log_e \frac{T_c}{T_c'} = C_p \log_e \frac{T_b}{T_a}$$

$$= C_p \log_e \frac{T_a r^{r-1}}{T_a \left(\frac{p_c}{p_a}\right)^{\frac{r-1}{r}}}.$$

But

$$r^{r-1} = \left(\frac{v_a}{v_b}\right)^{r-1} = \left(\frac{p_b}{p_a}\right)^{\frac{r-1}{r}};$$

$$\therefore \; \phi_c - \phi_c' = C_p \log_e \left(\frac{p_b}{p_a}\frac{p_a}{p_c}\right)^{\frac{r-1}{r}}$$

$$= C_p \log_e \left(\frac{p_b}{p_c}\right)^{\frac{r-1}{r}}.$$

And the amount of heat necessary for this isothermal expansion from B to C,

$$H_1 = T_b C_p \log_e \left(\frac{p_b}{p_c}\right)^{\frac{r-1}{r}}$$

But

$$\frac{\gamma-1}{\gamma} = \frac{\dfrac{C_p}{C_v}-1}{\dfrac{C_p}{C_v}} = \frac{C_p-C_v}{C_p};$$

$$\therefore H_1 = T_b(C_p-C_v) \log_e \frac{p_b}{p_c}$$

and

$$\log_e \frac{p_b}{p_c} = \frac{H_1}{T_b(C_p-C_v)}.$$

Put

$$\frac{H_1}{T_b(C_p-C_v)} = Z;$$

then will

$$\frac{p_b}{p_c} = e^Z,$$

$$p_c = \frac{p_b}{e^Z}.$$

That is to say, starting at the state B and adding a quantity of heat H_1, isothermally the resulting pressure is

$$p_c = \frac{p_b}{e^Z} = \frac{p_a \gamma^\gamma}{e^Z}. \quad \cdots \cdots \cdots \quad (3)$$

Since

$$\frac{p_b}{p_c} = \frac{v_c}{v_b},$$

$$\frac{v_c}{v_b} = e^Z,$$

$$v_c = v_b e^Z = \frac{v_a}{\gamma} e^Z, \quad \cdots \cdots \cdots \quad (4)$$

$$T_c = T_b = T_a \gamma^{\gamma-1}, \quad \cdots \cdots \cdots \quad (5)$$

$$p_d = p_a, \quad \cdots \cdots \cdots \cdots \quad (6)$$

$$v_d = v_c \left(\frac{p_c}{p_d}\right)^{\frac{1}{\gamma}} = \frac{v_a e^Z}{\gamma} \left(\frac{p_a \gamma^\gamma}{e^Z p_a}\right)^{\frac{1}{\gamma}} = v_a e^{Z\left(\frac{\gamma-1}{\gamma}\right)}.$$

Now

$$Z = \frac{H_1}{T_b(C_p - C_v)};$$

$$\therefore Z\left(\frac{\gamma - 1}{\gamma}\right) = \frac{H_1(C_p - C_v)}{T_b C_p(C_p - C_v)} = \frac{H_1}{T_b C_p}.$$

In Cycle III,

$$Y = 1 + \frac{H_1}{T_b C_p}.$$

Hence

$$Z\left(\frac{\gamma - 1}{\gamma}\right) = Y - 1.$$

Whence

$$v_d = v_a e^{Y-1}. \qquad \cdots \qquad \cdots \qquad (7)$$

Similarly

$$T_d = T_c\left(\frac{p_d}{p_c}\right)^{\frac{\gamma-1}{\gamma}} = T_b\left(\frac{p_a e^Z}{p_a \gamma^\gamma}\right)^{\frac{\gamma-1}{\gamma}} = T_a e^Z\left(\frac{\gamma-1}{\gamma}\right) = T_a e^{Y-1}. \quad (8)$$

Apply the perfect gas law,

$$\frac{p_a v_a}{T_a} = R,$$

$$\frac{p_b v_b}{T_b} = \frac{v_a p_a \gamma^\gamma}{\gamma T_a \gamma^{\gamma-1}} = R,$$

$$\frac{p_c v_c}{T_c} = \frac{p_a \gamma \ v_a e^Z}{e^Z \gamma T_a \gamma^{\gamma-1}} = R,$$

$$\frac{p_d v_d}{T_d} = \frac{p_a v_a e^{Y-1}}{T_a e^{Y-1}} = R,$$

verifying the formulæ.

$$H_2 = C_p(T_d - T_a) = C_p T_a(e^{Y-1} - 1), \quad \cdots \quad (9)$$

$$W = H_1 - H_2 = H_1 - C_p T_a(e^{Y-1} - 1), \quad \cdots \quad (10)$$

$$E = \frac{W}{H_1} = 1 - \frac{C_p T_a(e^{Y-1} - 1)}{H_1}, \quad \cdots \quad (11)$$

$$R_\phi = \frac{H_1}{T_b}, \quad \cdots \quad (12)$$

$$\text{M.E.T.} = \frac{1}{2}\left(\frac{H_1 + H_2}{R_\phi}\right) = \frac{1}{2}\left\{\frac{H_1 + C_p T_a(e^{Y-1} - 1)}{\frac{H_1}{T_b}}\right\};$$

$$\therefore \text{M.E.T.} = \frac{1}{2}\left[T_b + \frac{C_p T_a T_b(e^{Y-1})}{H_1}\right]$$

$$= \frac{1}{2}\left[T_a \gamma^{\gamma-1} + \frac{C_p(T_a)^2 \gamma^{\gamma-1}(e^{Y-1} - 1)}{H_1}\right], \quad (13)$$

$$R_v = v_d - v_b = v_a\left[e^{Y-1} - \frac{1}{\gamma}\right], \quad \cdots \quad (14)$$

$$\text{M.E.P.} = J\frac{W}{R_v} = J\frac{H_1 - C_p T_a(e^{Y-1} - 1)}{v_a\left(e^{Y-1} - \frac{1}{\gamma}\right)}, \quad \cdots \quad (15)$$

$$R_p = p_b - p_a = p_a(\gamma^\gamma - 1), \quad \cdots \quad (16)$$

$$\text{M.E.V.} = J\frac{W}{R_p} = J\frac{H_1 - C_p T_a(e^{Y-1} - 1)}{p_a(\gamma^\gamma - 1)}, \quad \cdots \quad (17)$$

$$R_T = T_b - T_a = T_a(\gamma_r^{-1} - 1). \quad \cdots \quad (18)$$

Tabulating for Cycle IV:

Symbol.	Formula as First Derived.	Formula Reduced.
p_b	$p_a\left(\dfrac{v_a}{v_b}\right)^\gamma$	$p_a\gamma^\gamma$
v_b	$\dfrac{v_a}{\gamma}$	$\dfrac{v_a}{\gamma}$
T_b	$T_a\left(\dfrac{v_a}{v_b}\right)^{\gamma-1}$	$T_a\gamma^{\gamma-1}$
Equation of condition.	$H_1 < T_b\log_e\dfrac{T_b}{T_a}$	$H_1 < T_a\gamma^{\gamma-1}\log_e\gamma^{\gamma-1}$
p_c	$\dfrac{p_b}{\dfrac{H_1}{e^{T_d(C_p - C_v)}}}$	$\dfrac{p_a\gamma^\gamma}{e^Z}$
v_c	$v_b\dfrac{p_b}{p_c}$	$\dfrac{v_a}{\gamma}e^Z$

Symbol.	Formula as First Derived.	Formula Reduced.
T_c	T_b	$T_a \gamma^{\gamma-1}$
p_d	p_a	p_a
v_d	$v_c \left(\dfrac{p_c}{p_d}\right)^{\frac{1}{\gamma}}$	$v_a e^{Y-1}$
T_d	$T_c \left(\dfrac{p_d}{p_c}\right)^{\frac{\gamma-1}{\gamma}}$	$T_a e^{Y-1}$
H_2	$C_p(T_d - T_a)$	$C_p T_a(e^{Y-1} - 1)$
W	$H_1 - H_2$	$H_1 - C_p T_a(e^{Y-1} - 1)$
E	$1 - \dfrac{H_2}{H_1}$	$1 - C_p T_a(e^{Y-1} - 1)$
R_ϕ	$\dfrac{H_1}{T_b}$	$\dfrac{H_1}{\gamma^{\gamma-1} T_a}$
M.E.T.	$\dfrac{1}{2}\left(\dfrac{H_1 + H_2}{R_\phi}\right)$	$\dfrac{\gamma^{\gamma-1} T_a}{2}\left[\dfrac{H_1 + C_p T_a(e^{Y-1} - 1)}{H_1}\right]$
R_v	$v_d - v_b$	$v_a\left(e^{Y-1} - \dfrac{1}{\gamma}\right)$
M.E.P.	$J\dfrac{W}{R_v}$	$J\left\{\dfrac{H_1 - C_p T_a(e^{Y-1} - 1)}{v_a\left(e^{Y-1} - \dfrac{1}{\gamma}\right)}\right\}$
R_p	$p_b - p_a$	$p_a(\gamma^\gamma - 1)$
M.E.V.	$J\dfrac{W}{R_p}$	$J\left[\dfrac{H_1 - C_p T_a(e^{Y-1} - 1)}{p_a(\gamma^{\gamma-1} - 1)}\right]$
R_T	$T_b - T_a$	$T_a(\gamma^{\gamma-1} - 1)$

CYCLE IV A.

In the first modification of the type form in Group IV the expansion is not complete, so that:

From *A* to B. Adiabatic compression from atmospheric pressure.

From *B* to *C*. Addition of heat isothermally.

From C to D. Adiabatic expansion to pressure above atmosphere.

From D to E. Cooling isometrically to atmospheric pressure.

From E to A. Cooling at atmospheric pressure.

Figs. 137 and 138 are its diagrams. This is the case for a

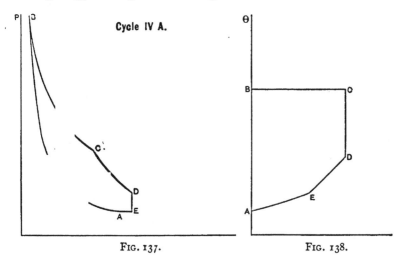

Cycle IV A.

FIG. 137.   FIG. 138.

Diesel engine in which the line of heating BC is too long for the size of the engine cylinder to permit of complete expansion.

The results of IV up to point C may be assumed. The point D lies somewhere on the adiabatic between C and atmosphere and is subject to the conditions

$$p_c > p_d > p_e, \quad \cdots \cdots \cdots \quad (1)$$

$$v_d > v_a. \quad \cdots \cdots \cdots \cdots \quad (2)$$

Then

$$v_d = v_c \left(\frac{p_c}{p_d}\right)^{\frac{1}{r}} = \frac{v_a}{r} e^z \left(\frac{p_a r^r}{p_a e^z}\right)^{\frac{1}{r}} = v_a e^z \left(\frac{r-1}{r}\right) \left(\frac{p_a}{p_d}\right)^{\frac{1}{r}};$$

$$\therefore v_d = v_a \left(\frac{p_a}{p_d}\right)^{\frac{1}{r}} e^{r-1}. \quad \cdots \cdots \quad (3)$$

Similarly

$$T_d = T_c\left(\frac{p_d}{p_c}\right)^{\frac{\gamma-1}{\gamma}} = T_a \gamma^{\gamma-1}\left(\frac{p_d e^Z}{p_a \gamma^\gamma}\right)^{\frac{\gamma-1}{\gamma}};$$

$$\therefore T_d = T_a\left(\frac{p_d}{p_a}\right)^{\frac{\gamma-1}{\gamma}} e^{Y-1}, \quad \cdots \cdots \quad (4)$$

$$v_e = v_d = v_a\left(\frac{p_a}{p_d}\right)^{\frac{1}{\gamma}} e^{Y-1}, \quad \cdots \cdots \quad (5)$$

$$p_e = p_a,$$

$$T_e = T_d\frac{p_e}{p_d} = T_a\left(\frac{p_d}{p_a}\right)^{\frac{\gamma-1}{\gamma}} e^{Y-1}\frac{p_a}{p_d} = T_a\left(\frac{p_a}{p_d}\right)^{\frac{1}{\gamma}} e^{Y-1}.$$

Apply the perfect gas law to D and E:

$$\frac{p_d v_d}{T_d} = \frac{p_d v_a p_a^{\frac{1}{\gamma}} e^{Y-1} p_a^{\frac{\gamma-1}{\gamma}}}{p_d^{\frac{1}{\gamma}} T_a p_d^{\frac{\gamma-1}{\gamma}} e^{Y-1}} = R.$$

$$\frac{p_e v_e}{T_e} = \frac{p_a v_a\left(\frac{p_a}{p_d}\right)^{\frac{1}{\gamma}} e^{Y-1}}{T_a\left(\frac{p_a}{p_d}\right)^{\frac{1}{\gamma}} e^{Y-1}} = R.$$

The heat abstracted is

$$H_2 = C_v(T_d - T_e) + C_p(T_e - T_a)$$

$$= C_v\left[T_a\left(\frac{p_d}{p_a}\right)^{\frac{\gamma-1}{\gamma}} e^{Y-1} - T_a\left(\frac{p_a}{p_d}\right)^{\frac{1}{\gamma}} e^{Y-1}\right]$$

$$+ C_p\left[T_a\left(\frac{p_a}{p_d}\right)^{\frac{1}{\gamma}} e^{Y-1} - T_a\right];$$

$$\therefore H_2 = C_v T_a e^{Y-1}\left(\frac{p_a}{p_d}\right)^{\frac{1}{\gamma}}\left[\frac{p_d}{p_a} - 1\right] - C_p T_a\left[\left(\frac{p_a}{p_d}\right)^{\frac{1}{\gamma}} e^{Y-} - 1\right], \quad (6)$$

$$W = H_1 - H_2, \quad \cdots \cdots \quad (7)$$

$$E = 1 - \frac{H_2}{H_1}, \quad \cdots \cdots \quad (8)$$

$$R_v = v_d - v_a = v_a\left[\left(\frac{p_a}{p_d}\right)^{\frac{1}{\gamma}} e^{Y-1} - \frac{1}{\gamma}\right], \quad \cdots \quad (9)$$

$$\text{M.E.P.} = J\frac{W}{v_a\left[\left(\dfrac{p_a}{p_d}\right)^{\frac{1}{r}}e^{r-1} - \dfrac{1}{r}\right]}, \quad \cdots \quad (10)$$

$$R_p = p_b - p_a = p_a(r^r - 1),$$

$$\text{M.E.V.} = J\frac{W}{p_a(r^r - 1)}, \quad \cdots \cdots \quad (11)$$

$$R_\phi = \frac{H_1}{T_b}, \quad \cdots \cdots \cdots \quad (12)$$

$$\text{M.E.T.} = \frac{1}{2}\left\{\frac{H_1 + H_2}{\dfrac{H_1}{T_b}}\right\} = \frac{T_b}{2}\left(1 - \frac{H_2}{H_1}\right), \quad \cdots \quad (13)$$

$$R_T = T_b - T_a = T_a(r^{r-1} - 1). \quad \cdots \cdots \quad (14)$$

Cycle IV B.

In the second modification of the fourth group the expansion goes below atmosphere, so that

From A to B. Adiabatic compression from atmospheric pressure.

From B to C. Addition of heat isothermally.

From C to D. Adiabatic expansion to pressure below atmosphere.

From D to E. Cooling isothermally to atmospheric pressure.

From E to A. Cooling at atmospheric pressure.

Let Figs. 139 and 140 be its diagrams.

The operations up to C are as in IV and those results may be assumed.

The point D is subject to the condition

$$p_d < p_a, \quad \cdots \cdots \cdots \quad (1)$$

and the point E to the condition

$$v_e > v_a. \quad \cdots \cdots \cdots \quad (2)$$

Then

$$v_d = v_a\left(\frac{p_a}{p_d}\right)^{\frac{1}{r}}e^{r-1}, \quad \cdots \cdots \quad (3)$$

and

$$T_d = T_a \left(\frac{p_d}{p_a}\right)^{\frac{r-1}{r}} e^{Y-1}; \quad \cdots \cdots \quad (4)$$

$$T_e = T_d = T_a \left(\frac{p_d}{p_a}\right)^{\frac{r-1}{r}} e^{Y-1}, \quad \cdots \cdots \quad (5)$$

$$p_e = p_a,$$

$$v_e = v_d \frac{p_d}{p_e} = v_a \left(\frac{p_d}{p_a}\right)^{\frac{r-1}{r}} e^{Y-1}. \quad \cdots \cdots \quad (6)$$

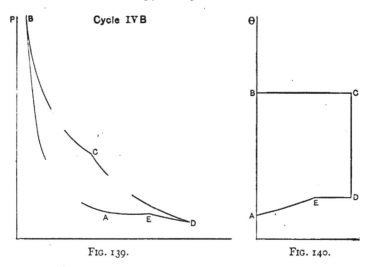

FIG. 139. FIG. 140.

Following the methods already adopted, it will be true to write

$$H_2 = T_d(\phi_d - \phi_c) + C_p(T_e - T_a).$$

But

$$\phi_d - \phi_e = (\phi_c - \phi_b) - (\phi_e - \phi_a),$$

$$\phi_c - \phi_b = \frac{H_1}{T_b} = \frac{H_1}{T_a r^{r-1}},$$

$$\phi_e - \phi_a = C_p \log_e \frac{T_e}{T_a}$$

$$= C_p \log_e \left[\left(\frac{p_d}{p_a}\right)^{\frac{r-1}{r}} e^{Y-1}\right].$$

$$\therefore\ H_2 = T_e\left[\frac{H_1}{T_a r^{r-1}} - C_p \log_e\left\{\left(\frac{p_d}{p_a}\right)^{\frac{r-1}{r}} e^{Y-1}\right\}\right]$$

$$+ C_p T_a\left[\left(\frac{p_d}{p_a}\right)^{\frac{r-1}{r}} e^{Y-1} - 1\right]$$

$$\text{(7)}$$

$$W = H_1 - H_2, \quad \cdot \quad \cdot \quad \cdot \quad \cdot \quad \cdot \quad \cdot \quad \text{(8)}$$

$$E = 1 - \frac{H_2}{H_1}, \quad \cdot \quad \cdot \quad \cdot \quad \cdot \quad \cdot \quad \cdot \quad \text{(9)}$$

$$R_v = v_d - v_b = v_a\left[\left(\frac{p_a}{p_d}\right)^{\frac{1}{r}} e^{Y-1} - \frac{1}{r}\right]. \quad \cdot \quad \cdot \quad \cdot \quad \text{(10)}$$

$$\therefore\ \text{M.E.P.} = J\frac{W}{v_a\left[\left(\frac{p_a}{p_d}\right)^{\frac{1}{r}} e^{Y-1} - \frac{1}{r}\right]}, \quad \cdot \quad \cdot \quad \cdot \quad \text{(11)}$$

$$R_p = p_b - p_d = p_a r^r - p_d. \quad \cdot \quad \cdot \quad \cdot \quad \cdot \quad \cdot \quad \text{(12)}$$

$$\therefore\ \text{M.E.V.} = J\frac{W}{p_a r^r - p_d}, \quad \cdot \quad \cdot \quad \cdot \quad \cdot \quad \text{(13)}$$

$$R_\phi = \frac{H_1}{T_b} = \frac{H_1}{T_a r^r}, \quad \cdot \quad \cdot \quad \cdot \quad \cdot \quad \cdot \quad \text{(14)}$$

$$\text{M.E.T.} = \frac{1}{2}\left(\frac{H_1 + H_2}{R_\phi}\right) = \frac{1}{2}\left\{\frac{H_1 + H_2}{\frac{H_1}{T_b}}\right\}$$

$$= \frac{1}{2}\left(T_b + T_b\frac{H_2}{H_1}\right) = \frac{T_b}{2}\left(1 + \frac{H_2}{H_1}\right), \quad \cdot \quad \cdot \quad \cdot \quad \text{(15)}$$

$$R_T = T_b - T_a = T_a(r^{r-1} - 1). \quad \cdot \quad \cdot \quad \cdot \quad \cdot \quad \text{(16)}$$

CYCLE IV C.

The third modification in Group IV is the cycle known as the Carnot Ideal Cycle, in which the temperature is carried down to the initial value, so that the isothermal will close it. Hence there is

From *A* to *B*. Adiabatic compression from atmospheric pressure.

From *B* to *C*. Addition of heat isothermally.

From C to D. Adiabatic expansion to pressure below atmosphere such that we get

From D to A. Cooling isothermally to original volume and atmospheric pressure.

Let Figs. 141 and 142 be its diagrams.

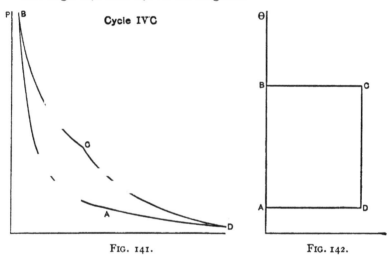

FIG. 141.   FIG. 142.

Assume results up to C as in IV.

The adiabatic through C must meet the isothermal through A to locate the point D.

From the adiabatic relations,

$$p_d = p_c \left(\frac{v_c}{v_d}\right)^r.$$

From the isothermal relation,

$$p_d = p_a \frac{v_a}{v_d}.$$

$$\therefore \ p_c \left(\frac{v_c}{v_d}\right)^r = p_a \frac{v_a}{v_d},$$

$$p_c v_c{}^r = p_a v_a v_d{}^{r-1}.$$

$$\therefore \ v_c = \left(\frac{p_c v_c{}^r}{p_a v_a}\right)^{\frac{1}{r-1}} = \left(\frac{p_a r^r v_a{}^r e^{rZ}}{e^Z r^r p_a v_a}\right)^{\frac{1}{r-1}}.$$

$$\therefore \ v_d = v_a e^Z = r v_c. \quad \cdots \quad \cdots \quad (\text{I})$$

By substitution,

$$p_d = \frac{p_a}{e^Z} = \frac{p_c}{\gamma^r}, \quad \cdots \cdots \quad (2)$$

$$T_d = T_a. \quad \cdots \cdots \quad (3)$$

By inspection it is easily seen the perfect gas law is satisfied.

$$H_2 = 1_a(\phi_d - \phi_a) = T_a(\phi_c - \phi_b) = T_a\frac{H_1}{T_b} = \frac{T_a H_1}{T_a \gamma^{r-1}}.$$

$$\therefore H_2 = \frac{H_1}{\gamma^{r-1}}, \quad \cdots \cdots \quad (4)$$

$$W = H_1 - H_2 = H_1\left(1 - \frac{1}{\gamma^{r-1}}\right), \quad \cdots \quad (5)$$

$$E = \frac{W}{H_1} = 1 - \frac{H_2}{H_1} = 1 - \frac{1}{\gamma^{r-1}}, \quad \cdots \quad (6)$$

$$R_v = v_d - v_b = v_a\left(e^Z - \frac{1}{\gamma}\right), \quad \cdots \quad (7)$$

$$\text{M.E.P.} = J\frac{H_1\left(1 - \frac{1}{\gamma^{r-1}}\right)}{v_a\left(e^Z - \frac{1}{\gamma}\right)}, \quad \cdots \quad (8)$$

$$R_p = p_b - p_d = p_a\left(\gamma^r - \frac{1}{e^Z}\right), \quad \cdots \quad (9)$$

$$\text{M.E.V.} = J\frac{H_1\left(1 - \frac{1}{\gamma^{r-1}}\right)}{p_a\left(\gamma^r - \frac{1}{e^Z}\right)}, \quad \cdots \quad (10)$$

$$\text{M.E.T.} = \tfrac{1}{2}(T_a + T_b) = \frac{T_a}{2}(1 + \gamma^{r-1}). \quad \cdots \quad (11)$$

Tabulating for Cycle IV C:

Symbol.	Formula as First Derived.	Formula Reduced.
p_b.	$p_a\left(\dfrac{v_a}{v_b}\right)^{\gamma}$	$p_a\gamma^{\gamma}$
v_b	$\dfrac{v_a}{\gamma}$	$\dfrac{v_a}{\gamma}$
T_b	$T_a\left(\dfrac{v_a}{v_b}\right)^{\gamma-1}$	$T_a\gamma^{\gamma-1}$.
Equation of condition	$H_1>0$	$H_1>0$
p_c	$\dfrac{p_b}{e^{\frac{H_1}{T_b(C_p-C_v)}}}$	$\dfrac{p_a\gamma^{\gamma}}{e^Z}$
v_c	$v_b\dfrac{p_b}{p_c}$	$\dfrac{v_a}{\gamma}e^Z$
T_c	T_b	$T_a\gamma^{\gamma-1}$
p_d		$\dfrac{p_a}{e^Z}$
v_d		v_ae^Z
T_d		T_a
H_2.	$T_a(\phi_d-\phi_a)$.	$\dfrac{H_1}{\gamma^{\gamma-1}}$
W	H_1-H_2.	$H_1\left(1-\dfrac{1}{\gamma^{\gamma-1}}\right)$
E	$1-\dfrac{H_2}{H_1}$	$\left(1-\dfrac{1}{\gamma^{\gamma-1}}\right)$
R_ϕ	$\dfrac{H_1}{T_b}$	$\dfrac{H_1}{T_a\gamma^{\gamma-1}}$
M.E.T.	$\dfrac{1}{2}\left(\dfrac{H_1+H_2}{R_\phi}\right)$	$\dfrac{T_a}{2}(1+\gamma^{\gamma-1})$
R_v	v_d-v_b.	$v_a\left(e^Z-\dfrac{1}{\gamma}\right)$

Symbol.	Formula as First Derived.	Formula Reduced.

$$\text{M.E.P}\ldots\ldots\ldots\ldots J\frac{W}{R_v}\ldots\ldots\ldots\ldots JH_1\left\{\frac{1-\dfrac{1}{r^{v-1}}}{v_a\left(e^z-\dfrac{1}{r}\right)}\right\}$$

$$R_p\ldots\ldots\ldots\ldots\ldots p_d-p_b\ldots\ldots\ldots\ldots p_a\left(r^r-\frac{1}{e^z}\right)$$

$$\text{M.E.V}\ldots\ldots\ldots\ldots J\frac{W}{R_p}\ldots\ldots\ldots\ldots JH_1\left\{\frac{1-\dfrac{1}{r^{r-1}}}{p_a\left(r^r-\dfrac{1}{e^z}\right)}\right\}$$

$$R_T\ldots\ldots\ldots\ldots T_b-T_a\ldots\ldots\ldots\ldots T_a(r^{r-1}-1)$$

If the adiabatic compression and expansion be replaced by isometric changes of temperature, while the heating and cooling phases remain isothermal, the cycle which results is that of the Stirling hot-air engine.

187. Compression Cycle with the Heating Process Arbitrary. —A fifth group may be formed from those cycles in which the heating process follows some arbitrary law, which does not fall into one of the normal types heretofore treated. That is, the volume pressure and temperature may all vary while heat is being added. Such variations will give the pv and $\theta\phi$ diagrams herewith in which

From A to B. Adiabatic compression from atmospheric pressure.

From B to C. Addition of heat at variable pvT.

From C to D. Adiabatic expansion to atmospheric pressure.

From D to A. Cooling at atmospheric pressure.

Cycles V, A, B, and C may have the same modification on Cycle V, as II, A, B, and C have on III, for example.

Let Figs. 143 and 144 be the diagrams of the cycles.

If heat be added at increasing p, v, and T the curves of states will lie somewhere between the isometric and isopiestic on both diagrams and the cycle is somewhere between III and II. If the heat addition took place at decreasing p, increasing v and T, the curve of states might lie between the isopiestic and the isothermal and the cycle lie between III and IV. It is impossible, however, to calculate the appropriate set of formulæ without knowing the law of variation of states. The number of ways of variation is infinite, and while any one might be assumed, nothing

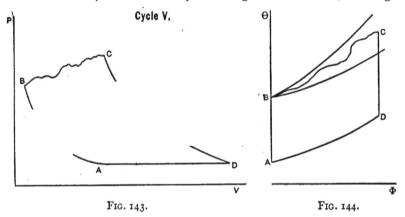

FIG. 143.   FIG. 144.

could be gained by the calculation unless the law of variation chosen was pre-eminently simple or maintains in practice. Whatever it may be, however, the previous discussion will enable it to be classed pretty well without entering much into details.

188. Cycles with Atmospheric Heating.—A group of cycles must be formed to include those in which, with or without compression, the gas is heated at or below atmospheric pressure. These form groups from VI to X. In Group VI there will be

From A to B. Addition of heat at atmospheric pressure

From B to C. Cooling isometrically.

From C to A. Adiabatic compression.

Let Figs. 145 and 146 be the diagrams of the cycle.

Heat being added isopiestically,

$$T_b - T_a = \frac{H_1}{C_p};$$

$$\therefore\ T_b = T_a\left(1 + \frac{H_1}{C_p T_a}\right) = T_a x, \quad \cdots \cdots \quad (1)$$

$$v_b = v_a \frac{T_b}{T_a} = v_a x, \quad \cdots \cdots \cdots \quad (2)$$

$$v_c = v_b = v_a x. \quad \cdots \cdots \cdots \quad (3)$$

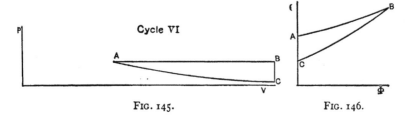

FIG. 145.   FIG. 146.

The point C lies on the adiabatic through A, hence

$$p_c = p_a\left(\frac{v_a}{v_c}\right)^\gamma = p_a\left(\frac{v_a}{v_a x}\right)^\gamma = \frac{p_a}{x^\gamma}, \quad \cdots \cdots \quad (4)$$

$$T_c = T_a\left(\frac{v_a}{v_c}\right)^{\gamma-1};$$

$$\therefore\ T_c = \frac{T_a}{x^{\gamma-1}}. \quad \cdots \cdots \cdots \quad (5)$$

The perfect gas law is seen by inspection to be satisfied:

$$H_2 = C_v(T_b - T_c) = C_v\left(T_a x - \frac{T_a}{x^{\gamma-1}}\right);$$

$$\therefore\ H_2 = C_v T_a x\left(1 - \frac{1}{x^\gamma}\right), \quad \cdots \cdots \quad (6)$$

$$W = H_1 - C_v T_a x\left(1 - \frac{1}{x^\gamma}\right), \quad \cdots \cdots \quad (7)$$

$$E = 1 - \frac{C_v T_a x\left(1 - \frac{1}{x^\gamma}\right)}{H_1}, \quad \cdots \cdots \quad (8)$$

$$R\phi = \phi_b - \phi_a = C_p \log_e \frac{T_b}{T_a};$$

$$\therefore R\phi = C_p \log_e x. \quad \cdot \cdot \cdot \cdot \cdot \cdot \cdot \cdot \cdot \quad (9)$$

$$\text{M.E.T.} = \frac{1}{2} \left\{ \frac{1 + C_v T_a x \left(1 - \frac{1}{x^\gamma}\right)}{C_p \log_e x} \right\}, \quad \cdot \cdot \cdot \quad (10)$$

$$R_v = v_c - v_a = v_a(x - 1). \quad \cdot \cdot \cdot \cdot \cdot \cdot \quad (12)$$

$$\text{M.E.P.} = J\frac{H_1 - C_v T_a x \left(1 - \frac{1}{x^\gamma}\right)}{v_a(x-1)}, \quad \cdot \cdot \cdot \cdot \quad (13)$$

$$R_p = p_a - p_c = p_a\left(1 - \frac{1}{x^\gamma}\right), \quad \cdot \cdot \cdot \cdot \cdot \cdot \quad (14)$$

$$\text{M.E.V.} = J\frac{H_1 - C_v T_a x \left(1 - \frac{1}{x^\gamma}\right)}{p_a\left(1 - \frac{1}{x^\gamma}\right)} = J\left\{ \frac{H_1}{p_a\left(1 - \frac{1}{x^\gamma}\right)} - \frac{C_v T_a x}{p_a} \right\} \quad (15)$$

$$R\phi = T_b - T_a = T_a(x - 1). \quad \cdot \cdot \cdot \cdot \cdot \quad (16)$$

CYCLE VII.

In Cycle VII there will be

From *A* to *B*.   Addition of heat at atmospheric pressure.

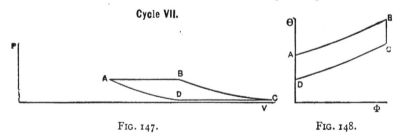

Cycle VII.

FIG. 147.   FIG. 148.

From **B** to *C*.   Adiabatic expansion.
From *C* to *D*.   Cooling isopiestically.
From *D* to *A*.   Adiabatic compression.
Let Figs. 147 and 148 be its diagrams.

For **B** as before,

$$v_b = v_a x, \quad \text{. . . . .} \quad \textbf{(1)}$$

$$p_b = p_a, \quad \text{. . . . .} \quad \textbf{(2)}$$

$$T_b = T_a x. \quad \text{. . . . .} \quad \textbf{(3)}$$

The point **C** lies on an adiabatic through **B** and is subject to the condition

$$p_a > p_c > o, \quad \text{. . . . . . .} \quad \textbf{(4)}$$

$$v_c = v_b \left(\frac{p_b}{p_c}\right)^{\frac{1}{r}} = v_a x \left(\frac{p_a}{p_c}\right)^{\frac{1}{r}}, \quad \text{. . . . .} \quad \textbf{(5)}$$

$$T_c = T_b \left(\frac{p_c}{p_a}\right)^{\frac{r-1}{r}} = T_a x \left(\frac{p_c}{p_a}\right)^{\frac{r-1}{r}}, \quad \text{. . . .} \quad \textbf{(6)}$$

$$v_d = v_a \left(\frac{p_d}{p_a}\right)^{\frac{1}{r}} = v_a \left(\frac{p_a}{p_c}\right)^{\frac{1}{r}} = \frac{v_c}{x}. \quad \text{. . . . .} \quad \textbf{(7)}$$

But

$$v_b = v_a x.$$

Hence

$$\frac{v_a}{v_d} = \frac{v_b}{v_c}. \quad \text{. . . . .} \quad \textbf{(8)}$$

Similarly

$$\frac{T_a}{T_d} = \frac{T_b}{T_c} \quad \text{. . . . .} \quad \textbf{(9)}$$

and

$$T_d = \frac{T_c}{x};$$

$$H_2 = C_p(T_c - T_d) = C_p T_c \left(1 - \frac{1}{x}\right)$$

$$= C_p T_a x \left(\frac{p_c}{p_a}\right)^{\frac{r-1}{r}} \left(1 - \frac{1}{x}\right), \quad \text{. . . . .} \quad \textbf{(10)}$$

$$W = H_1 - H_2 = H_1 - C_p T_a x \left(\frac{p_c}{p_a}\right)^{\frac{r-1}{r}} \left(1 - \frac{1}{x}\right), \quad \text{. .} \quad \textbf{(11)}$$

$$E = 1 - \frac{H_2}{H_1}, \quad \cdots \cdots \cdots \quad (12)$$

$$R\phi = C_p \log_e x \text{ as in VI.} \quad \cdots \quad (13)$$

$$\therefore \text{M.E.T.} = \frac{1}{2}\left(\frac{H_1 + H_2}{C_p \log_e x}\right), \quad \cdots \cdots \quad (14)$$

$$R_v = v_c - v_a = v_a\left[\left(\frac{p_a}{p_c}\right)^{\frac{1}{r}} x - 1\right], \quad \cdots \quad (15)$$

$$\text{M.E.P.} = J\frac{W}{R_v} = J\frac{W}{v_a\left[\left(\frac{p_a}{p_c}\right)^{\frac{1}{r}} x - 1\right]}, \quad \cdots \quad (16)$$

$$R_p = p_a - p_c = p_a - p_d, \quad \cdots \cdots \quad (17)$$

$$\text{M.E.V.} = J\frac{W}{p_a - p_d}, \quad \cdots \cdots \quad (18)$$

$$R_T = T_b - T_a = T_a(x-1) \text{ as in VI.} \quad \cdots \quad (19)$$

Cycle VIII.

In Cycle VIII there are three steps only, viz.:

From A to B. Addition of heat at atmospheric pressure.

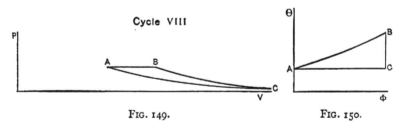

FIG. 149. FIG. 150.

From B to C. Adiabatic compression to such a pressure that we get

From C to D. Isothermal compression to original state.

Figs. 149 and 150 are its diagrams.

It will be true for B that

$$v_b = v_a x, \quad \cdots \quad \cdots \quad (1)$$

$$p_b = p_a, \quad \cdots \quad \cdots \quad (2)$$

$$T_b = T_a x. \quad \cdots \quad \cdots \quad (3)$$

The isothermal through A intersects the adiabatic through B to determine C.

From the adiabatic

$$p_c = p_b \left(\frac{v_b}{v_c}\right)^r.$$

From the isothermal

$$p_c = p_a \frac{v_a}{v_c};$$

$$\therefore p_b \left(\frac{v_b}{v_c}\right)^r = p_a \frac{v_a}{v_c}.$$

But

$$v_b{}^r = v_a{}^r x^r.$$

$$\therefore \frac{v_a{}^r x^r}{v_a} = \frac{v_c{}^r}{v_c};$$

$$\therefore v_c = v_a x^{\frac{r}{r-1}}. \quad \cdots \quad \cdots \quad (4)$$

By substitution

$$p_c = \frac{p_a}{x^{\frac{r}{r-1}}}, \quad \cdots \quad \cdots \quad (5)$$

$$H_{\cdot} = T_a(\phi_b - \phi_a) = T_a C_p \log_e x, \quad \cdots \quad (6)$$

$$W = H_1 - T_a C_p \log_e x, \quad \cdots \quad \cdots \quad (7)$$

$$E = 1 - \frac{T_a C_p \log_e x}{H_1}, \quad \cdots \quad \cdots \quad (8)$$

$$R_v = v_c - v_a = v_a(x^{\frac{r}{r-1}} - 1), \quad \cdots \quad \cdots \quad (9)$$

$$\text{M.E.P.} = J\frac{H_1 - T_aC_p\log_e x}{v_a(x^{\frac{r}{r-1}} - 1)}, \quad \ldots \quad (10)$$

$$R_p = p_a - p_c = p_a\left(1 - \frac{1}{x^{\frac{r}{r-1}}}\right), \quad \ldots \quad (11)$$

$$\text{M.E.V.} = J\frac{H_1 - T_aC_p\log_e x}{v_a(x^{\frac{r}{r-1}} - 1)}, \quad \ldots \quad (12)$$

$$\text{M.E.T.} = \frac{1}{2}\left(\frac{H_1 + T_aC_p\log_e x}{C_p\log_e x}\right) = \frac{1}{2}\left(\frac{H_1}{C_p\log_e x} + T_a\right). \quad (13)$$

CYCLE IX.

In Cycle IX the expansion is incomplete, calling for a cooling at constant volume, so that

From *A* to B.   Addition of heat at atmospheric pressure.
From *B* to *C*.   Adiabatic expansion.
From *C* to *D*.   Cooling isometrically.
From *D* to *A*.   Compression adiabatically.
Let Figs. 151 and 152 be its diagrams.

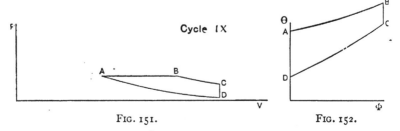

FIG. 151.   FIG. 152.

Up to the point *C* the results of VII may be assumed.

The point *D* lies on an adiabatic through *A* and is subject to the conditions

$$v_d = v_c, \quad \ldots \quad \ldots \quad \ldots \quad (1)$$

$$p_c > p_d > 0, \quad \ldots \quad \ldots \quad \ldots \quad (2)$$

$$v_d > v_b, \quad \cdots \quad \cdots \quad \cdots \quad (3)$$

$$p_d = p_a\left(\frac{v_a}{v_d}\right)^r = p_a\left(\frac{v_a}{v_c}\right)^r = p_a\left\{\frac{v_a}{v_a x\left(\frac{p_a}{p_c}\right)^{\frac{1}{r}}}\right\}^r = \frac{p_a p_c}{p_a x^r} = \frac{p_c}{x^r}, \quad (4)$$

$$T_d = T_c\frac{p_d}{p_c} = \frac{T_c}{x^r}, \quad \cdots \quad \cdots \quad (5)$$

$$H_2 = C_v(T_c - T_d) = C_v T_c\left(1 - \frac{1}{x^r}\right)$$

$$(6)$$

$$= C_v T_a x\left(\frac{p_c}{p_a}\right)^{\frac{r-1}{r}}\left(1 - \frac{1}{x^r}\right),$$

$$W = H_1 - H_2 = H_1 - C_v T_a x\left(\frac{p_c}{p_a}\right)^{\frac{r-1}{r}}\left(1 - \frac{1}{x^r}\right), \quad \cdots \quad (7)$$

$$E = 1 - \frac{H_2}{H_1} = 1 - \frac{C_v T_a x\left(\frac{p_c}{p_a}\right)^{\frac{r-1}{r}}\left(1 - \frac{1}{x^r}\right)}{H_1}, \quad \cdots \quad (8)$$

$$R_\phi = C_p\log_e x \quad \text{as before.} \quad \cdots \quad \cdots \quad (9)$$

$$\text{M.E.T.} = \frac{1}{2}\left(\frac{H_1 + H_2}{C_p\log_e x}\right), \quad \cdots \quad \cdots \quad (10)$$

$$R_v = v_c - v_a = v_a\left[x\left(\frac{p_a}{p_c}\right)^{\frac{1}{r}} - 1\right], \quad \cdots \quad \cdots \quad (11)$$

$$\text{M.E.P.} = J\frac{W}{v_a\left[x\left(\frac{p_a}{p_c}\right)^{\frac{1}{r}} - 1\right]}, \quad \cdots \quad \cdots \quad (12)$$

$$R_p = p_a - p_d = p_a - \frac{p_c}{x^r}, \quad \cdots \quad \cdots \quad (13)$$

$$\text{M.E.V.} = J\frac{W}{p_a - \frac{p_c}{x^r}}, \quad \cdots \quad \cdots \quad (14)$$

$$R_T = T_b - T_a = T_a(x - 1). \quad \cdots \quad \cdots \quad (15)$$

CYCLE X.

In this cycle, as in the last four, heat is added at atmospheric pressure, then follows adiabatic expansion, after which heat is abstracted according to some law as yet undefined. Adiabatic compression completes the cycle. As the law of abstraction of heat is as yet undefined formulæ cannot be derived for the cycle and its discussion will be left as with Cycle V.

Formulæ might have been derived for the imperfect carrying out of Cycles VI, VII, VIII, and IX, but they are of such slight importance in practice that it did not seem desirable.

Besides the twenty-two cycles considered there may be others due to the combination or differentiation of these typical ones, but the object of this analysis will be best accomplished by a study of types, the non-typical or synthetic cycles being omitted. The method of study here set forth, being of universal application to all possible cycles, will furnish means of reaching a clear understanding of any of the unconsidered cycles should need arise.

Referring now to the quantitative graphical presentation of the *PV* diagrams for the most important first four cycles and reproduced in Figs. 153 to 161, it should be observed that these areas are all derived from the following data and were plotted to scale of twenty atmospheres to one inch for pressures and 200 cubic feet to the inch for volumes. The illustrations have been made by reducing the full-size drawings to one-half size, which has therefore doubled the scale. The data for plotting were:

A. Initial condition:

Pressure.................... one atmosphere
Temperature............... 492° F. absolute
Volume.................... 12.4 cubic feet (approx.

B. Compression, final equal to $\frac{1}{10}$ initial volume and also $\frac{1}{2}$, making two cases.

The B.T.U. added per pound of air were 500 for all cycles except IV and IV *C*, in which 250 only were added, because in Cycle IV a maximum of 278 B.T.U. brings the isothermal

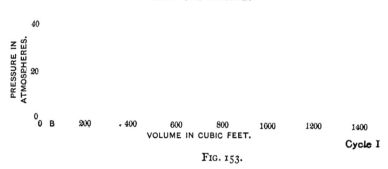

FIG. 153.

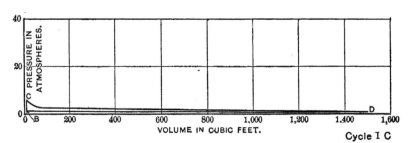

FIG. 154

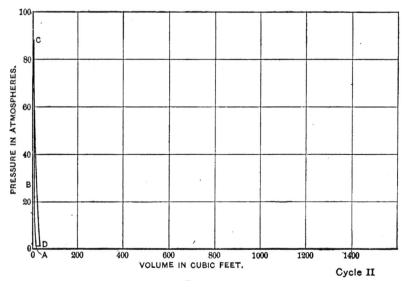

FIG. 155.

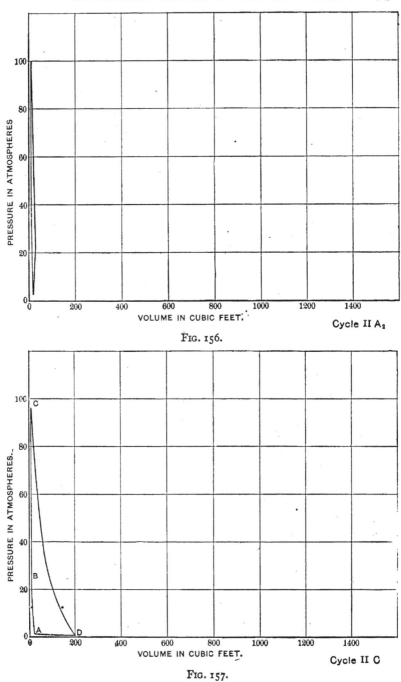

FIG. 156.

FIG. 157.

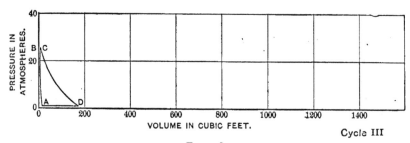

FIG. 158.

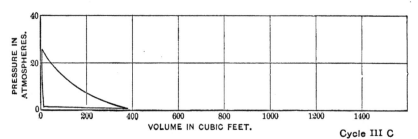

FIG. 159.

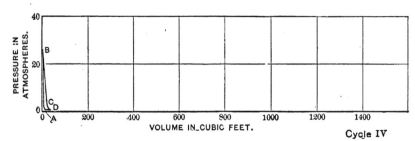

FIG. 160.

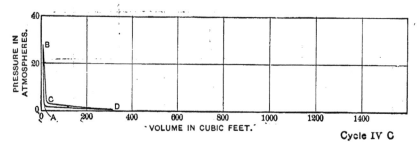

FIG. 161.

down to atmosphere, and the last phase, the adiabatic expansion, becomes impossible.

189. Comparison of Cycles with Respect to Temperatures before Expansion.—Of the many cycles considered for comparison only those will be chosen that might be called the perfect cycles, because accurately defined, and these are Cycles I, I C, II, II A_2, II C, III, III C, IV, IV C. The atmospheric cycles are of comparatively little importance and will be neglected in the discussion. Each variable will be taken up separately, beginning with temperatures, and its value examined in the different cases by formula and by calculated examples expressed in curves which are then the graphical formulæ. The curves given are approximately correct, and as the same approximation will probably maintain for all the cases the curves will serve as well for comparison as if absolutely exact. Two cases of each are given, one with compression 2:1 and one with 10:1 (volume ratios). Call the atmospheric values p_a, v_a, T_a.

TEMPERATURES AFTER ADDITION OF H_1 B.T.U.

Cycle.

I, I C. $$T_c = T_a X = T_a \left(1 + \frac{H_1}{C_v T_a} \right) \quad \text{(Fig. 162)} \quad . \quad (1)$$

II, II A, II C. $$T_c = T_b X = T_b \left(1 + \frac{H_1}{C_v T_b} \right) \quad \text{(Fig. 163)} \quad . \quad (2)$$

III, III C. $$T_c = T_b X = T_b \left(1 + \frac{H_1}{C_v T_b} \right) \quad \text{(Fig. 164)} \quad (3)$$

IV, IV C. $$T_c = T_b. \quad \text{(Fig. 165)} \quad . \quad . \quad . \quad (4)$$

Using axes of T_c and H_1 it will be observed that these are all straight lines passing through the axis of temperatures at T_b above the origin except in cycles (I, I C) where the intersection is at T_a. These lines are inclined to the axis of H and make with it an angle α such that in

I, I C, II, II A, II C $$\tan \alpha = \frac{1}{C_v}, \quad . \quad . \quad . \quad . \quad . \quad . \quad (5)$$

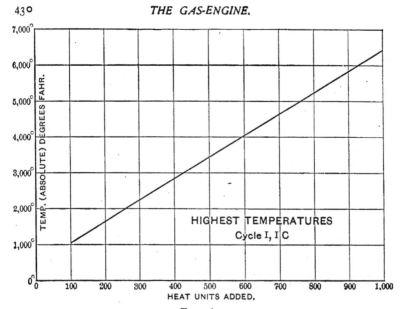

FIG. 162.

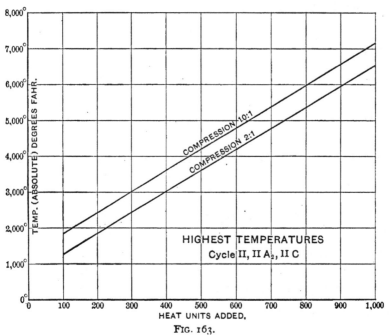

FIG. 163.

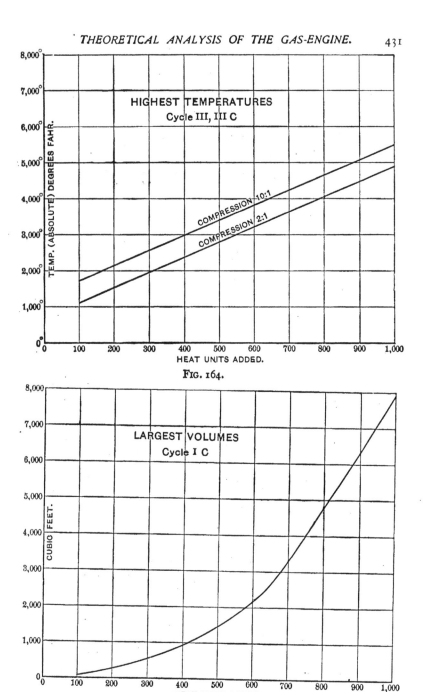

FIG. 164.

FIG. 165.

and in

III, III C $\qquad \tan \alpha' = \dfrac{1}{C_p},$ . . . . (6)

while IV, IV C are lines parallel to axis H_1.

190. Comparison of Cycles with Respect to Temperatures after Expansion.—A treatment similar to the foregoing respecting temperatures after expansion gives the following comparison diagram.

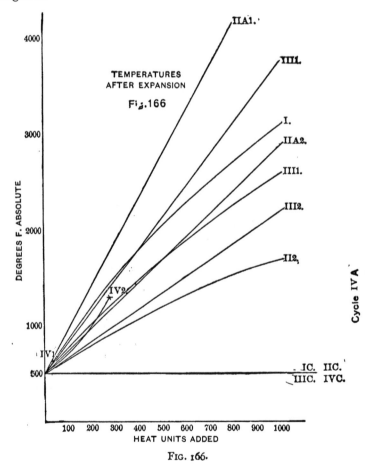

FIG. 166.

Cycle.

I. $$T_d = T_a(X)^{\frac{1}{\gamma}} = T_b\left(1 + \frac{H_1}{C_v T_a}\right)^{\frac{1}{\gamma}}, \quad \cdots \quad (19)$$

I C. $$T_d = T_a, \quad \cdots \quad (20)$$

II. $$T_d = T_a(X)^{\frac{1}{\gamma}} = T_a\left(1 + \frac{H_1}{C_v T_b}\right)^{\frac{1}{\gamma}}, \quad \cdots \quad (21)$$

II A. $$T_d = T_a X = T_a\left(1 + \frac{H_1}{C_v T_b}\right), \quad \cdots \quad (22)$$

II C. $$T_d = T_a, \quad \cdots \quad (23)$$

III. $$T_d = T_a Y = T_a\left(1 + \frac{H_1}{C_v T_b}\right), \quad \cdots \quad (24)$$

III C. $$T_d = T_a, \quad \cdots \quad (25)$$

IV. $$T_1 = T_a e^{Y-1} = T_a e^{\frac{H_1}{C_p T_b}}, \quad \cdots \quad (26)$$

IV C. $$T_d = T_a. \quad \cdots \quad (27)$$

Curves (19) and (21) are similar in form, cutting axis T_d at different points, however, and having different slopes. It is easily seen that (21) is always greater than (19), also that (22) is greater than (21), since

$$\frac{1}{\gamma} < 1.$$

Both (22) and (24) are straight lines, but they have different slopes through intersecting axis T_d at the same point:

$$(\tan \delta)_{\text{II. A.}} = \frac{T_a}{C_v T_b} = \frac{1}{\gamma^{\gamma-1} C_v}, \quad \cdots \quad (28)$$

$$(\tan \delta)_{\text{III.}} = \frac{1}{\gamma^{\gamma-1} C_p}, \quad \cdots \quad (29)$$

whence (22) is always greater than (24). Equation (26) is an exponential cutting T_d axis at T_a. It is concave up and slopes up to the right, since

$$\frac{dT_d}{dH_1} = \frac{1}{r^{\gamma-1}C_p}e^{\frac{H_1}{C_pT}}, \quad \cdots \cdots \quad (30)$$

$$\frac{d^2T_d}{dH_1{}^2} = \frac{1}{r^{\gamma-1}C_p{}^2T_b}e^{\frac{H_1}{C_pT_b}}. \quad \cdots \cdots \quad (31)$$

These curves are shown in Fig. 166 for the two cases.

191. Deductions from the Comparisons of Temperature.— Translating and analyzing these equations, the following deductions seem unavoidable:

1. For the same previous compression the temperature resulting in each cycle from heat addition, and which is the maximum for the cycle, will be different. That is, the addition of the same amount of heat will result in a different temperature for each group of cycles.

2. Gases passing through Cycle I may, on addition of a certain amount of heat, H_1, have a temperature equal to what the same gas would have passing through Cycle III. However, for more heat added the temperature for I will become higher than that for III, while for less heat added III will be higher.

3. Increase of compression before heating changes the temperature after heating by only so much numerically as the varied compression has resulted in changing the temperature before heating begins.

4. The temperature increase due to heating is proportional to the amount of heat added H_1, and the constant of proportionality involves the reciprocal of the specific heat for the process and the weight of the gas present.

5. After the gas has expanded to the greatest volume possible in the cycle, no two cycles will leave the gas with the same temperature except in a few special cases.

6. Cycle I C, II C, III C, IV C by definition have the same temperatures at the end of expansion, and this is moreover constant no matter what H may be and is equal to the initial temperature of the cycle.

7. There will be a value of H_1 for a limited range of compressions for which Cycle III may give to the gas the same final expansion temperature as Cycle I.

8. Similarly II for one compression may coincide in final temperature with II A for some other compression.

9. The temperature after expansion for Cycle II A_1 will always be higher than for III and III higher than for II.

10. In round numbers II A may be 25 per cent higher than III, and may even be 100 per cent higher than II for the same compression for possible values of H_1.

11. With variation of compression the temperature at the termination of expansion will vary, always becoming lower, but the extent of the lowering will depend on how much heat was added before expansion and in case II A and III is exactly proportional to H_1.

12. A change of compression $\frac{1}{2}$ to $\frac{1}{10}$ may change the temperature at the end of expansion in the case of Cycle II A and III as much as 80 per cent for possible values of H_1.

13. Mean effective ·temperatures, Fig. 167, are different for different cycles and for different compressions in the same cycle.

14. Cycle IV C is the only cycle with constant mean effective temperature.

15. Mean effective temperatures of all other cycles increase with H_1.

16. For large values of H_1 the order of magnitude of mean effective temperatures will be: Lowest, IV C, III C, I C, II C, III, I, II, highest, II A.

17. For lower values of H_1 this order may be somewhat changed, and there will be points at which two different cycles will have simultaneous values of M.E.T. and H_1.

The following graphical comparison of mean effective temperatures in the various cycles is also instructive (Fig. 167):

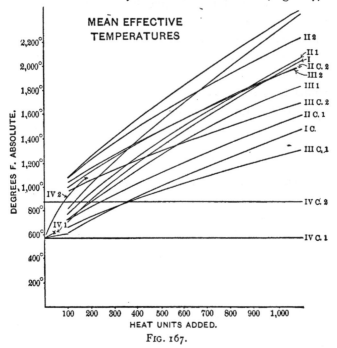

Fig. 167.

192. Comparison of Cycles with Respect to Pressures after Addition of Heat before Expansion.—A similar treatment of the equations and plotting of the pressures after addition of H_1 give the following comparison (Figs. 168–170):

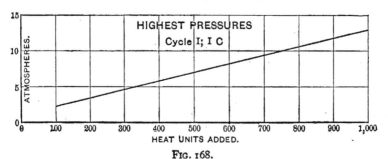

Fig. 168.

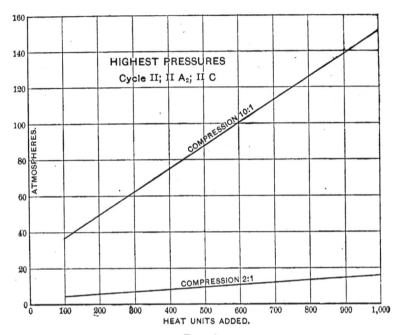

FIG. 169.

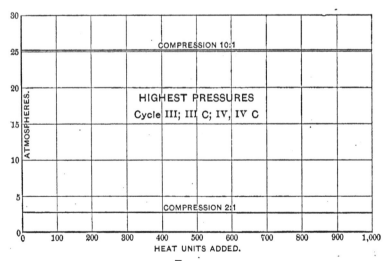

FIG. 170.

Cycle.

I, I C.
$$p_c = p_a X = p_a\left(1 + \frac{H_1}{C_v T_b}\right) \quad \text{(Fig. 168)} \quad . \quad (7)$$

II, II A, II C.
$$p_c = p_b X = p_b\left(1 + \frac{H_1}{C_v T_b}\right) \quad \text{(Fig. 169)} \quad . \quad (8)$$

III, III C.
$$p_c = p_b \quad \text{(Fig. 170)} . \; . \; . \; . \; . \; . \quad (9)$$

IV, IV C.
$$p_c = \frac{p_b}{e^Z} = \frac{p_b}{e^{\frac{H_1}{(C_p - C_v) T_b}}} \quad \text{(Fig. 170)}. \quad . \; . \quad (10)$$

Equations (7), (8), and (9) are all straight lines, (9) being parallel to axis H_1, while (7) and (8) are inclined. Equation (10) is an exponential curve sloping down to the right and concave up and asymptotic to axis of H, as can be seen from the derivatives

$$\frac{dp}{dH_1} = -\frac{p_b}{(C_p - C_v) T_b e^{\frac{H_1}{(C_p - C_v) T_b}}}, \quad . \; . \; . \quad (11)$$

$$\frac{d^2 p}{dH_1{}^2} = \frac{p_b}{(C_p - C_v)^2 T_b{}^2 e^{\frac{H_1}{(C_p - C_v) T_b}}}. \quad . \; . \; . \quad (12)$$

193. Comparison of Cycles with Respect to Pressures after Expansion.—A similar treatment for the pressures after expansion gives the curves and equations which follow:

Cycle.

I.
$$p_d = p_a, \quad . \; . \; . \; . \; . \; . \; . \quad (32)$$

I C.
$$p_d = \frac{p_a}{X^{\frac{1}{r-1}}} = \frac{p_a}{\left(1 + \frac{H_1}{C_v T_a}\right)^{\frac{1}{r-1}}}, \quad . \; . \; . \; . \quad (33)$$

II.
$$p_d = p_a, \quad . \; . \; . \; . \; . \; . \quad (34)$$

II A.
$$p_d = p_a X = p_a\left(1 + \frac{H_1}{C_v T_b}\right); \quad . \; . \; . \; . \quad (35)$$

II C.
$$p_d = \frac{p_a}{X^{\frac{1}{\gamma-1}}} = \frac{p_a}{\left(1 + \dfrac{H_1}{C_v T_b}\right)^{\frac{1}{\gamma-1}}}, \quad \cdots \cdot \quad (36)$$

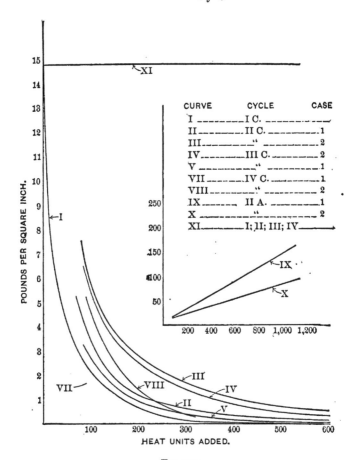

FIG. 171.

III.
$$p_d = p_a, \quad \cdots \cdots \cdot \quad (37)$$

III C.
$$p_d = \frac{p_a}{X^{\frac{1}{\gamma-1}}} = \frac{p_a}{\left(1 + \dfrac{H_1}{C_p T_b}\right)^{\frac{1}{\gamma-1}}}, \quad \cdots \cdot \quad (38)$$

IV.
$$p_d = p_a, \quad \cdot \quad \cdot \quad \cdot \quad \cdot \quad \cdot \quad \cdot \quad (39)$$

IV C.
$$p_d = \frac{p_a}{e^Z} = \frac{p_a}{e^{\frac{H_1}{(C_p - C_v) T_b}}} \cdot \quad \cdot \quad \cdot \quad \cdot \quad \cdot \quad (40)$$

Equations (32), (34), (37), (39) are identical and represent a straight line parallel to axis H_1. Curve (55) is a straight line inclined to H_1. All the others are concave up, sloping down to the right; their relative positions are seen in Fig. 171 for two compressions.

194. Comparison of Mean Effective Pressures in the Various Cycles.—The mean effective pressure during the working-stroke of the engine is one of the most important practical data concerning the cycle under which it is working. The larger this value the smaller the volume of the cylinder need to be for a given power, and the lighter the weight of the engine. The following graphical presentation of the comparison of cycles in this respect is most instructive (Figs. 172 to 178). The superiority of Cycle II A_2 (the Otto) is manifest. A statement of the conclusion with respect to the pressures in the various cycles must include the following conclusions on page 387.

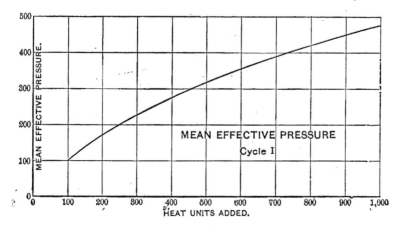

Fig. 172.

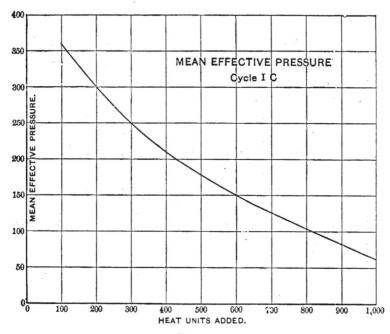

FIG. 173.

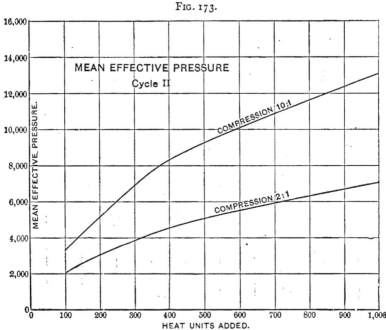

FIG 174.

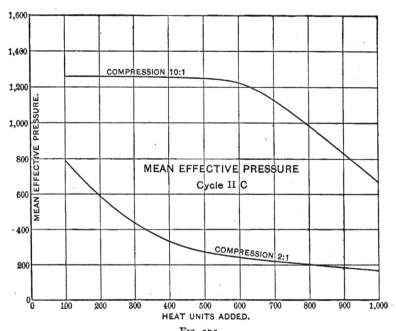

FIG. 175.

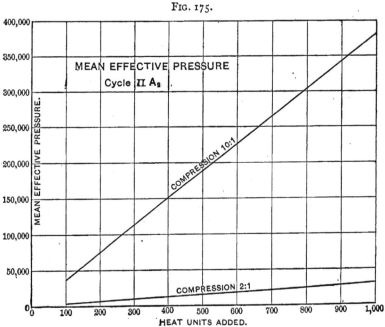

FIG. 176.

100   200   300   400   500   600   700   800   900   1000

HEAT UNITS ADDED.

FIG. 177.

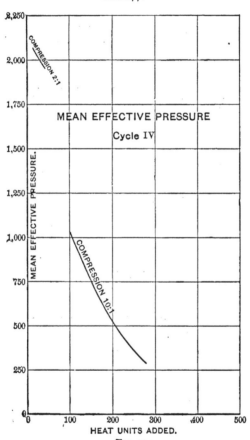

FIG. 179.

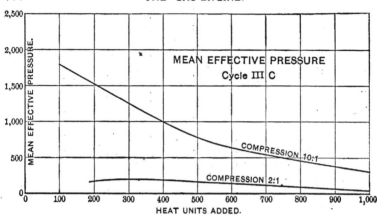

Fig. 178.

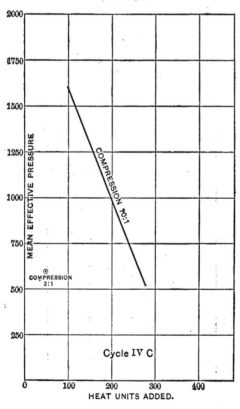

Fig. 180.

PRESSURES (Figs. 168 to 180).

1. The pressures resulting from heat addition are different for cycles with different numerals, but the same in any one group. Thus II, II A, II B, II C or Group II will all have the same pressures, whereas those of Group II will differ from those of Groups III and IV.

2. Lines representing pressures as functions of the heat supplied, H_1, will cross as these functions are different for different groups, and it will hence be possible for the different groups of cycles to have the same pressures for certain values of H_1.

3. Groups I and II have pressures after heating that increase with H_1, while in Group III the pressure is constant and in IV decreasing with increase of H_1.

4. For same compressions Group II will always have the highest pressure after heating, and III, IV, and I come in the order named for moderate H_1, while for large H_1 IV cannot exist.

5. Increase of compression will change the pressure after heating in Group III only so much as results from the changed compression before heating. In Groups II and I the change is such as to keep the pressure ratio before and after heating constant; so that for a given change in H_1 the resulting pressure change in II will be greatest for high compressions, less for moderate compressions, and least for no compression, *i.e.*, for Group I.

6. After expansion by definition the pressures of I, II, III, and IV are all atmospheric and equal.

7. The pressure which II A_2 will reach when the gas has expanded to original volume increases with H_1 and is such that the ratio of this pressure to atmospheric is the same as the ratio of pressure after heating to that before.

8. Cycles with letter C all go below atmosphere in expanding to such a pressure as will bring the temperature down to that originally existing in the gas. These resulting pressures after expansion are different for each cycle, but the lines representing them as functions of H_1 may intersect.

9. The lines for IV C may cross others, but I C, II C, III C cannot intersect, and these will always be in the order of magnitude II C, III C, I C, and all asymptotic to axis of H_1, so that the terminal pressure can never be zero.

10. An increase of compression will cause an increase in final pressure for same H_1.

11. Mean effective pressure expressed as a function of H_1 will give for every cycle and every different compression a different M.E.P. curve, but as before these may intersect.

12. For all cycles except those ending with isothermal return to the original state, the M.E.P. increases with H_1, but for those bearing the letter C the M.E.P. decreases and for no cycle is it constant.

13. For the same previous compression the cycles have M.E.P. of about the following order of magnitude when H_1 is large enough.

Greatest M.E.P., II A_2, 200; II, 40; I, 25; III, 15; II C, 1.5; I C, 0.3; III C, 0.2. When H_1 is small IV will probably come between III and II C.

14. A change in compression from $\frac{1}{2}$ to $\frac{1}{10}$ (vols.) may cause a change in II A_2 of 35 per cent, II of 100 per cent, III of 300 per cent for the same possible values of H_1.

15. The effect of changed compression before heating is the more marked on M.E.P. resulting when M.E.P. is lowest and the extent of the increase is greater with H_1.

195. Comparison of Cycles with Respect to Volumes After Heating.—Following the analysis of the effects of temperatures and pressures comes naturally a comparison of the volumes filled when a unit of heat is added in the various cycles. The following lines show this relation graphically for the phase of heating the gas:

VOLUMES AFTER HEATING BY H_1, B.T.U.

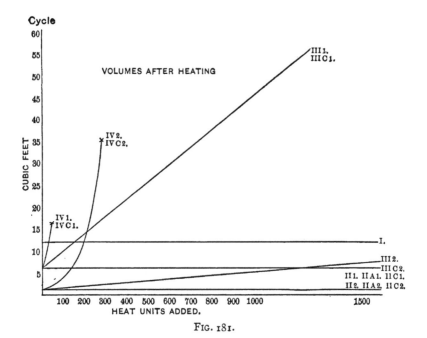

FIG. 181.

I, I C. $\qquad$ $v_c = v_a,$ . . . . (13)

II, II A, II C. $\qquad$ $v_c = v_b,$ . . . . (14)

III, III C. $\qquad$ $v_c = v_b Y = v_b \left(1 + \dfrac{H_1}{C_v T_b} \right),$ . . . . . (15)

IV, IV C. $\qquad$ $v_c = v_b e^Z = v_b e^{\frac{H_1}{(C_p - C_v) T_b}}.$ . . . . . (16)

Formula (13) is a straight line parallel to H_1 and is always less than (14), which is similar, but cuts axis of V_c at a point v_b higher than v_a. Equation (15) is a straight line inclined to H_1. Equation (16) is an exponential curve cutting axis V_c at point V_b; it is concave up and slopes up to the right as is shown by the derivatives

$$-\frac{dv_c}{dH_1}=\frac{v_b}{(C_p-C_v)T_b}e^{\overline{(C_p-C_v)T_b}}, \quad \cdot \ \cdot \ \cdot \ (17)$$

$$\frac{d^2v_c}{dH_1}=\frac{v_b}{(C_p-C_v)^2T_b^2}e^{\overline{(C_p-C_v)T_b}}. \quad \cdot \ \cdot \ \cdot \ (18)$$

These curves are shown in Fig. 179 for the two cases.

196. Comparison of Cycles with Respect to Volumes After Expansion.—A similar treatment gives the following plotted curves,(Figs. 182–189) for volumes after expansion:

VOLUMES AFTER EXPANSION.

Cycle.

I. $v_d=v_aX^{\frac{1}{r}}=v_a\left(1+\frac{H_1}{C_vT_a}\right)^{\frac{1}{r}}$ (Fig. 182)   . . (41)

I C. $v_d=v_aX^{\frac{1}{r-1}}=v_a\left(1+\frac{H_1}{C_vT_a}\right)^{\frac{1}{r-1}}$ (Fig. 183)   . (42)

II. $v_d=v_aX^{\frac{1}{r}}=v_a\left(1+\frac{H_1}{C_vT_b}\right)^{\frac{1}{r}}$ (Fig. 184)   . . (43)

II A. $v_d=v_a$ (Const.)   . . . . . (44)

II C. $v_d=v_aX^{\frac{1}{r-1}}=v_a\left(1+\frac{H_1}{C_vT_b}\right)^{\frac{1}{r-1}}$ (Fig. 185)   . (45)

III. $v_d=v_aY$ (Fig. 186)   . . . . (46)

III C. $v_d=v_aY^{\frac{1}{r-1}}=v_a\left(1+\frac{H_1}{C_pT_b}\right)^{\frac{r}{r-1}}$ (Fig. 187). . (47)

IV. $v_d=v_ae^{Y-1}=v_ae^{\frac{H_1}{C_pT_b}}$ (Fig. 188). . (48)

IV C. $v_d=v_ae^{Z}=v_ae^{\frac{H_1}{(C_p-C_v)T_b}}$ (Fig. 189)   . (49)

These curves will admit of considerable discussion, but the curves of Figs. 182 to 189 show at a glance all which it is necessary to know in general.

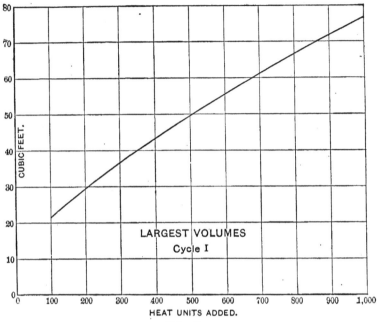

FIG. 182.

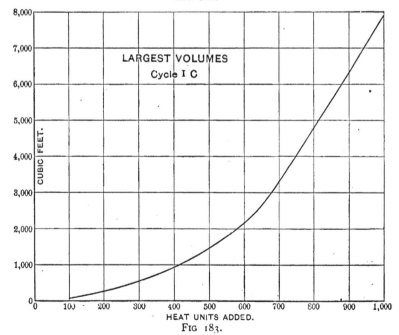

FIG 183.

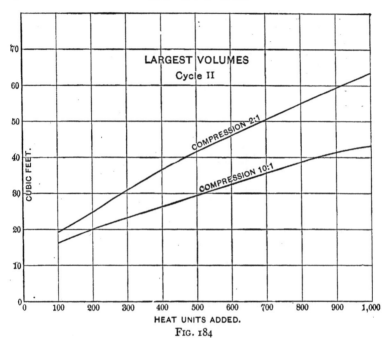

LARGEST VOLUMES
Cycle II

COMPRESSION 2:1

COMPRESSION 10:1

CUBIC FEET.

HEAT UNITS ADDED.

FIG. 184

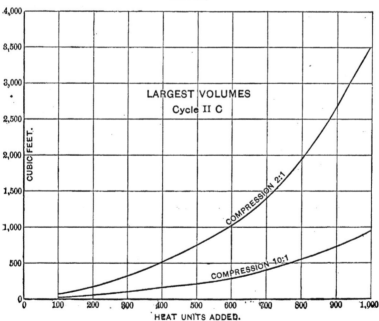

LARGEST VOLUMES
Cycle II C

COMPRESSION 2:1

COMPRESSION 10:1

CUBIC FEET.

HEAT UNITS ADDED.

FIG. 185.

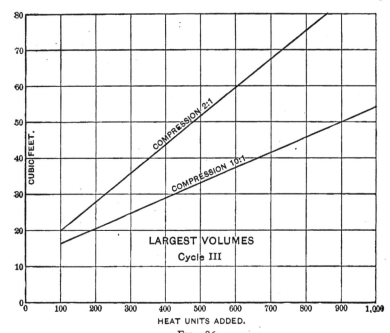

FIG. 186.

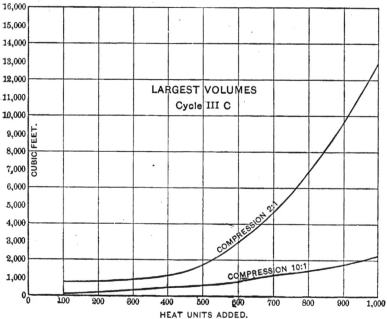

FIG. 187.

197. **Deductions from Comparisons of Cycles with Respect to Volumes.**—In Fig. 188 is presented a graphical comparison of mean effective volumes for certain cases of Cycles I, II, and III. A statement of the conclusions capable of being drawn from the curves would give (Figs. 188 to 190):

1. The volumes after heating are the same for cycles of the same group, and for all groups increase with H_1 except in Groups

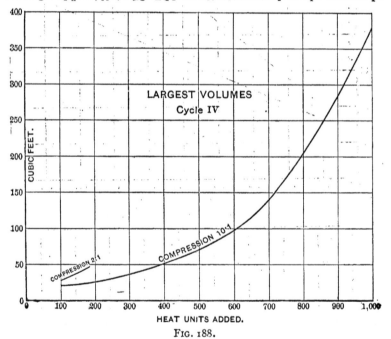

FIG. 188.

I and II, where by definition they are constant and equal to the volumes existing before heating.

2. In Group III the volumes after heating are proportional to H_1 with the same constant of proportionality for the same compression. Increase of compression decreases this constant of proportionality.

3. In Group IV the volumes increase rapidly with H_1, but are not proportional to H_1 so long as H_1 is small; with large H_1 Group IV cannot exist.

4. Lines of volumes after heating ɪepresented as functions of H_1 may cross in some cases. II, IV, and III may cross I, i.e., the compression cycles may cross the non-compression ones. But for the same compression II, III, and IV can never have the same volumes after heating. Lines of III and IV for high compres-

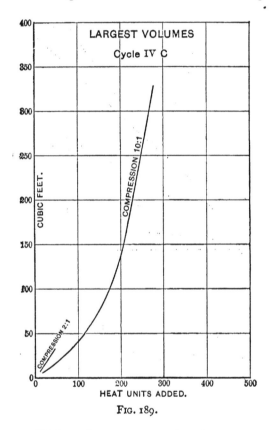

FIG. 189.

sion may cross II for a lower compression, but cannot cross each other.

5. For possible values of H_1 the volumes after heating for the different groups may have the following order of magnitude if H_1 is large enough: Group III, 55.00; Group I, 12.38; Group II, 6.00.

6. After expansion is completed the volume occupied by the gas in the different cycles will vary through very wide limits, increasing with H_1.

7. The volume occupied by Cycle III will be such as to keep the ratio between this final volume and the volume before com-

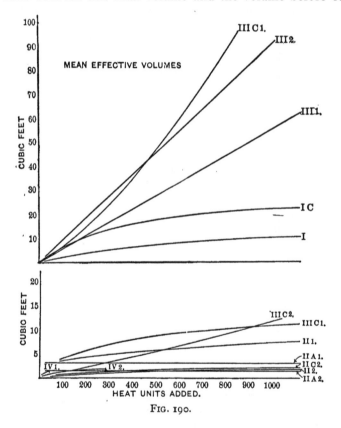

FIG. 190.

pression the same as the ratio of volume after heating to that before, and the final volume is proportional to H_1. The constant of proportionality is decreased by compression increase.

8. The final volume of Cycle II A is least and equal to that existing before compression.

9. When H_1 is large enough there may be a value for which the final volume may exist in the following order of magnitude:

III C_1, 7000.00; I C, 4200.00; II C, 2300.00; III, 75.00; I, 65.00; II, 51.00; II A_2, 12.38. A change of compression by which the volume after compression is one fifth that for the previous case may change this list to the following: III C, 1000.00; I C, 4200.00; II C, 500.00; III, 40.00; I, 65.00; II, 34.00; II A_2, 12.38.

10. The mean effective volumes increase with H_1 for all cycles except II A_2, in which this variable is constant.

11. For Cycle III the M.E.V. is proportional to H_1, and increase of compression increases the constant of proportionality.

198. Comparison of Cycles with Respect to Heat Discharged or Abstracted. Work Done. Efficiencies.—A comparison of the several cycles from this point of view leads at once to the deduction concerning their relative efficiency. Plotting the curves from the following equations:

Cycle.

I.
$$H_2 = C_p T_a (X^{\frac{1}{\gamma}} - 1), \quad \ldots \ldots \ldots \ldots \quad (50)$$

I C.
$$H_2 = T_a C_v \log_e X = T_a C_v \log_e \left(1 + \frac{H_1}{C_v T_b}\right), \quad \ldots \quad (51)$$

II.
$$H_2 = C_p T_a (X^{\frac{1}{\gamma}} - 1), \quad \ldots \ldots \ldots \ldots \quad (52)$$

II A_2.
$$H_2 = C_v T_a (X - 1) = \frac{H_1}{\gamma^{\gamma - 1}}, \quad \ldots \ldots \ldots \quad (53)$$

II C.
$$H_2 = C_v T_a \log_e \left(1 - \frac{H_1}{C_v T_b}\right), \quad \ldots \ldots \ldots \quad (54)$$

III.
$$H_2 = C_p T_a (Y - 1) = \frac{H_1}{\gamma^{\gamma - 1}}, \quad \ldots \ldots \ldots \quad (55)$$

III C.
$$H_2 = C_p T_a \log_e \left(1 + \frac{H_1}{C_p T_b}\right), \quad \ldots \ldots \quad (56)$$

IV.
$$H_2 = C_p T_a (e^{Y - 1} - 1) = C_p T_a (e^{\frac{H_1}{C_p T_b}} - 1), \quad \ldots \quad (57)$$

IV C.
$$H_2 = \frac{H_1}{\gamma^{\gamma - 1}}, \quad \ldots \ldots \ldots \ldots \ldots \quad (58)$$

the following curves result.

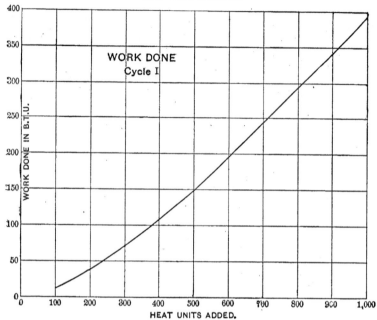

FIG. 192.

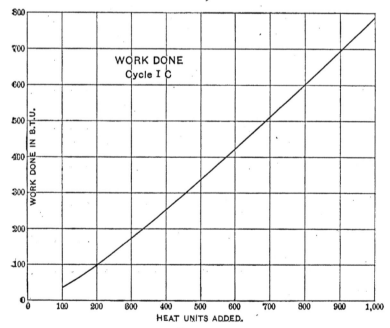

FIG. 193.

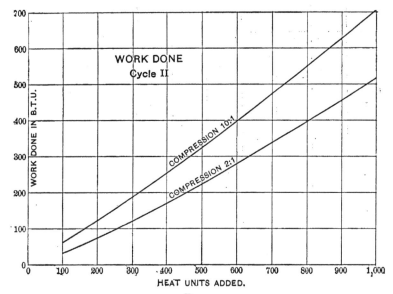

FIG. 194.

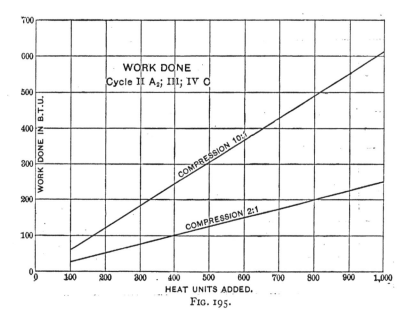

FIG. 195.

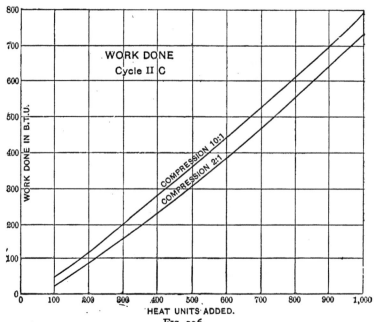

FIG. 196.

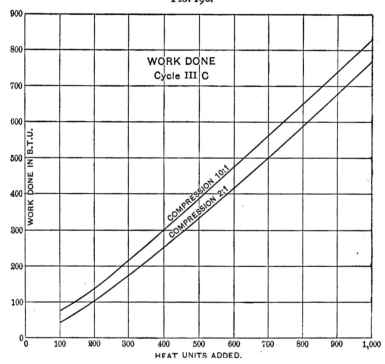

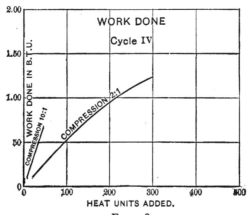

FIG. 198.

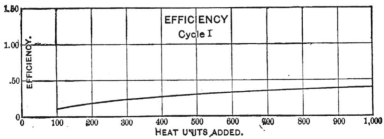

FIG. 199.

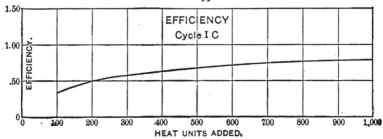

FIG. 200.

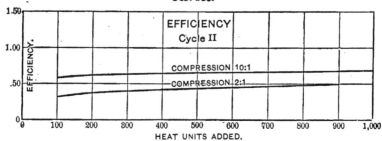

FIG. 201.

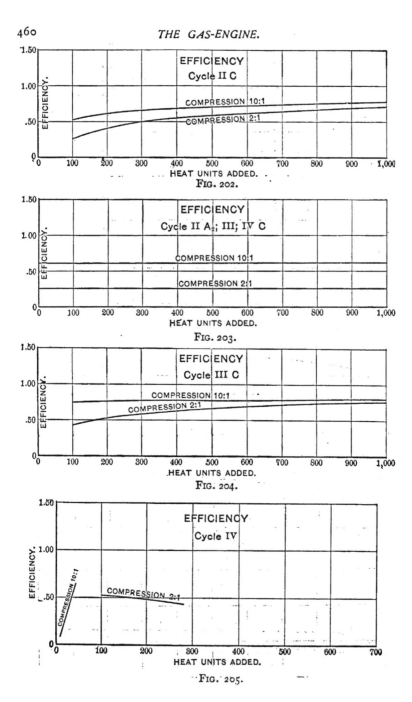

FIG. 202.

FIG. 203.

FIG. 204.

FIG. 205.

The work done in the various cycles in B.T.U. is shown by Figs. 192 to 198, and the efficiencies in Figs. 199 to 205.

Equations (53), (55), and (58) are identical, that is, these three cycles will discharge the same amount of heat and have the same efficiency; moreover, this efficiency will be independent of everything but the compression. These three cycles have, further, a common property not seen by the formula, but from their definitions each receives and discharges all its heat according to the same law.

Cycle II A receives all heat at constant volume and discharges all at constant volume.

Cycle III receives all heat at constant pressure and discharges all at constant pressure.

Cycle IV C receives all heat at constant temperature and discharges all at constant temperature.

A consideration of the above would seem to warrant the proposition:

When all the heat is discharged according to the same law under which it was received, then the cycle will have an efficiency independent of everything but the previous compression and will be given by

$$E = 1 - \frac{1}{r^{\gamma-1}}.$$

We may remark here that as IV C is the Carnot Cycle we can state that Cycles II A$_2$ and III have the same efficiency as the Carnot Cycle with same previous compression. This is an important supplementary to the old theorem that the Carnot Cycle has the highest efficiency for its temperature range.

The relation between the other values of H_2 are best shown by the curves of Figs. 199 to 205 by implication.

The following comparisons will be interesting and useful:

6. For Cycles II A$_2$, III, IV C the efficiency is a function of the adiabatic compression only and the same function for each.

It is independent of the amount of heat supplied, i.e., is not a function of H.

7. For all cycles the efficiency increases with the compression, but not according to the same law.

8. For Cycles IV, IV A, IV B the efficiency decreases with increase of heat added to the same mass of gas.

9. For all other cycles except II A_2, III, IV C the efficiency increases with H_1, but according to different laws, so that the distance between efficiency curves will vary.

10. For these cases a change in H_1 will produce more effect when H_1 is small than when it is large.

11. After heat has been added the efficiency will vary with the degree of expansion. Cycle II, therefore, will have an efficiency always higher than II A and lower than II B or II C.

12. Cycles in which an adiabatic compression precedes heating will always have a higher efficiency than those lacking this compression, other things being equal.

13. For the same initial conditions and same heat added, if H_1 is large enough Cycle II C will always have the highest efficiency, and then come in order III C; I C; II $\left\{ \begin{array}{l} \text{II A} \\ \text{III} \\ \text{IV C} \end{array} \right\}$, always remembering that IV, IV A, IV B, IV C cannot exist if H_1 be large.

14. The difference in efficiency between the curtailed expansion of Cycle II A_2 and that of II increases with the amount of heat, the difference being small when H_1 is small, and greater as H_1 increases, the greatest possible being about 12 per cent.

15. Expanding Cycle II to original temperature, making Cycle II C, may increase the efficiency from 5 to 15 per cent approximately for possible values of H_1.

16. Cycle III may add by expansion to original temperature as much as 25 per cent to the efficiency for possible values of H_1.

17. Cycles IV, IV A, IV B have an efficiency decreasing with increase of H provided H remain small; when H passes a certain limit the cycle ceases to be possible.

ɩ8. A change in the volume ratio of compression from $\frac{1}{2}$ to $\frac{1}{10}$ will increase the efficiency of the cycles as follows for possible values of H_1:

Cycle II........30–20 per cent approximately, depending on H_1.

" II A$_2$
" III $\Bigg\}$ ......35 per cent approximately, depending on H_1.
" IV C

ʻ II C.......40–5 per cent approximately, depending on H_1.

199. General Conclusions from the Analysis of Cycles.—Certain useful general conclusions may be drawn from the foregoing analysis, beside the specific ones referred to under their appropriate titles:

If the cycle consists of a series of operations, or pressure-volume-temperature changes resulting in a return to the original state of pressure, volume, and temperature, then:

1. The P.V.T. at any point of a cycle depends on: (*a*) The cycle itself qualitatively considered, i.e., the nature and order of succession of the processes or phases already completed; (*b*) the extent or intensity of each phase of the cycle quantitatively; (*c*) the amount of heat, H, added before reaching the point considered. For example, the temperature at the end of combustion will be different for different cycles, and will vary with the compression before heating, the law of compression, and the amount of heat added.

2. The part of the total heat transformed into work is a function of the cycle, and will vary with the order, nature, and extent of the cyclic phases, except when all the heat is added and all abstracted according to the same law.

3. When the laws of heating and of cooling are identical, then the part of the total heat supplied which becomes transformed into work is constant for the same previous compression, and this resulting efficiency is a function of the previous compression only when these other two phases, compression and expansion, completing the cycle, have likewise the same law.

4. The range of changes in pressure, volume, and temperature

is different for different cycles, and in any one cycle will depend on the amount of heat added.

5. While the variations noted do in general hold, yet in the different cycles each variable may be a different function of H_1, so that two or more curves may intersect, and for that particular value of H_1 the variable will have the same value in two or more different cycles simultaneously.

From the data here set down the selection of a cycle on purely ideal grounds can be made with a full knowledge of all the conditions surrounding the selection; that is, knowing what results are desired, the cycle which, theoretically, ideally, or mathematically considered, gives the results can be found, and in addition it is easy to see what accompanying circumstances are inevitable. If that cycle which transforms the greatest amount of heat into work ideally is wanted, it is readily seen that II C with as high compression as possible must be selected, but it is also evident that a very large volume range must be submitted to. If that cycle with the lowest temperature range is wanted, then any of Group IV must be taken.

If a cycle is desired which will convert of any amount of heat the same proportion into work, then any one of II A, III, or IV C may be chosen, but of these one has the lowest pressure range, another the lowest temperature range, and the last the lowest volume range.

For example, it is from a comparison such as this that the following suggestions are derived:

(A) Cycle I and its variations, by reason of its poor showing in efficiency and mean effective pressure as compared with the previous-compression Cycle II, must be set aside.

(B) The atmospheric cycles, by reason of their low mean effective pressure and consequent large volume range, are useless for power purposes as compared with the other cycles.

(C) This leaves as the only cycles worthy of application II, III, IV, and their variations.

(D) Of the last mentioned, the three which are peculiar, Cycle II A_2, Otto, heating and cooling the gas at constant volume; Cycle III, Brayton, heating and cooling the gas at constant pressure; and Cycle IV C, Carnot, heating and cooling the gas at constant temperature, have the same efficiency for the same compression, and should consequently, with the same heat supplied, do the same work.

The efficiency of each is given by

$$E = 1 - \left(\frac{V_b}{V_a}\right)^{r-1},$$

where V_a is the volume before compression,

V_b " " " after "

r " " ratio of specific heats, and for air $r = 1.406$.

(E). The other cycles, II, II B and C; III A and B; IV, IV A and B, can easily be given their proper comparative position by remembering that each is a more or less complete expansion of one of the above three. For example, if in the Otto the expansion were carried to atmospheric pressure, the efficiency would be greater than for the Otto. Similarly with the Carnot, if the expansion were stopped at atmospheric pressure, as was first suggested by Diesel, the resulting Cycle IV would have an efficiency less than the Carnot, and hence less than either the Otto or Brayton cycles.

(F) If, as respects the other variables entering each of the cycles adopted for comparison, there be assumed   .

The same { mass of gas, heat supplied after, compression,

there will result for

Cycle II A, Otto,
    " III, Brayton,
    " IV C, Carnot,   } same work done, and hence same efficiency.

And, further,

	Lowest.	Intermediate.	Highest
Maximum temperature.	Carnot	Brayton	Otto
Pressure range...............	Brayton	Carnot	Otto
Volume range.	Otto	Brayton	Carnot
Temperature range...........	Carnot	Brayton	Otto
Mean effective pressure.	Carnot	Brayton	Otto
$\dfrac{\text{Pressure range}}{\text{Mean effective pressure}}$	Brayton	Carnot	Otto
Mean effective temperature. ...	Carnot	Brayton	**Otto**

The relation of the Diesel to the Otto and Brayton is easily seen if it be recalled that it is an imperfect Carnot.

11. Some of these variables should be a maximum and some a minimum. For the maximum temperature the Carnot holds first place, but its impracticability yields the place to Brayton. Neither pressure range nor mean effective pressure is wanted by itself, but only the ratio between them, for it is to this ratio that the weight of the engine must be approximately proportional; here Brayton holds the most favorable place.

Volume range should be low, and here first place is held by the Otto. The mean effective temperature should be low, and the Brayton is exceeded only by the Carnot.

The low mean effective pressure of the Carnot, and all other isothermal combustion cycles, is sufficient warrant for cutting them out of consideration in comparison with the Cycles II, III, and their variations.

The conclusion is thus reached that, theoretically, the last-named cycles only are worthy of further consideration.

12. In the above, the hypothesis that heat could be added to the gas has been assumed, and no account taken of the means of so doing, but this point needs consideration. If heat be added through walls from a source of known supply, of which we can control and use as much or as little as we please, there will be no alteration in the formulæ or results of the analytical comparison; but the internal-combustion method of heating presents some new questions for solution. First, the air and fuel become carbonic acid, steam, etc., and as to what value of the specific heat should be used, who can say? (Par. 55.) Second, the chemical change is accompanied by an intrinsic volume change. (Par. 14.) Third, there may be reasons why the fuel should give out more heat when burned in one way than when burned in another.

13. The only ways of heating by internal combustion which are worth anything for power are the constant-volume and constant-pressure methods. On theoretical grounds there is no reason

for saying that, for any particular system of combustion, more heat can be developed one way than the other. The evidence that heat has been added to a mass of gas in an engine is, for the two cases, (A) an increase of pressure, and (B) an increase of volume. This pressure increase on the one hand and volume increase on the other can be readily observed by indicators, and the results of these observations on a large number of indicator-cards show that the increase is not what it should be if all the calorific value of the fuel had developed.

In short, there is in practice abundant evidence of heat suppression, and whether this be due to radiation, conduction, dissociation, or an increase of specific heat, or to an actual non-production of heat is unknown. What is known and can be asserted is that the effects on pressure and volume are such as they would be if only a part of the heat supposed to be generated had appeared. The result might be worked up to give a new value to the heating power of the fuel, to be called its *effective* calorific value, or a new value given to the specific heat, to be called the *effective* specific heat of the process.

14. For constant-volume combustion the value for H_1, the British thermal units per pound of mixture, will be derived from the equation

$$\frac{p_2}{p_1} = \frac{T_2}{T_1} = 1 + \frac{H_1}{C_v T_1},$$

where p_1 = pressure before compression;
T_1 = temperature before combustion;
p_2 = pressure after combustion;
T_2 = temperature after combustion;
C_v = specific heat at constant volume.

This ratio in the general run of gas-engines will average about 3.5. In some cases it may reach 4, but it seldom has reached 6. Some values are given below:

Engine.	$\dfrac{p_2}{p_1}$	Remarks.
Westinghouse.	3	On gas
Otto.	4.5.	N. Y. gas
Hornsby-Ackroyd.	3.5	Kerosene
Nash.	4	N. Y. gas
Clerk.	4	Glasgow gas
Crossley.	3	Dowson gas
Priestman.	3.5	Kerosene
Crossley oil.	3.5	Kerosene

A general statement, very nearly true, would give these pressure and temperature ratios about 50 per cent of what the usual values of H_1 and C_v would produce. These figures, while not strictly true for any one case, give a fair average value.

15. The other system of combustion—that at constant pressure—may be observed in the same way. The only indicator-card available from this type of engine was taken from a Brayton oil-engine with its smoky fire. The volume ratio, in this case, is quite well given by the relative lengths of the delivery line of the compressor and the admission line of the power cylinder, and is given by

$$\frac{v_2}{v_1} = 3.2.$$

To compare this with the pressure ratios given. Theoretically,

$$\frac{v_2}{v_1} = \frac{T_2}{T_1} = 1 + \frac{H_1}{C_p T_1},$$

where C_p is the specific heat at constant pressure and the other symbols are as heretofore; combining this with the similar one for the other type, we get

$$\frac{H_1}{T_1} = C_v\left(\frac{p_2}{p_1} - 1\right) = C_p\left(\frac{v_2}{v_1} - 1\right),$$

or

$$\frac{p_2}{p_1} = 1 + \gamma\left(\frac{v_2}{v_1} - 1\right).$$

$$\frac{v_2}{v_1} = 3.2, \quad \text{``} \quad \frac{p_2}{p_1} = 4.44.$$

16. This shows that when a Brayton engine gives a volume ratio in combustion of 3.2 there is evidence of as much heat as would cause a pressure ratio of 4.44 in an explosion engine; hence it would seem that, for the combustion process alone, the Brayton engine, even with its poor fire, was giving evidence of as much heat as the very best explosion engine, and more than can most of them. This point is very striking, and, in order to verify or disprove it, a large mass of data is necessary, which can be collected only after considerable time.

The above point bears strongly on the formulæ of cyclic comparison. The analysis showed that the Otto and Brayton cycles must have the same efficiency for the same heat added; but if one, by reason of its system of combustion, can take from the fuel more heat than the other, then that one must have the higher efficiency in practice, assuming equal subsequent heat losses and equal friction losses in the mechanism.

200. Formula for Theoretical Mean Effective Pressure. Otto Cycle.—A most serviceable deduction can be made from the analysis of the cycle for Groups II in which the expansion and compression curves are similar between two terminal verticals

(Fig. 206). The mean effective pressure will be the area from the diagram under the expansion curve diminished by the area under the compression curve and divided by the length of the diagram between verticals. From paragraph **51** the mean effect-

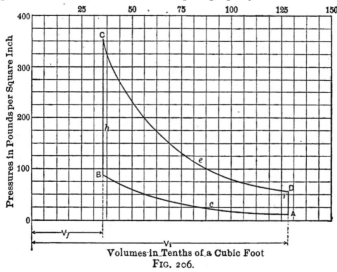

FIG. 206.

ive pressure will be for the two adiabatics *CD* and *AD*, with the same value for *n* in each,

$$\text{M.E.P.}_{CD} = \frac{p_c}{n-1} \times \frac{1}{r-1}\left[1 - \frac{1}{r^{n-1}}\right].$$

$$\text{M.E.P.}_{AB} = \frac{p_b}{n-1} \times \frac{1}{r-1}\left[1 - \frac{1}{r^{n-1}}\right].$$

The difference will be the mean working pressure on the piston, or

$$\text{M.E.P.} = \frac{p_c - p_b}{n-1} \times \frac{1}{r-1}\left[1 - \frac{1}{r^{n-1}}\right]$$

$$= \left(\frac{p_b}{n-1}\right)\left(\frac{p_c}{p_b} - 1\right)\frac{1}{r-1}\left[1 - \frac{1}{r^{n-1}}\right].$$

But the line $p_c - p_b$ measures the increase of pressure due to the ignition of the charge, and the ratio $\dfrac{p_c}{p_b}$ is the ratio of the compression pressure to the explosion pressure and must, therefore, depend on the fuel varying with it, and having a definite relation to the calorific power. From paragraphs 14 and 20 it became clear that for a fuel of Q calorific power in B.T.U. the burning of y pounds in x pounds of air gave a temperature increase of

$$T_c - T_b = \frac{y}{x+y} \times \frac{Q}{C_v}.$$

If this be divided by T_b,

$$\frac{T_c}{T_b} - 1 = \frac{y}{x+y} \times \frac{Q}{C_v T_b}.$$

But $\dfrac{T_c}{T_b} = \dfrac{p_c}{p_b}$; whence

$$\frac{p_c}{p_b} - 1 = \frac{y}{x+y} \times \frac{Q}{C_v T_b}.$$

Substituting this, there results the formula for M.E.P. first suggested by Lucke,

$$\text{M.E.P.} = \frac{p_b}{n-1} \times \frac{yQ}{(x+y)C_v T_b} \times \frac{1}{r-1}\left[1 - \frac{1}{r^{n-1}}\right],$$

which should be the mean effective pressure resulting when one pound of a mixture having a fuel value Q is compressed from a volume v_i to a volume v_f and it is ignited and then allowed to expand down to the original volume without losses in the process.

If instead of weights as in the foregoing, it be preferred to use volumes, so as to make the data of the table in paragraph 29 more immediately available, the formula for mean effective pressure used in paragraphs 29 and 40 may be used involving the calorific value of the mixture of fuel and air per cubic foot taken into the cylinder in the aspiration stroke. This formula is

$$\text{M.E.P.} = \frac{778}{144}\left[1 - \left(\frac{P_a}{P_b}\right)^{.29}\right]\frac{H}{a+1}$$

$$= \frac{H}{a+1}\left[\begin{matrix}\text{Tabular Value com-}\\ \text{puted in Par. 152.}\end{matrix}\right]$$

The diagram in Fig. 205 is the standard type reference diagram of the Otto cycle (pars. 184, 40, 50, and 152), using the constants for air to give quantitative results. The points A and B are found from the data of paragraphs 11, 47, and 152, and the tabular values computed in the table of the latter paragraph. The length of the vertical BC or the line of ignition causing

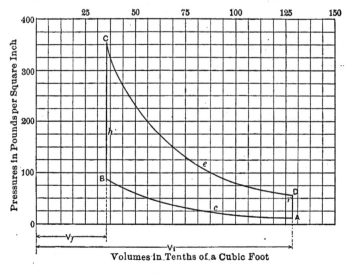

FIG. 205.

increase of pressure and temperature at a constant volume is deduced from the data of paragraph 46. If the quantity of heat given to one pound of air by the ignition of the fuel be:

$$H_t = C_v (T_1 - T_3),$$

then by the Gay-Lussac-Mariotte law (pars. 44, 45):

$$\frac{p_1 v_1}{T_1} = \frac{p_3 v_3}{T_3}$$

whence

$$\frac{p_1}{p_3} = \frac{T_1}{T_3} = \frac{H_t}{C_v T_3} + 1$$

since

$$\frac{V_3}{V_1} = 1 \text{ and } \frac{T_1}{T_3} - 1 = \frac{H_i}{C_v}.$$

Multiplying through by p_3 it is also true that

$$p_1 - p_3 = \frac{H_1}{C_v} \times \frac{p_3}{T_3}.$$

But since

$$p_3 v_3 = RT_3$$

and

$$R = 778 \, (C_p - C_v)$$

and

$$\frac{V_3}{V_0} = \left(\frac{p_0}{p_3}\right)^{\frac{1}{n}}$$

it follows that

$$\frac{p_3}{T_3 C_v} = \frac{778 \, (C_p - C_v)}{C_v \left(\frac{p_0}{p_3}\right)^{\frac{1}{n}}}$$

$$= \frac{778 \, (n - 1)}{V_0 \left(\frac{p_0}{p_3}\right)^{\frac{1}{n}}}$$

where the pressures are in pounds per square foot. Dividing by 144 to reduce to pressures in pounds per square inch and substituting for $\frac{p_3}{T_3 C_v}$ its value $\frac{p_1 - p_3}{H_i}$ it follows that

$$p_1 - p_3 = 2.21 \, \frac{H_i}{V_0} \times \frac{1}{\left(\frac{p_0}{p_3}\right)^{.71}}.$$

But the ratio

$$\frac{H_i}{V_o} = \frac{\text{B.T.U. per pound of the mixture}}{\text{Volume of mixture in cubic feet}}$$

and hence

$$= \text{B.T.U. per cubic foot of mixture}$$

which is the quantity tabulated in columns 11 and 12 of the table in paragraph 29, and which is there called $\dfrac{H}{a + 1}$ if there are no neutrals present. When neutrals are presented to dilute the mixture the quotient should be less heat per cubic foot, or the denominator should be $\dfrac{H}{n + a + 1}$. Substituting this, the pressure rise from ignition should be

$$p_1 - p_3 = 2.21 \left[\frac{H}{n + a + 1} \right] \times \frac{1}{\left(\dfrac{p_0}{p_3} \right)^{.71}}$$

If there is no compression of the mixture so that $\dfrac{p_0}{p_3}$ is unity the rise of temperature is less as has been elsewhere established.

If the volume of neutrals be taken as filling the clearance volume completely and at the temperature (62° F.) of the incoming mixture to eliminate the temperature factor, no essential error will be made. In this case n will be $N = V_f$, and

$$\frac{N}{a + 1} = \frac{V_3}{V_1 - V_f}$$

whence

$$p_1 - p_3 = \frac{2.21}{\left(\dfrac{p_0}{p_3} \right)^{.71}} \times \frac{H}{\left(1 + \dfrac{v_f}{v_1 - v_f} \right) a + 1}.$$

In this by substituting for $\dfrac{v_f}{v_1 - v_f}$ its equivalent $1 - \left(\dfrac{p_0}{p_3} \right)^{.71}$ the expression becomes

$$p_1 - p_3 = 2.21 \frac{H}{a + 1} \times \left(\frac{1}{\left(\dfrac{p_0}{p_3} \right)^{.71}} - 1 \right)$$

in which the tabular values may be substituted from paragraphs 29 and 152. When the position of the point C is thus found the location of the theoretical expansion line is fixed and the curve drawn by the relation $p_x v_x^{1.41} = p_3 v_3^{1.41}$ in which the second member is given, and each pressure found by dividing this quantity by the assumed volume. Such a standard diagram can then be taken as a standard or as unity of possible performance. The actual performance will be some fraction of the possible air-card standard; or the air-card area or performance should be multiplied by a factor less than unity, to give the actual performance. Such a multiplier may be called as in steam-engine design, the "diagram factor," and some values for it will be given in the next paragraph after investigating the causes which increase the size of its denominator.

If the first formula above for M.E.P. be applied to an example with the following data:

$$T_a = 62° \text{ F.} + 460 = 522°;$$
$$p_a = 14.7 \text{ lbs. per sq. inch or } 2116.4 \text{ per sq. foot;}$$
$$1 \text{ cu. ft. gas weighing } .032 \text{ lbs.;}$$
$$1 \text{ cu. ft. air weighing } .078 \text{ lbs.,}$$

so that 5.6 cubic feet of air, the best volume for one volume of gas weighs .437 pound, and 6.6 cubic feet of the mixture will weigh .437 + .032 = 0.469 pound, whence 1 cubic foot of the mixture will weigh $\frac{.4690}{6.6} = .071$ pound, or the volume of one pound of the mixture $v_a = 14$ cubic feet. Let the value for n be taken the same as for air $1.4, C_v = 0.1689$, and suppose the ratio $\frac{v_a}{v_b} = 5$. Then

$$p_b = p_a(5)^{1.4} = 2116.4 \times 9.5 = 20,106 \text{ lbs. per sq. ft.}$$
$$= 14.7 \times 9.5 = 140 \text{ lbs. per sq. in.;}$$
$$T_b = T_a(5)^{.4} = 522 \times 1.9 = 992° \text{ absolute;}$$
$$r^{n-1} = (5)^{.4} = 1.9.$$

For this gas $\frac{yQ}{x+y} = 1600$; whence

$$\text{M.E.P} = \frac{140}{.4} \times \frac{1600}{.1689 \times 992} \times \frac{1}{4}\left[1 - \frac{1}{1.9}\right] = 392.$$

If the specific heat be taken as that for a composition of the products of combustion made up of one-third CO_2 and the rest steam-gas, the M.E.P. will be less as this greater value for the specific heat comes in the denominator. Or again, if the experimental or effective values for the specific heat be taken as given in paragraph 55, then the M.E.P. comes lower, or a little over 100 pounds per square inch, which, while still higher than praetice values, is much nearer than the computation above will give. What are the factors explaining this loss of heat?

201. Factors Reducing Computed Mean Effective Pressure.— Diagram Factor.—Some of these have been already referred to in other connections, but are here recapitulated and supplemented.

1. The pressure in the cylinder at the time when the volume is v_a is not that of the atmosphere, but is below it. Hence the full weight of the mixture is not really present. The causes for this diminished weight include:

(a) Friction in the suction valve and the piping and passages, if tortuous or small.

(b) The delay in opening of the valve if an automatically lifting one. It will not open until the pressure on top is below that underneath it.

(c) The inertia of the valve and of the column of air and fuel outside the valve.

2. Loss of weight in the charge by heating as it enters while the valve is still open to atmospheric pressure. This heating may be effected by the hot metal, or by the hot products of combustion trapped in the clearance volume. High compression and lessened clearance volume diminish this loss.

3. The compression may be nearer isothermal than adiabatic by reason of low jacket temperature taking off heat in compression and lowering the value of p_b. This loss will be less with hot jacket water and at high speeds of the piston, but the hotter walls will increase the loss in No. 2.

4. The ignition line instead of being vertical may be inclined by a retarding of the time of firing the charge. Compare the discussion of governing by ignition in paragraph 115. The indicator-diagram with retarded ignition (Figs. 86 and 87) shows a lower pressure at its maximum point because the mixture had partly lost its compression pressure before it was ignited.

5. The presence of diluent neutral gases from a previous stroke, or an impoverished or excessively rich mixture, will delay the propagation of the flame in the mixture, increasing loss No. 4.

6. The degree of the compression before explosion. The higher the compression pressure the higher the value for p_c when other things are equal. But the qualities of the fuel cannot be disregarded here with respect to the temperatures at which the heat caused by compression will cause them to ignite, perhaps before the completion of the compression stroke. With common illuminating-gas from coal or with gasoline a compression to 90 pounds or 6 atmospheres should not be exceeded; with kerosene vapor the ignition or pressure limit is at $3\frac{1}{2}$ to 4 atmospheres with hot cylinder-walls, while the weaker gases from producers or from blast-furnaces may be compressed to 15 atmospheres without pre-ignition.

7. The length of the explosion line BC (Fig. 205) will be fixed by the character of the fuel, other things being equal. With the compression pressure p_b fixed by the limits set in No. 7 the ratio of the explosion pressure p_c to it will be:

$\dfrac{p_c}{p_b}$ for weak producer and furnace gas $= 2$;

" " illuminating- or coal-gas $= 2.5$ to 4;

" " natural gas and carbureted gas $= 3$ to 4.5;

" " gasoline $= 3$ to 5;

" " kerosene $\begin{Bmatrix} \text{quite variable in en-} \\ \text{gines using injection} \\ \text{and other systems} \end{Bmatrix} = 3$ to 6.

The ratio can be made anything less than this by incomplete combustion from any cause. For instance, gas- or gasoline-engines which should normally give a ratio of 4 can easily be brought down to an actual relation of 1.5, as revealed by applying an indicator.

8. The expansion line may have its pressure ordinates lowered by conduction of heat to the walls and jacket.

9. The exhaust-valve is almost universally made to lead the piston slightly so as to open before the stroke ends and relieve the resistance against the return or exhausting stroke from the very start. This causes a loss of work area to the diagram.

10. The exhaust stroke should be made against atmospheric pressure only. When the exhaust area is small at the valve or in the passages an unnecessary resistance acts like a brake to diminish the power of the engine. Mufflers may produce this effect, but the loss here is not so much the direct consumption of power as it is the effect produced on the succeeding charge as treated in loss No. 5.

These causes of loss are of importance not only in operating, but as affecting design of new work. The necessary disagreement between theory and practice at this point opens a promising field for research and experiment, so as to ascertain **for** each class of engine the factor or coefficient by which the theoretical mean effective pressure is to be multiplied, so that the formula shall represent actual output from assumed values of heat energy supplied. An empirical formula of this sort (Grover) makes

$$\text{M.E.P.} = 2p_b - 0.01 p_b^2,$$

in which as before p_b is the pressure at end of compression. (See par. 40.)

It will be plain in reviewing these occasions of loss that some will be inherent in the motor by reason of lack of skill or experience in the designer, and others from lack of these same qualities in the operator. If the valves are responsible for the failure to receive the full weight of charge, as in No. 1, it may be because they are:

(*a*) Of too small diameter.

(*b*) Not caused to lift enough.

(*c*) Fitted with too heavy or stiff a spring.

(*d*) Fitted with badly designed cams.

(*e*) Opened by cams improperly set.

(*f*) Controlled by a governor poorly designed, or working badly from lack of care.

Defective valve design or adjustment may also cause:

(*g*) One cylinder with exhaust at high pressure, because just beginning, flowing back into another just finishing its exhaust stroke.

(*h*) Excess of back-pressure preventing the entry of new mixture under its low pressure.

(*i*) In multiple-cylinder engines, as in the motor-car, the shape of the inlet passages may preclude equable distribution of fuel mixture to all of the cylinders, so that some get less than others. The passages may be tortuous or crooked.

(*j*) The speed of the piston may be too high in relation to areas, or for the inertia of the mixture itself to be overcome and permit motion of air and fuel.

But, granting the charge is of full weight, the events in the cylinder or adjustments of the carburetor may preclude the charge getting its chance to work. Leakage past the piston is of all others the most fatal of these defects. It not only lowers the mean effective pressure directly, but by the loss of compression pressure it acts indirectly also. Defective mixture in relation to the piston speed and too early release by opening the exhaust prematurely, add to the effect of incorrect timing of the ignition. If the breech end, or the volume of the clearance for combustion of the mixture is unskilfully moulded, the ignition pressure possible for that mixture is not reached, and the mean pressure is less than it might be if the maximum is lowered. If the charge is partial at the outset, by reason of throttling or governing action to reduce the power of the stroke, the proportionate effect is increased and the mean effective pressure falls, therefore, faster in this case. What allowances should, therefore, be made quantitatively for these actions, and what values should be given to the diagram factor? The following table (Lucke) brings together some accepted comparisons of theoretical air-card values with the observed values from the actual card. The average seems to lie between 50 per cent and 65 per cent, provided the engine pistons and valves are fairly tight, and the mixture, and ignition are properly adjusted, and neither back-pressure nor suction throttling are present.

TABLE OF VALUES OF THE DIAGRAM FACTOR.*

1	2	3	4	5
Fuel used in motor.	Compressive Pressure Absolute. Pounds per sq. in.	Ratio of Ob. Pressure / A. c. Pressure	Range of Average Compressions.	Range of Average Ratios.
		Per cent		
Gasoline vapor...............	40	58	80–130	50–30
" "	75	56	...	...
" carburetor	75	40	...	...
Kerosene vapor	65	60	45–75	30–40
" injected	50	40	...	20
Carbureted water-gas........	90	58	...	...
" "	60	59	60–90	45
Coal gas	60	60	...	...
" "	100	52	80	45
Natural gas	115	61	90–140	52–40
Mond gas	120	51	100–160	56–40
Blast-furnace gas...........	184	55	130–180	48–30

202. Design of Cylinder Volumes. — The formula for the mean effective pressure leads directly to the choice of the cylinder volume for a required horse-power to be developed. The accepted formula for a piston motor (par. 40) takes the form

$$\text{H.P.} = \frac{PLAN}{33,000},$$

in which for the gas-engine P is the mean effective pressure just discussed, and N is the number of working strokes or ignitions occurring in one minute. The ratio of d, the diameter of the cylinder $(\pi \frac{d_2}{4} = A)$, to the length of the stroke L has been generally conceded to lie between $d = L$ and $L = 2d$. A stroke of twice the diameter, however, which is quite usual with other media, is rarely encountered in small gas-engine design, and

* The diagram factor for two-cycle power cylinders may be taken at 0.8, that of four-cycle cylinders to which the above table refers. The rate of M.E.P. increase seems to decrease as the compression increases.

preference seems to centre around $L = 1.25d$ and $L = 1.3d$. With P and N assumed, and L expressed in terms of d, the equation can be solved for d, which will give the piston displacement. For example, if the piston speed be assumed to be 500 feet per minute, $p_b = 80$ pounds, and $S = 1.5d$, the computation for a 25-H.P. engine will be

$$\text{R.p.m.} = \frac{\text{piston speed}}{2 \times \text{stroke}} = \frac{500}{3d},$$

giving for an Otto cycle,

$$\text{Explosions per minute} = \frac{\text{R.p.m.}}{2} = \frac{500}{6d}$$

Using the empirical formula for M.E.P.,

$$\text{M.E.P.} = 2p_b - 0.01 p_b{}^2$$
$$= 160 - [0.01 \times 6400]$$
$$= 96 \text{ pounds per sq. in.}$$

Whence, since $\text{H.P.} = \dfrac{PLAN}{33,000}$,

$$25 = 96 \times \frac{3D}{2} \times \frac{D^2\pi}{4} \times \frac{500}{6D},$$

or

$$D^2 = 87.5 \text{ about,}$$

and

$$D = 9.5 \text{ nearly,}$$

and

$$L = 9.5 \times 1.5 = 14 + .$$

203. Volume of the Clearance.—For the volume of the clearance, the practice of the present day has settled upon a ratio of initial to final pressures and volumes expressed by the equation

$$p_a v_a{}^{1.35} = p_b v_b{}^{1.35},$$

when the subscripts a belong to the state at the beginning of compression, and the subscripts b belong to the higher values just before ignition. If these be written

$$p_b = p_a \left(\frac{v_a}{v_b}\right)^{1.35},$$

the compression pressure in the clearance is given when the initial pressure is known (usually the atmospheric) and the ratio of volumes. (See previous treatment in paragraph 152, from which this is repeated.) Ordinarily, however, the compression pressure limit is fixed by the condition that pre-ignition is not to be caused, and the formula will be used in the form

$$v_b = v_a \left(\frac{p_a}{p_b}\right)^{1.35}.$$

The usual compression values in pounds per sq. in. for the various kinds of gas are about as follows:

In Engines of	Blast-fur-nace Gas.	Producer-gas.	Weak Illum. Gas	Rich Illum. Gas	Natural Gas.
Small and medium sizes........	....	125	90	65	55
Large sizes.................	150	140	80	60	60

Taking from this the ratio of pressures desired (probably about 4 atmospheres pressure for ordinary gas for p_b will give the ratio $\dfrac{p_a}{p_b} = \dfrac{1}{4}$) and the ratio of the clearance volume to the piston displacement follows directly when this assumption is made. Hence

$$v_a = \text{piston displacement} + \text{clearance};$$
$$v_b = [\text{stroke} \times \text{area} + \text{clearance}] \left(\frac{p_a}{p_b}\right)^{1.35},$$

which can be transposed into

$$[\text{clearance}]\left[1 - \left(\frac{p_a}{p_b}\right)^{1.35} \right] = [\text{piston displacement}]\left(\frac{p_a}{p_b}\right)^{1.35}.$$

Applying this to the $9\frac{1}{2} \times 14$ cylinder of the 25-H.P. engine schemed in the foregoing paragraph, with 80 pounds compression pressure, and assuming the value for p_a to be 14 pounds, we have

$$v_a = v_b + 14 \text{ inches,}$$

substituting lengths for volumes, to which they are proportional when the area is constant as in a cylinder. Then

$$14v_a{}^{1.3} = 95(v_a - 14)^{1.3};$$

whence

$$v_a = \sqrt[1.3]{6.78(v_a - 14)^{1.3}}$$
$$= (v_a - 14)(3.36) \text{ nearly}$$
$$= 18.1;$$

whence the actual volume of the clearance

$$v_b = (18.1 - 14) \times \frac{\pi D^2}{4}$$
$$= 283 \text{ cubic inches.}$$

204. Velocity through Valves, Ports, and Passages.— The discussion in paragraph 98 indicates that considerable loss of power both direct and indirect will follow if the flow of gas and air through the valves is made so rapid as to entail excessive friction. Such loss is both of pressure and effective volume. This trouble is worse with automatic than with mechanically operated valves. Present good practice in small engines at very high speed, for automobile uses, favors keeping the velocity of inlet flow at or below 60 lineal feet per second for automatic valves, while permitting 90 feet per second for mechanically lifted valves. The exhaust flow out of the cylinder will be rapid enough at 75 feet per second.

205. Mechanical Design of Gas-engines Regarded as Machines.—The design of the gas-engine from the structural point of view, with respect to weight of fly-wheel, diameter of shaft, bearing surfaces, cross-sectional area of parts, and the like, belongs to a separate department from that before the student throughout this treatise. It has also been so well and completely worked out by others in forms accessible to every one interested that it does not seem desirable to expand the treatment of this section to include it.*   The gas-engine as an achievement in machine design falls under the same laws and principles which apply to the steam-engine, regard being paid to the special character of the impulses originating in the cylinder.   The foregoing treatment has shown how to arrive at the effort in the cylinder, and the rest is machine designing along lines of accepted ge:eral practice in that art.

---

* A most complete and exhaustive discussion of this topic will be found in "Gas Engine Design," by Prof. Chas. E. Lucke, 1905.   Parts II and III, Forces Due to Gas Pressure and Inertia, and Dimensions of Engine Parts.

CHAPTER XVIII.

INTERNAL-COMBUSTION ENGINES WITH HEATING AT CONSTANT
PRESSURE.

210. Introductory.—The previous chapters have been concerned principally with the so-called explosive engines, in which a mixture of fuel and air was ignited at a constant volume with a resulting increase in pressure which was utilized to impel a piston. Such engines operate upon the Otto cycle, in its two-stroke or four-stroke form, and are widely used and familiar.

It is quite possible, however, to burn the same explosive mixtures so that the result of heating them by the ignition or combustion process shall be an increase of their volume at a constant pressure, and if free to expand against a movable piston, such expansion of volume will take place at a constant pressure exerted on the piston while the volume increase takes place. This action is that of Cycle III, and its analysis appears in paragraph 185 et seq., as attaching to the design of the Brayton engine in America and the Simon engine in England. One difficulty at the time of their first presentation was that due to the difficulty of handling explosive mixtures with a continuous or intermittent release of heat energy to act in a motor. The work of Charles E. Lucke on combustion of such explosive mixtures in motion has removed this obstacle, and has brought this cycle within the scope of practical realization either in reciprocating motors or in those of continuous type, such as the gas-turbine principle presents. The method for securing such continuous, manageable, and complete

485

combustion and some principles underlying it form the subject of this chapter, together with some forms of apparatus using the principle of constant-pressure heating.

211. Lucke Apparatus for Continuous Combustion of Explosive Mixtures.—Referring to the general statements of paragraph 10, let it be assumed that a mass of explosive mixture is passing through a non-conducting tube with a uniform velocity v. Then, if inflammation be started at some point, the surface of combustion may remain at rest or move with or against the current. Denote the rate of propagation by r. Then, when $v>r$, the surface of combustion will move with the current, and if the tube has an end, the flame will "blow off" and combustion cease; if $v=r$, the surface of combustion will remain at rest, other influences being inoperative; if $v<r$, the surface of combustion will move back toward the source, or "back-flash."

Of course, a small tube of heat-conducting material will exert considerable cooling effect, but for the present such tubes need not be considered.

In a practicable system of burning an explosive mixture continuously the following are desiderata:

I. "Back-flashing" must be prevented.

II. "Blow-off" must be prevented.

III. Combustion surface must be localized.

VI. It must remain localized for wide ranges of feed or velocity of flow of the mixture.

V. The localization must be unaffected by changes of temperature.

VI. Large or small quantities must be burned without affecting the above, and the transition from very small quantities to very large, or *vice versa*, however sudden, should be easy.

The first requirement might be accomplished in three ways:

(*a*) By using a long tube of some conducting material and so small in diameter as to prevent the passage of the flame-cap under any circumstances.

(*b*) By using wire-gauze screens.

(*c*) By causing the mixture to flow at some point with a velocity always greater than the rate of propagation.

The first (*a*) is impracticable, as it permits of only small quantities being burned; the second (*b*) will not work when the wire gauze gets hot; this leaves (*c*), which is practicable, as a valve in a pipe will answer for the necessary contraction and consequent increase of velocity. Hence the first requirement in the desired method of combustion will be the following: At some point before the combustion surface is reached the velocity of feed must be such that $v > r$.

Requirement II might be accomplished in three ways:

(*a*) By so reducing the velocity after passing the high-speed point that at some surface $v = r$.

(*b*) By suddenly increasing the temperature of the mixture so as to increase the rate of propagation while v remains constant; or,

(*c*) By both reducing v, by spreading the current, and increasing r by heating.

All of these ways are practicable; but, as a reduction of velocity alone or a sufficient heating alone would not produce the desired result so well as both operating together, there will be introduced as the second requirement in the desired method the following: After passing the point where $v > r$, the velocity of the mixture should be so reduced and its temperature increased as to make $v' = r'$.

Let the mixture issue from an orifice into the air. By properly regulating the velocity of exit, the flame-cap can be maintained at the orifice—the only device successful for this purpose in certain experiments was to cause water to drip into the supply-chamber. The position of the flame-cap is so extremely sensitive to changes of flow that all other methods which were tried for obtaining a constant velocity of exit, variable at will, failed. Increase the velocity of flow slightly, and the flame-cap will lift off. This may be done until the flame-cap is as much as 2 inches (with illuminating-gas and air) from the orifice before extinction takes place. It would seem that the impinging of the jet on the atmosphere should spread it and so reduce its velocity, but no appreciable increase of diameter could be observed. When the cap is close to the orifice it is of a deep blue color, uniform in shade over the disk, and the edges are sharply defined; whereas, as it lifts off some distance, it becomes indistinct and unsteady at the edges until, at the moment of extinction, it fades and disappears. When the cap is away from the orifice, while there is no visible connection with the source of supply, there really exists a column of mixture extending from the orifice to the cap and passing through the atmosphere. Naturally, at the surface of this column, diffusion will take place, and the longer the column the greater will be this diffusion effect, thus affecting the composition of the advancing column of mixture and causing partial loss of gas. This is the real cause of extinction.

From these experiments can be drawn the conclusion that the current cannot be sufficiently reduced in velocity by issuing into an atmosphere of lower pressure to prevent "blow-off" before diffusion with the atmosphere so alters the character of the mixture as to cause extinction or loss of fuel by dilution before reaching the surface of combustion. This calls for a new condition besides those noted in the requirements for combustion. The reduction of velocity of the mixture, after passing the place where $v > r$, must be accomplished in such a way as to prevent diffusion with any other gas.

To prevent this diffusion, there naturally suggests itself the expedient of surrounding the issuing jet with a shield of larger diameter, to still permit of the desired expansion. This is shown in Fig. 210, and is essentially the same as proposed by Ladd, Schmid, Beckfeld, and others. The mixture must issue from orifice a with a velocity $v_a > r$; this will prevent "back-flash." If the distance from a to b is long enough to allow the gas to spread and reduce velocity, "blow-off" will not occur until $v_b > r$, and within these limits the flame-cap should remain within the shield. A trial shows that when $(Dia)_b$ is but slightly larger than $(Dia)_a$, the feed velocity may be varied in about the proportions noted, but this means that action is confined within very narrow working limits. The flame-cap seems to lose its flat, volumeless character for some reason not at first clear. When $(Dia)_b$ is much larger, say four or five times $(Dia)_a$, a slow increase of feed velocity above r reveals the advancing flame-cap just as if the shield were not there. Later a slight spreading is noted, and then the

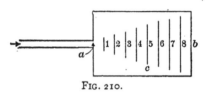

FIG. 210.

flame actually begins to show volume, as if there were no longer an explosive mixture present; this heats up the shield. A little consideration will show this to be due to the diffusion of the advancing and slightly spreading column with the products of combustion within the shield, and the high temperature of the shield helps to maintain a combustion of what is now a diluted explosive mixture beyond a point where that combustion would be possible if cold. An increase of velocity will cause extinction by "blow-off."

Here the results are somewhat better than in the last case without the shield. The principles operating, with the results, are: back-flash prevented by sufficiently great initial velocity at a; a spreading to reduce velocity, but very slight and insufficient, as proved by the narrow working limits; diffusion is not prevented; gas is partly heated before burning by the shield, which helps to continue combustion. If the advancing column did increase in cross-section and decrease in velocity while advancing, successive possible positions of the flame-cap would be as shown at 1, 2, 3, 4, etc., of Fig. 210.

It is obvious that at any point between a and 7, such as 4, the cap is surrounded by products of combustion, and the advancing column of mixture is passing through an atmosphere chiefly composed of the same, resulting in disastrous diffusion. This at once suggests giving the shielding envelope the form of a cone, supposing the orifice circular, so that the flame-cap at any instant may entirely fill up the space between the walls.

Apparatus with this end in view was tried and gave some interesting results. Fig. 211 shows a cone of 45 degrees angle,

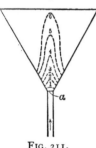

FIG. 211.

with a $\frac{1}{8}$-inch orifice such as was used. The velocity of feed was so adjusted as to cause the flame-cap to advance slowly from a, with the expectation stated above. The flame-caps at successive positions took the forms shown at the lines 1, 2, 3, 4, 5, 6, etc., and finally "blow-off" occurred. Since the only place where the combustion surface can remain at rest is on a surface where $v = r$, and since, secondly, r is here constant, the curves indicating the intersection of the combustion surfaces by meridian planes give us graphical values of the velocity of the advancing column of mixture. It is seen that the expected spreading did not take place, and that at any circular cross-section of the cone the velocity was greatest at the centre, decreasing toward the edges.

The curves 1, 2, 3, etc., are really cross-sections of successive constant-velocity surfaces in the advancing column, and the surface of combustion will lie on that surface of constant-transmission velocity where $v = r$.

A constant-velocity surface may be defined as a surface at every point of which the moving particles of gas have equal instantaneous velocities. If these successive surfaces had remained flat or nearly so, the proper sort of spreading of current and uniform decrease of velocity would be indicated. This gives an accurate definition of how the velocity is to be reduced after passing the point where $v > r$. The velocity of the advancing mixture must be reduced without diffusion, so as to keep the surfaces of constant velocity of such form that adjacent points on any one will be at approximately the same distance from the point where spreading begins. Reducing the angle of the cone, while helping matters considerably, reduces the range of feed velocities within impracticable limits.

Many ways of bringing about the above were tried, but only one seemed pre-eminently good both by reason of its simplicity and effectiveness, for it fulfils almost perfectly the requirements proposed for the desired method; this is, to fill the cone with fragments of refractory material, such as pottery, broken crucibles, bits of magnesite, or any other rock which will stand the high temperature without fusing. In cones of 60 degrees, and with a $\frac{1}{4}$-inch orifice, pieces about $\frac{3}{8}$-inch diameter seem to answer well.

These separate pieces of solid matter interpose many reflecting surfaces without materially hindering the advance of the mixture, and cause it to spread in the way desired, keeping the surface of combustion spherical and preventing diffusion. A variation of velocity causes the spherical surface of combustion to vary only in diameter, and the limits of feed are determined only by the size of the cone.

A cone of given altitude will give the greatest range of variation of diameter of cross-section when its angle is 180 degrees. This is a plane surface which, with the orifice and broken rock, should appear as in Fig. 212. Here the surface of combustion

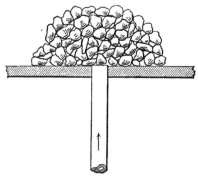

FIG. 212.

is approximately a hemisphere. Trial shows that this arrangement works perfectly, and the limits of feed are determined only by the size of the pile of rock surrounding the opening. A cone of 360 degrees, or no cone at all, suggests the surrounding of the nozzle by broken rock without any enclosing walls (Fig. 213). This arrangement also works remarkably well.

FIG. 213.

The surface of combustion is here approximately a sphere, giving the greatest possible increase in area of the surface of combustion for the distance travelled from the nozzle.

·If d denote the distance from the point where spreading begins to the surface of combustion, and S the area of the surface, we have:

For a cone, $\qquad S = \pi d^2 \tan^2 \alpha.$
· For no walls (Fig. 213), $S' = 4\pi d^2.$

Not only is the greatest possible range of action by velocity reduction thus obtained, enabling the greatest possible amount of mixture to be burned in a given volume, but this amount is further augmented by reason of the increase of the rate of propagation caused by the passage of the mixture between the hot fragments. Hence both principles operate simultaneously toward the desired end.

Hence a method of continuously burning explosive mixtures of all sorts, whether in the chemical proportion or not, as classified in Classes IV and V of paragraph, 10 seems to have been found which fulfils all the conditions set down as necessary, and which may be stated as follows:

I. Cause the mixture to pass a point where its velocity of transmission shall always be greater than the rate of propagation of inflammation through the mixture. This may be done by a valve in the feed-pipe.

II. So spread the current of mixture after it passes this point of high velocity that surfaces of constant-transmission velocity shall be of such form as to keep adjacent points on any one at approximately the same distance from the point where spreading begins. The whole spreading must take place so that the advancing unburned mixture cannot diffuse with any other gas. This can be accomplished by surrounding the orifice with solid fragments, introducing numerous reflecting surfaces which accomplish the spreading; also, by the passage through the interstices between this solid matter, the mixture is heated and the rate of propagation increased, making possible the burning of more mixture in unit volume.

When a chemical proportion is maintained in the mixture, all the combustion takes place on the *combustion surface,* giving absolutely neutral products of combustion; but when an excess of gas is present within certain limits, all gas which can find oxygen burns explosively between the solids, while the excess acts merely as a neutral diluent to be burned when it meets an oxygen atmosphere later on. By properly placing the oxygen atmosphere to burn the excess gas, the hot products can be made either *reducing* or *oxidizing*—reducing after leaving the explosive-combustion surface and before meeting the excess of oxygen in the atmosphere, oxidizing after that meeting.

It might be here noted that the principle well known in explosive combustion at constant volume, and constantly operating in the gas-engine, that "to a chemical mixture of air and gas there may be added large quantities of gas without altering the explosive properties of the mixture," is, by these experiments, extended. It appears that in explosive combustion at constant pressure, or "continuous combustion of explosive mixtures," the same principle applies, and, though no real proportion measurements have yet been made, it seems to a wider degree. That is to say, mixtures of air and gas, with gas in excess of the amount the air present can support, will burn explosively. The excess gas present acts merely as a neutral diluent, such as the nitrogen of the air. It is a fact, also, that as the solid fragments heat up, the excess may be greater than when they are cold.

212. Engines which have Operated with Constant-pressure Heating.—To carry out this principle in the past, the engine of Stephen Wilcox in 1865 was designed. The central burner between the working cylinders receives gas centrally from the feed-pump G, and air is delivered from the pump H around the gauze surrounding the gas-jet (Fig. 214). The heating of the air used for combustion, and some additional air entering through the valve $\tilde{M}$, causes the expansion which produces the working stroke.

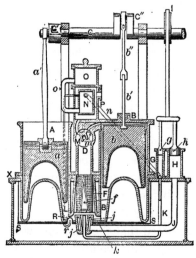

FIG. 214.

When the jet of fuel is projected into the mass of air to be heated when the latter is in the cylinder, the engine resembles the Diesel (1892) or the Gibbs (1897) in form. In the Diesel the air is heated to the ignition-point of the fuel by the compression of the previous stroke; in the Gibbs the air is not heated, but the gas is admitted after the compressed air has expanded after cut-off to reach the lower pressure in the gas-reservoir. The gas then flows in and is ignited electrically (Fig. 215).

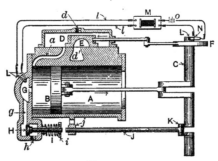

FIG. 215.

Explosive mixtures burning without any atmosphere of air around them are the feature of the Brayton, the Reeve, and the Schmid and Beckfeld. The Schmid and Beckfeld design in Fig. 216 shows a supply of gas entering through the pipe *E* and

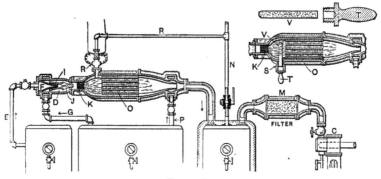

FIG. 216.

air through *G*, and mixing in the nozzle *F*. The ignition takes place in the long fusiform chamber in which a perforated brick tuyere helps to raise the temperature and act as a re-igniter. An igniting-plug of coke or carbon is pre-heated and inserted by the handle *T* to start combustion.

In the burner principle of Sidney **A. R**eeve of 1897 both air and fuel were supplied by separate pumps, and the proportions

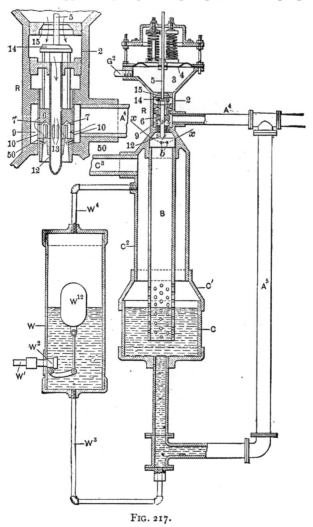

Fig. 217.

maintained by maintaining equal pressures in two receivers which these pumps supplied. A loaded check-valve maintained a pressure in the receivers higher than that in the combustion-chamber. In the enlarged detail (Fig. 217) the pressure is equalized by a diaphragm 4 with springs adjusted on its back. The diaphragm actuates a plunger 6 in a perforated sleeve. Gas enters through the central tube 10, and air through 10 around the central gas-current. The ignition takes place at *b*. The water-seal is kept automatically at a desired level by a float in

FIG. 218.

an auxiliary chamber, and by its presence the hot gases are kept at a constant pressure corresponding to that of saturated steam at that temperature, and with a constant difference of pressure at the discharge point from that at the supply point.

213. The Brayton Engine.—This engine (Fig. 218) had the air compressed in the pump *B*, whose volume was one-half that

of the power cylinder A. The two constant-pressure tanks at
the base of the frame delivered the air through the pipe O in

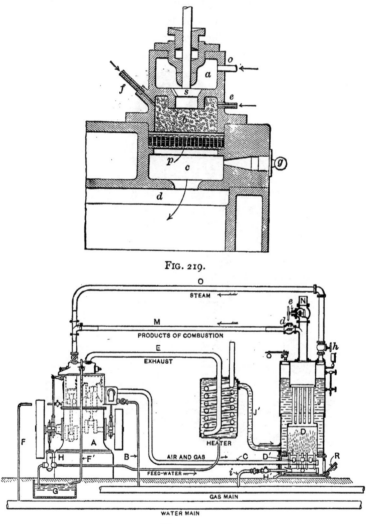

FIG. 219.

FIG. 220.

Fig. 219, which is the burner for the oil-engine, and passed it
through the absorbent material or wick b to which the oil was

fed by a pump. The air and fuel combine here so that the air becomes carburated (par. 107) and passes through a wire-gauze grating p into the cylinder d, where it burns on meeting the flame on the bottom of the gauze. The air is never completely shut off by the admission-valve on its nominal closure, but enough always flows through f to keep a small flame alight. This flame increases in volume for the power stroke. Combinations of steam and products-of-combustion engines belong in this class. In Fig. 220 is shown a steam-boiler with a feed-water heater utilizing waste heat in the escaping gases. The air and gas burn first in an open fire-box to start the engine with steam, and later, when the engine itself can compress the air and gas mixture, the fire-box and chimney are closed, and the engine works on a circuit of gas and steam which escape together from the exhaust-pipe after going through the coil.

214. Apparatus for Observing Increase in Volume with Constant-pressure Heating.—The inconvenient limits of size for an apparatus in which the maximum heating effect could be ob-

Fig. 221.

served as producing a maximum increase of volume at constant pressure resulted in the design of the apparatus by Dr. C. E. Lucke which is shown in Fig. 221. It depends for its action on the principles of gas flow through an orifice. The rate of flow of a gas though an orifice is proportional to the form of orifice and to the pressure drop through the orifice. Now if the gas be caused to pass through a hole in a plate before combustion, and later, after combustion, pass through a similar hole in a similar plate, the constant due to the form of orifice would be eliminated in comparing velocities through the two holes. Secondly, when the fall in pressure through each hole is the same the velocity of flow through each plate will be equal, and the volume passing will be proportional to the area of the orifice only if the pressures used be small enough to make correction for compression vanishingly small. Gas and air are mixed in any proportion desired at the compressor intake and delivered, mixed, to the chamber AB, from which the mixture will pass to the upper chamber C through a hole in the plate secured between the flanges. In chamber C there is placed a cone of brick to keep the lower plate cool, and in the cone is placed broken rock to permit of the combustion of the explosive mixture. The top plate between the flange D is provided with asbestos sheets to keep the hot gases from chilling just before issuing.

Both the brick cone for the lower and the asbestos sheet protection for the upper plates can be removed for the taking of observations, and a one-inch lining of fire-clay can be supplied to prevent radiation. Mercury manometers to both chambers indicate the interior pressures, and hence the drop in pressure through each plate.

215. The Future of the Engine which uses Constant-pressure Heating of the Working Medium. The Gas-turbine. — The author believes that in the future there will be increasing attention given to the engine which operates under the constant-pressure heating cycle. The motor may be either a reciprocating or a continuous rotary one, of the turbine order. If a reciprocating engine, it will be a compromise between the original Brayton and the present Diesel types, avoiding the inconveniently high compression of the latter, and improving on the mechanical design of the former. Among the advantages accruing from the use of this cycle other than the theoretical ones already referred to in the foregoing may be listed:

(1) The avoidance of the sudden changes of pressure in the cylinder. The suddenness of these changes in the constant-volume cycle is the cause of the difficulty in obtaining a uniform turning effort in these motors.

(2) The work diagram becomes more flexible, and the control of the effort more perfect in its adjustment to the resistance. The turning effort on the crank is more nearly uniform throughout the cycle.

(3) The motor becomes more easily reversible, and approaches the flexibility of the steam-engine.

(4) Low-grade liquid fuels are easily used.

(5) High compression and its good economy are easily secured without danger of pre-ignitions.

With respect to the development of this cycle for application in engines of the turbine class, which are particularly adapted to work under just these conditions, it must be said as yet that the uncertainties concerning the transformation of heat energy into kinetic energy in a free expansion, and the problems as to suitable structural materials at high temperatures, make it premature at present to pursue this attractive path beyond this point. Success will be likely to follow only from considerable further expenditure of time in research and capital in experiment. The gas-turbine would be specially valuable if it could be made to operate under pressures not greatly above atmosphere, while utilizing high initial temperatures and large increases of volume.

CHAPTER XIX.

TESTS ON EXPLOSIVE MIXTURES.

220. Introductory.—The theoretical computations made in connection with paragraph 20 have indicated the accepted method of arriving at the rise in temperature which follows from an ignition of a fuel which contains the necessary amount of oxygen to produce so rapid a combustion as to be designated as an explosion, which again is the result of the practically instantaneous propagation of a flame introduced into the mixture at one point. The condition of such ignition is that present in every explosive gas-engine where the mixture of gas and air is ignited in the constant volume of the combustion-chamber, which is filled by the explosive mixture at the pressure resulting from the previous compression and which is ignited by the electric spark or whatever igniting device is used. It becomes a matter of considerable interest to compare the theoretical pressures and temperatures with those which are actually realized in real engines or under the conditions which prevail with respect to the presence of varying volumes of fuel and air, or varying volumes of neutral or incombustible gases in the explosive mixture itself.

A diagram may be drawn presenting graphically the temperature value as computed theoretically from the formula

$$T = 32 + \frac{Q}{[(Rx) + y]C_v}$$

(par. 55) for a gas in which Q is 649 B.T.U. per cubic foot, whose weight is 0.03348 pound per cubic foot, while x is the weight of a cubic foot of air, or 0.08073. It would give the curve below (Fig. 222).

It will be apparent that the better the explosion process in any engine, the smaller the cylinder volume which will be required to overcome a given mechanical resistance. Hence an explosive mixture will be called the best which produces the greatest initial pressure for a unit volume; and secondly, that which maintains

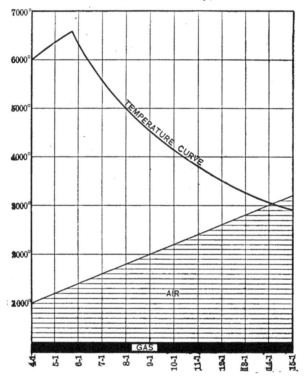

FIG. 222.

the highest pressure for the longest time when exposed to the cooling action of the cylinder walls and jackets. It is a matter of common observation in gas-engines that if the piston be blocked so that it cannot move, and the compressed charge behind it be ignited, the pressure due to the ignition will fall very rapidly by reason of the absorption of heat from the conducting walls. It becomes significant, therefore, to investigate the behavior of such explosive mixtures.

221. Clerk's Explosion Experiments.—The standard English experiments upon this question are those of Mr. Dugald Clerk. His apparatus is presented in Fig. 225, and consisted of a closed

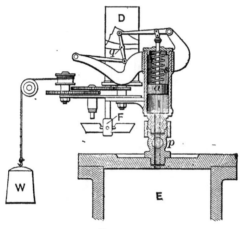

FIG. 225.

cylindrical vessel seven inches in diameter and eight and one-quarter inches on the inside and holding, therefore, 317 cubic inches. Upon one cover was attached a steam-engine indicator whose drum was made to revolve by a weight through multiplying gearing with a fan governor to maintain a uniform speed. The cylinder was filled with the mixture to be tested, the revolving drum set in motion with the pencil-point bearing against it, and an electric spark was passed between terminals at the bottom of the vessel. The rate of the revolution of the indicator-drum being known, the interval of time elapsing between any two points of the explosion curve or the cooling curve are at once given, and the position of the highest pressure gives both the value of that maximum and the time taken to reach it. The diagrams from such an apparatus appear as in Fig. 226, when mixtures of ordinary illuminating-gas and air were tested. The following table gives the results of this investigation. It is unfortunate for the value of these results for subsequent comparison with other tests, that their author did not give in some fulness the composition of the gas which he used. The presence of more or less percentages of neutrals or diluents in their analysis makes the mixtures behave

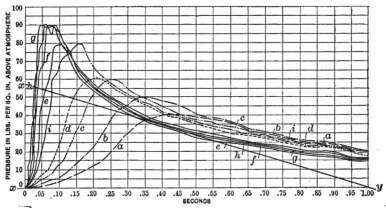

FIG. 226.

quite differently at different times. Mr. Clerk simply calls the gas Glasgow and Oldham gas, and published analyses taken at some other time cannot always be connected safely to these data.

1	2	3	4	5	6	7
Proportions.		Maximum Observed Pressure, Lbs. per Sq. In.	Time to Reach Maximum Pressure, Seconds.	Pressure Calculated Due to 1 Cu. In. Gas.	Pressure after 0.2 Second after Maximum.	Col. 5+Col. 6 2
Gas.	Air.					
1	14	40	0.45	728	602	665
1	13	51.5	0.31	....	....	....
1	12	60	0.24	756	576	666
1	11	61	0.17	....	....	....
1	10	....	....	690	470	580
1	9	78	0.08	....	....	....
1	8	....	....	712	440	556
1	7	87	0.06	....	....	....
1	6	90	0.04	576	342	459
1	5	91	0.055	....	....	....
1	4	80	0.16	....	....	....

Temperature before explosion 60° F. Pressure atmospheric.

With a mixture of hydrogen and air instead of gas and air at 55° F. the following results showed that it was inferior to coal-gas, as follows:

1	2	3	4
Proportions.		Maximum Observed Pressure, Pounds per Square Inch.	Time to Reach Maximum Pressure, Seconds.
Hydrogen.	Air.		
1	6	41	0.15
1	4	68	0.026
2	5	80	0.01

This investigation as presented in the curves seems to indicate that the best results with respect to pressure are given with one volume of gas to five volumes of air, and that as the dilution with air increases, the area of work under the curve would diminish. The rapidity of the ignition in these proportions makes an inconvenient shock in the cylinder. If, now, the pressures in the third column resulting from explosion are reduced in common ratio so as to give the volume which they occupy with equal pressures from a uniform volume of gas ignited, such as one cubic inch, the fifth column of the table results.

If now from the diagram of pressures the pressure existing after one-fifth of a second be scaled off, a measure is found for the cooling effect of the walls in an engine making 300 revolutions per minute. This pressure tabulated in the 6th column can be used to compute the 7th column, which should give the highest arithmetical mean of the explosion effect and cooling effect with five strokes to the second. This column indicates that the best results with respect to mean pressure are given when the ratio of gas to air is between one-twelfth and one-fourteenth. Both with less gas and with more the mean pressure falls off. Clerk's experiments with hydrogen, reported in the second table, showed that it was not equal to the ordinary illuminating-gas volume for volume. Two volumes of hydrogen to five volumes of air gave a pressure of 80 pounds in one one-hundredth of a second, while with illuminating-gas the same pressure resulted from a mixture of one of gas to ten of air, with a combustion which was not so inconveniently rapid as to cause undue shock, jar, or vibration in the motor.

222. Lucke's Explosion Experiments.—The apparatus used by Dr. Charles E. Lucke in determining the pressures caused by explosion or by combustion at constant volume is shown in **Fig. 227.** To a tee *A* were attached two nipples *B* closed with

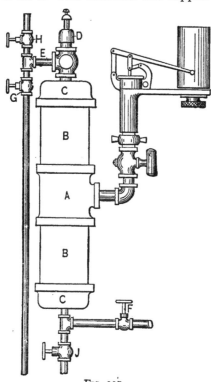

Fig. 227.

caps *C* above and below. To the branch of the tee was connected an indicator, and the igniting arrangement was attached in one branch of a three-way cock on the upper cap. The apparatus was first filled with water through the connection controlled by the valve *F* at the bottom until the water overflowed through the valve *G*, with the three-way cock in position to isolate the spark-plug but fill the apparatus completely and expel the air. When completely filled the water-valves *G* and *F* are closed and *H* and *J* open. The mixture to be tested in a closed tank under pressure flows through *H* and the opened three-way cock, and the water is expelled. When the three-way cock is reversed and the outlet *J* closed the mixture in the explosion vessel is at atmospheric pressure and can be fired by the spark-points, while the pressure on the indicator draws a diagram giving pressure and time values as in the Clerk apparatus.

While this investigation covered research into the heat developed by combustion at constant pressure, yet for the present purpose attention is mainly directed to the effect on the pressures at constant-volume combustion, as these are affected by varying constitution of the mixture from the presence of neutral gases, which are inert so far as producing temperature and pressure are concerned. If the gas be assumed, for example, to have a standard composition, such as:

CO_2.	3.8
C_2H_4.	14.6
CO.	28.0
H.	35.6
CH_4.	16.7
N.	1.3
Total	100.0

of this there will be

NEUTRAL.

CO_2.	3.8
N.	1.3
Total	5.1

Such a gas, moreover, will yield 691.59 B.T.U. per cubic foot products condensed, and in its combustion will call for 5.21 parts of air per one part of gas.

A *chemical mixture* then would have these characteristics:

$$\text{Air} \dots \dots \dots \dots \dots \dots \quad 5.21 \text{ volumes}$$
$$\text{Gas} \dots \dots \dots \dots \dots \dots \quad 1.00 \quad \text{``}$$

$$\text{Total} \dots \dots \dots \dots \dots \quad 6.21$$

of which

$$\text{Neutral} \begin{cases} \text{Neutral in gas..} & .051 \\ \text{Nitrogen in air..} & 4.120 \end{cases}$$

$$\text{Total neutral..} \quad 4.17$$

in 6.21 parts, or 67 per cent neutral.

Various mixtures of this gas and air will give

Gas	1	1	1	1	1	1	1	1	1
Air	3.0	3.5	4.0	4.5	5.0	5.5	6.0	6.5	7.0
Inactive air, i.e., excess						.29	.79	1.29	1.79
Active air	3.0	3.5	4.0	4.5	5.0	5.21	5.21	5.21	5.21
Inactive gas, i.e., excess	.427	.328	.232	.135	.050				
Active gas	.573	.672	.768	.865	.950	1.000	1.000	1.000	1.000
Neutral in active air	2.372	2.768	3.163	3.559	3.954	4.120	4.120	4.120	4.120
Neutral in active gas	.029	.034	.039	.044	.048	.051	.051	.051	.051
Total inactive or excess	2.828	3.130	3.434	3.735	4.052	4.461	4.961	5.461	5.961
Per cent inactive or excess	.701	.696	.687	.680	.675	.686	.704	.729	.746

It should be noted how very slightly the increase in percentage of dilution increases with the excesses of air and gas; though the proportions may vary over 100 per cent, the dilution varies through but little more than 5 per cent. There can be little doubt that the limits of combustibility are intimately associated with the per cent of neutral or inactive gases present.

On examination of the calorific values of some of these mixtures, i.e., the amount of heat which one cubic foot of gas can deliver when burned explosively in mixtures within the limits of explosive combustion, it will appear that the heat developed by a cubic foot of the gas in question is 691.59 B.T.U. when completely burned, i.e., in a chemical mixture, or in a mixture in which air is in excess, within of course the limits of combustibility. Hence,

Gas	1	1	1	1	1	1	1	1	1
Air	3.0	3.5	4.0	4.5	5.0	5.5	6.0	6.5	7.0
Gas burnt, i.e., gas that could find air enough to burn it	.573	.672	.768	.865	.950	1.000	1.000	1.000	1.000
B.T.U. available	396.3	464.8	531.2	598.2	657.0	691.6	691.6	691.6	691.6

These results are shown graphically on the curve *A* of Fig 228. The results are as follows, reduced to cubic feet of gas:

B.T.U. PER CUBIC FOOT OF GAS WHEN MIXED WITH AIR.

Gas.	Air.	B.T U. per cu. ft. Gas.
I	3.0	275.1
ᵥ	3.5	347.82
	4.0	401.57
	4.5	471.00
	5.0	541.70
	5.5	616.59
	6.0	600.78
	6.5	?

Each mixture of air to gas within the range of combustibility was fired, and then to each was added in turn successively in-

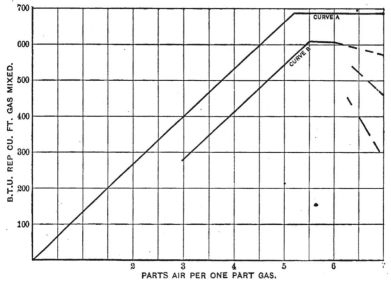

FIG. 228.

creasing amounts of neutral gases obtained by burning completely an explosive mixture.

It appeared that the resulting pressures were intimately connected with the percentage of dilution of neutral or excess gases, and as the gas used has already exhibited some agreement with what is possible with the typical water-gas chosen in comparison, it will be well to work out a table of percentage of dilution of different mixtures, and these figures will be placed on the curves of Figs. 229–234. The agreement and evident existence of a law is apparent.

WATER-GAS OF OBSERVED COMPOSITION.

Mixture, $\left\{ \begin{array}{l} \text{Air, } 3 \\ \text{Gas, } 1 \end{array} \right\}$ diluted.

Gas.	Air.	Added Neutral.	Primary Neutral.	Total Neutral.	Per Cent Neutral.
1	3	0	2.83	2.83	70.1
1	3	1	"	3.83	76.0
1	3	2	"	4.83	80.5
1	3	3	"	5.83	83.0

Mixture, $\left\{ \begin{array}{l} \text{Air, } 4 \\ \text{Gas, } 1 \end{array} \right\}$ diluted.

Gas.	Air.	Added Neutral.	Primary Neutral.	Total Neutral.	Per Cent Neutral.
1	4	0	3.43	3.43	68.7
1	4	1	"	4.43	74.0
1	4	2	"	5.43	77.6
1	4	3	"	6.43	80.6
1	4	4	"	7.43	82.8

Mixture, $\left\{ \begin{array}{l} \text{Air, } 5 \\ \text{Gas, } 1 \end{array} \right\}$ diluted.

Gas.	Air.	Added Neutral.	Primary Neutral.	Total Neutral.	Per Cent Neutral.
1	5	0	4.05	4.05	67.5
1	5	1	"	5.05	72.1
1	5	2	"	6.05	75.7
1	5	3	"	7.05	78.4
1	5	4	"	8.05	80.5

Mixture, $\left\{\begin{matrix} \text{Air, } 6 \\ \text{Gas, } 1 \end{matrix}\right\}$ diluted.

Gas.	Air.	Added Neutral.	Primary Neutral.	Total Neutral.	Per Cent. Neutral.
1	6	0	4.96	4.96	70.4
1	6	1	"	5.96	74.4
1	6	2	"	6.96	77.3
1	6	3	"	7.96	79.6
1	6	4	"	8.96	81.3
1	6	5	"	9.96	84.0

Mixture, $\left\{\begin{matrix} \text{Air, } 7 \\ \text{Gas, } 1 \end{matrix}\right\}$ diluted.

Gas.	Air.	Added Neutral.	Primary Neutral.	Total Neutral.	Per Cent Neutral.
1	7	0	5.96	5.96	74.6
1	7	1	"	6.96	77.3
1	7	2	"	7.96	79.6
1	7	3	"	8.96	81.3
1	7	4	"	9.96	83.0

The curves of Figs. 229–233 show the pressures given by the indicator for each mixture and are the mean values from a large number of lines drawn by the indicator. These curves are combined in Fig. 234, which is, therefore, a curve of pressures for all mixtures diluted or not within the range of explosive combustibility. The numbers on the curves show the percentage of dilution of the typical water-gas. The results are most remarkable and can be accounted for only by assuming that the presence of a large amount of dilution hinders combustion. The limits at which combustion ceases to be possible on too great a dilution are here indicated, whether that dilution be due to excess gas, excess air, or neutral gases. It will also be observed that the character of the diluent has an appreciable effect, but that when the dilution is least the pressure is greatest, about 60 pounds above atmosphere or a ratio of 5; and the presence of a constant per cent of neutral will make combustion impossible no matter

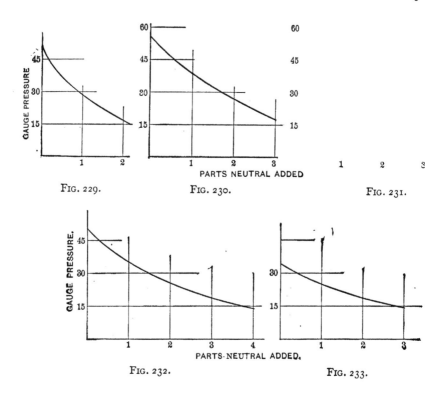

FIG. 229.

PARTS NEUTRAL ADDED

FIG. 230.

60
45
30
15
1 2 3

FIG. 231.

PARTS NEUTRAL ADDED.

FIG. 232.

FIG. 233.

3 3.5 4 4.5 5 5.5 6 6.5

PARTS AIR PER ONE PART GAS.

FIG. 234.

what the mixture of air and gas. The greatest neutral dilution gives the least pressure—about 15 pounds above atmosphere, or a ratio of about 2. These results give a reason for the decreased pressure in exploding gas-engines in which the mixture is always diluted by burnt products to an extent of 20–40 per cent of the volume of neutral addition to the gas mixture, which may already have neutral gas present to the extent of 65–70 per cent.

Neutral additions to the gases sent to the calorimeter and to the other apparatus showed, besides a corresponding and proper heat value for the resulting mixture, a decreased rate of propagation accompanied by a difficulty in ignition and constant tendency to incomplete combustion, i.e., tendency to cease burning after inflammation had been started and before the mass had been entirely burnt.

223. The Massachusetts Institute of Technology Experiments on Explosive Mixtures.—In 1898 an experimental apparatus was fitted up in the laboratories of the above Institute in Boston, and results from it have been published from time to time. The apparatus is a cast-iron cylinder with a flanged top to which is bolted the cover. The mixture is introduced and proportioned by the plan of exhausting the test-chamber by a pump to remove all previous charge, and is then scavenged by admitting fresh clean air. It is then exhausted again by the pump until a desired pressure lower than atmosphere is reached, so computed that when gas at atmospheric pressure is introduced the rise of pressure in the chamber to atmospheric pressure shall draw in just the desired volume of gas. Pressures above atmosphere can be used if desired by having suitable pumps and computed volumes adjusted to the higher pressures. This method of proportioning is not believed to be as reliable as that followed in the preceding paragraph, in view of the fact that the volumes are not directly measured.

The indicator makes its record upon a card on a power-driven disk, upon which the time record is simultaneously made by the vibrations of a tuning-fork, kept moving by an electromagnet. A pointer on one arm of the fork traces the time line. The mixtures are fired electrically. The same fifth-of-a-second limit was· used as chosen by Clerk, in correspondence with an engine running at 300 r.p.m. The analysis of the gas was:

CO. . . . . 25·3
Illuminants 12.0
CO$_2$. . . . . 1·9
CH$_4$. 28.9
N. 3
H. . . . . 27·9
O. 1

———

100.0

RESULTS OF TESTS ON EXPLOSIVE MIXTURES OF ILLUMINATING-
GAS AND AIR.

Mixture. (By Parts.)	Maximum Pressure. (Lbs. per Sq. In.)	Time of Explosion. (Seconds.)	First ⅕ Second.				⅕ Second after Maximum Pressure.				
			Area. (Square Inches.)	Mean Pressure (Lbs. per Sq. In.)	Mean Pressure ÷ Proportion of Gas.	Final Pressure.	Area. (Square Inches.)	Mean Pressure. (Lbs. per Sq. In.)	Mean Pressure ÷ Proportion of Gas.	Final Pressure.	Final Pressure ÷ Proportion of Gas.
1	2	3	4	5	6	7	8	9	10	11	12
Gas-Air.											
1–3	45	.49	0.32	11	44	26	1.30	43	172	40	160
1–4	86	.08	1.77	59	295	61	1.88	62	310	46	230
1–5	96	.05	1.86	62	372	52	1.93	64	384	44	264
1–6	88	.05	1.80	60	420	54	1.93	64	448	46	322
1–7	86	.06	1.97	66	528	58	1.93	64	512	48	384
1–8	87	.06	1.71	57	513	53	1.83	61	549	46	414
1–9	77	.08	1.60	53	530	57	1.86	62	620	46	460
1–10	71	.11	1.36	45	495	56	1.69	56	616	45	495
1–11	68	.14	1.21	40	480	60	1.66	55	660	43	516
1–12	39	.33	0.35	12	156	29	0.98	33	429	30	390
1–13	32	.42	0.18	6	84	16	0.79	26	364	24	336
1–14	9	.42	0.05	2	30	4	0.24	8	120	8	120

The diagrams developed on a straight base-line give the results in Fig. 235.

From these it appears that while the 15.4 percentage of gas gives the greatest initial pressure, the rate of cooling is also greater, so that, as in the other experiments, a mixture with 9 to 10 per cent

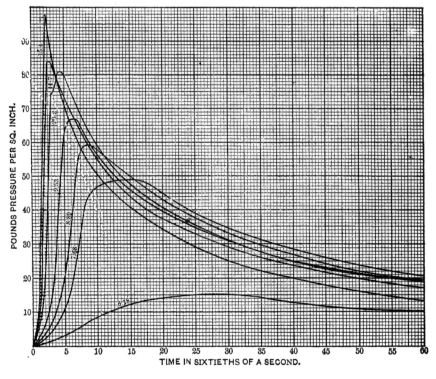

FIG. 235.

of gas gives the greatest average pressure for the first fifth of a second.

Fig. 236 shows the effect of varying the composition of the mixture both as to the time taken to reach the maximum pressure and the value of that maximum.

The quickest explosion, as elsewhere noted, gives the maximum pressure. The two following tables give interesting data from the same source concerning mixtures of air and gasoline.

GASOLINE (86°, Sp. gr. 0.648) AND AIR.

Proportion of Gas in Mixture.	Time of Explosion. (Sec.)	First 0.2 Second.				Maximum Pressure.	0.2 Sec. after Maximum Pressure.			
		Area. (Sq. In.)	Mean Pressure. (Lbs. per Sq. In.)	Mean Pressure ÷ Proportion of Gas.	Final Pressure.		Area. (Sq. In.)	Mean Pressure. (Lbs. per Sq. In.)	Mean Pressure ÷ Proportion of Gas.	Final Pressure.
.0151	.083	1.40	46.7	3080	46	70	1.48	49.4	3260	34
.0164	.100	1.41	47.0	2865	49	73	1.53	51.0	3110	36
.0179	.090	1.33	44.3	2480	44	71	1.43	47.7	2670	33
.0196	.083	1.36	45.3	2309	46	76	1.55	51.7	2634	35
.0217	.058	1.34	44.7	2055	37	70	1.45	48.4	2225	30
.0244	.067	1.50	50.0	2050	44	80	1.60	53.4	2190	36
.0256	.075	1.60	53.4	2081	50	84	1.69	56.4	2200	40
.0263	.059	1.65	55.0	2089	46	86	1.71	57.0	2164	38
.0278	.083	1.41	47.0	1691	48	78	1.62	54.0	1945	36
.0303	.091	1.40	46.7	1541	52	76	1.60	53.4	1760	38
.0323	.083	1.45	48.4	1500	48	77	1.62	54.0	1675	37
.0345	.083	1.51	50.3	1467	48	77	1.64	54.7	1587	37
.0385	.075	1.37	45.7	1188	47	66	1.50	50.0	1300	38
.0417	.066	1.25	42.7	1025	41	60	1.38	46.0	1104	35
.0476	.066	1.15	38.4	608	39	56	1.32	44.0	925	33

GASOLINE (76°, Sp. gr. 0.680) AND AIR.

Proportion of Gas in Mixture.	Time of Explosion. (Sec.)	First 0.2 Second.				Maximum Pressure.	0.2 Sec. after Maximum Pressure.			
		Area. (Sq. In.)	Mean Pressure. (Lbs. per Sq. In.)	Mean Pressure ÷ Proportion of Gas.	Final Pressure.		Area. (Sq. In.)	Mean Pressure. (Lbs. per Sq. In.)	Mean Pressure ÷ Proportion of Gas.	Final Pressure.
.0132	.167	.76	25.3	1925	52	52	1.28	42.7	3240	33
.0141	.117	1.15	38.4	2720	49	62	1.42	47.3	3360	35
.0151	.109	1.26	42.0	2770	48	64	1.45	48.6	2950	35
.0164	.182	.81	27.0	1650	50	51	1.25	41.7	2540	32
.0179	.109	1.27	42.3	2368	50	67	1.53	51.0	2855	36
.0196	.091	1.44	48.0	2441	48	73	1.53	51.0	2600	36
.0217	.082	1.43	47.7	2180	48	76	1.56	52.0	2391	37
.0244	.060	1.62	54.0	2213	45	85	1.63	54.3	2225	36
.0263	.058	1.61	53.7	2040	45	85	1.62	54.0	2052	36
.0278	.058	1.62	54.0	1943	46	84	1.64	54.7	1970	38
.0303	.066	1.49	49.7	1640	45	78	1.60	53.4	1760	37
.0323	.067	1.55	51.7	1602	48	83	1.70	56.7	1760	38
.0345	.100	1.34	44.7	1297	52	75	1.59	53.0	1536	38
.0385	.117	1.10	36.7	955	52	62	1.42	47.3	1230	35
.0417	.133	.98	32.7	761	52	55	1.40	46.7	1121	38
.0476	.210	.39	13.0	273	35	35	1.02	34.0	714	32

The greatest value for the mean pressure during the explosion period is found for both cases in the neighborhood of 25 parts in 1000, or 2½ per cent of gas in one hundred of the mixture.

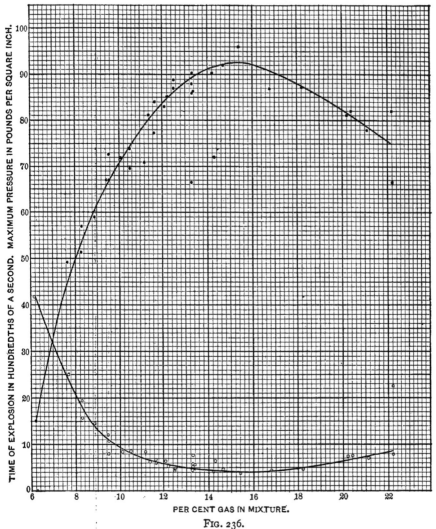

FIG. 236.

224. Grover's Experiments with Acetylene.—A third group of explosion experiments include those of Mr. Frederick Grover of

Leeds, England, between 1899 and 1901. The explosion-chamber was a piece of cast-iron flanged pipe. An indicator outfit gave the pressures due to the ignition, effected by electric spark. The timing for the speed of the diagram under the indicator pencil was done in a simple and elegant manner, by mounting a stop-watch ticking fractions of seconds upon a gear which was driven through a worm on the axis of the paper drum. By revolving the watch under the second hand and in the opposite direction, it is plain that at the speed of the second hand the hand would remain stationary in space. A mirror on the axis of the hand would show by steady reflection of a fixed object when the needle stood still in space, and the speed of the paper on the drum was easily computed when the reduction of the gear driving the watch-case was known. The paper would always be moving at the same rate for all experiments.

The tests gave results both of pressures on ignition and with

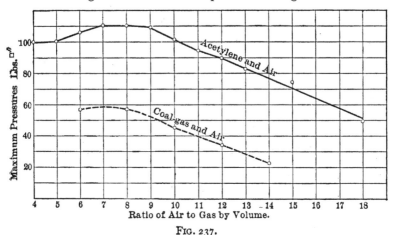

FIG. 237.

various degrees of previous compression and the time results with various mixtures of acetylene and air. Compressions of one, two, and three atmospheres were tried, and Figs. 237, 238, and 239 show the pressures with acetylene compared with the same mixtures of ordinary coal-gas and air. In Figs. 240, 241, and 242

are the pressures and the times required to reach them. Acetylene gives higher pressures than coal-gas, although less high

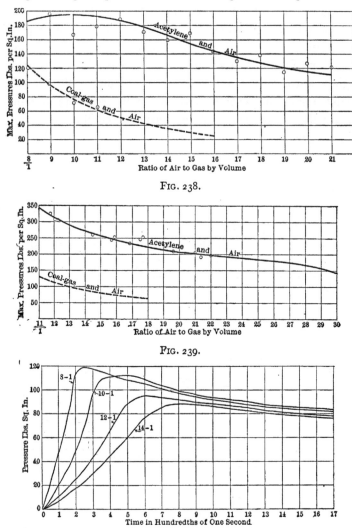

FIG. 238.

FIG. 239.

FIG. 240.

in proportion as the compressions increase. Coal-gas requires
5.7 volumes of air for its combustion, while acetylene requires

12.5 volumes. Acetylene fires more quickly, as the times ranged from 0.1 to 0.018 of a second for the proportions which reached their maximum with coal-gas in 0.5 and 0.25 of a second re-

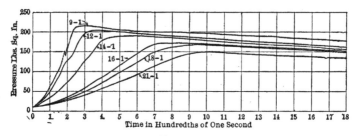

FIG. 241.

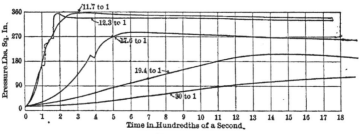

FIG. 242.

spectively. The weakest proportion of acetylene to air which would ignite is 1 to 18, while with coal-gas the limit is 1 to 15. An analysis of the products of combustion when the previous compression was three atmospheres gave the following table:

Mixture.		Constituent in Percentage.				
Air.	Acetylene.	CO_2.	CO.	O.	N.	Steam.
11.7	1	13.0	3.2	0.0	79.0	6.5
12.3	1	15.1	0.0	0.0	78.0	8.2
14.5	1	12.4	0.0	2.5	78.9	6.2
16.1	1	11.8	0.0	4.2	79.0	5.9
17.5	1	10.0	0.0	5.1	79.6	5.0
21	1	8.6	0.0	8.1	78.9	4.3
22	1	9.0	0.0	8.6	79.0	4.5
30	1	6.8	0.0	12.0	78.5	3.4

225. Grover's Experiments on **Effect of Neutrals in Explosive Mixtures.**—A series of experiments in the laboratories of the Yorkshire College at Leeds, England, instigated by Mr. Grover has added much to the knowledge concerning the effect of products of combustion in the explosive charge. The apparatus was essentially the same as in the previously described research. The charge consisted of:

1. The volume of neutral products of combustion from a previous ignition which was desired in proportion to the total volume.

2. Half the volume of pure air required to make the combustible mixture.

3. The full volume of coal-gas to make the combustible mixture.

4. The rest of the volume of pure air to complete the new charge.

Without waiting for diffusion the charge was then fired at atmospheric pressure.

The annexed diagram presents the results of this research in graphic form (Fig. 243). The experiments themselves and the diagram may be interpreted as suggesting:

1. The presence of products of combustion does not diminish the actual pressures as much as that same excess of air would do, filling the same volume of the clearance in the cylinder. This is not inconsistent with the observed fact that a scavenging action with pure air has diminished gas consumption, since the effect of that scavenging in cooling the cylinder is to increase the weight of explosive mixture which fills a given volume at that lower temperature, both by removing the heating effect of the hot gases, and by making the jacket effect prove pronounced.

2. The highest pressures are obtained when the volume of air in the charge is only slightly in excess of that required for complete combustion.

3. The time of explosion is reduced when the hot products of combustion take the place of cool fresh air.

4. The mixture will ignite in all cases where the volume of air exceeds the minimum of 5.5 times that of the gas as required for complete combustion, provided the proportion of neutrals present does not exceed 58 per cent of the combined mixture of

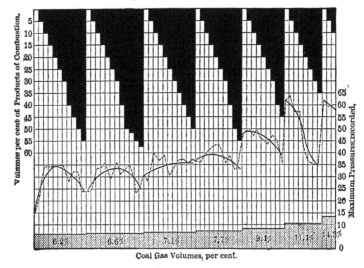

FIG. 243.

combustible, necessary air for combustion, and the neutral products taken together.

226. Temperature of Ignition or Inflammation.—The researches of Sir Humphry Davy showed that for ignition of gaseous explosive mixtures it was necessary that they be brought to a certain temperature at one point before they would ignite. The best modern work along this line has been done by Berthelot and Vielle, and particularly by Mallard and Le Chatelier. Students are referred to the monograph of the latter experimenters [*] for details of method and result under this and the following sections. To determine the temperature of inflammation, mixtures were admitted rapidly into a chamber previously heated

---

[*] Under the title "Recherches Expérimentales et Théoriques sur la Combustion des Mélanges Gazeux Explosifs," par MM. Mallard et Le Chatelier, Ingénieurs au Corps des Mines. Commission de Grison, 1883.

to a known temperature. It is then observed whether the mixture unites or not, and two limits are observed between which the temperature of ignition must lie. Testing with hydrogen, carbonic oxide, and marsh-gas, the limits were found to be for all mixtures:

H. . . . . $517-595°$, mixed with air, O and CO_2.
CO. . . . . $630-725°$, mixed with air, O and CO_2.
C_2H_4. . . . . $640-760°$, mixed with air and O.

Experiments on slow combustion show a discontinuity between it and that accompanied by light and heat changes.

In general the temperature of inflammation can be fixed at

$555°$ for explosive mixtures of H and O.
$655°$ "   "   "   " CO and O.
$656°$ "   "   "   " C_2H_4 and O.

The addition to explosive gas of even a considerable volume of inert gas modifies little or not at all the temperature of inflammation.

However, with mixtures of CO and O the addition of a notable quantity of CO_2 seems to elevate that temperature to a sensible degree. One volume of CO_2 added to explosive mixtures $CO + O$ raises the temperature from $655°$ to $700°$.

For mixtures in which H and O are the elements, the combustion takes place as soon as the temperature of inflammation is reached. It is entirely otherwise for marsh-gas, which may be likened to fire-damp. The mixtures formed by this gas with air or oxygen do not burn except after having been brought to a temperature equal or superior to that of inflammation and kept there for perhaps ten seconds. The retarding of inflammation increases with difference of temperature of gas and that of inflammation, and with the increase of the proportion of inert gas. This latter reason explains why, according to Davy, a bar of red-hot iron, though above $650°$, will not ignite a mixture of fire-damp. By opposing circulation one may easily provoke inflammation, because when it circulates freely the gas does not remain long enough exposed to the temperature of inflammation.

227. The Rate of Propagation of Flame.—Previous to Mallard and Le Chatelier this question had been attacked by Sir Humphry Davy and by MM. Bunsen, Schloesing and De Mondésir, Fonesca and Gouy, Berthelot and Vielle.

A summary of the work by Mallard and Le Chatelier on propagation of inflammation brings out the following facts:

There are two modes of propagation: (1) normal, which is that by conductivity, and (2) explosive, which takes place by the transmission through the mixture of a pressure sufficiently high to cause propagation by explosive wave. These correspond to deflagration and explosion of dynamite, etc. Each has a fixed velocity for a given mixture at a given pressure.

The phenomena of the explosive wave are of notable interest and will be referred to in the next paragraph. The rate of normal propagation, denoted by R, never exceeds 20 m. per sec.

For H and air the maximum is 4.30 m. per sec. for a 40 per cent H, i.e., an excess (30 per cent).

For C_2H_4 and air the maximum is 0.62 m. per sec. for a 12.2 per cent, i.e., an excess (9.4 per cent).

For illuminating-gas and air the maximum is 1.25 m. per sec. for a 17.0 per cent, i.e., an excess (15 per cent).

For CO and O and air the maximum is 2.00 m. per sec. always.

R increases with I, the rate of ignition, and when the tube is large is independent of diameter, but a tube small enough may cause extinction.

Agitation increases R. Combustion in the tube with slow R sets up oscillation which may cause extinction.

When for any reason of vibration or explosion of burnt gas the pressure transmitted to a layer next is equal to that which would elevate it to the temperature of inflammation, the combustion propagates with the same velocity as the compressive wave, resulting in the explosive wave.

228. The Propagation of an Explosive Wave. — Messrs. Schloesing and De Mondésir, in experiments upon ignition and propagation of flame in a long glass tube open at one end and closed at the other, noticed that in mixtures in which the normal rate of propagation was slow it was possible to produce true explosions of instantaneous character. These explosive projections of flame seemed to be the result of interior agitation in the mixture, probably of a vibratory order, and similar to the result of projecting a jet of gas at high velocity into a mass of gas at rest. If these agitations are produced in any given case, the usual rate of propagation becomes disturbed, and pressures will result much in excess of the expected pressure. The causes of such abnormal propagations as are traceable to disturbance of the mixture will be present:

1. When different parts of the mixture are of differing densities from temperature or other causes;

2. When an expansion caused by the increase in volume of a part of the gas on burning produces local compressions which are not instantly relieved by the yielding of the mobile mass. Pockets or confined volumes are particularly subject to this.

3. When a vibratory motion in the gas itself results from the process of propagation. When this synchronizes with the normal propagation rate the combination may result in a superposition of pressures beyond that which the confining vessel can withstand.

Messrs. Berthelot and Vielle carried observation of these phenomena farther, but the most conclusive demonstrations are those of MM. Mallard and Le Chatelier. These investigators not only produced the phenomena, but devised a photographic automatic record of the explosive process to which Berthelot and Vielle had given the name of "explosive wave" when generated in a tube. The pressures caused by this wave in tubes of glass of two millimetres in thickness were sufficient to reduce them to powder, although previously tested by static pressures above 150 pounds to the square inch. The pressure was sufficiently instantaneous in character to behave as high explosives do, in that a discharge in a funnel, immersed in water, so that only the resistance due to the inertia of the latter opposed free escape of the pressure, yet resulted in shivering the funnel to atoms. As observed by Mallard and Le Chatelier, and graphically recorded, the action might be analyzed into four phases, which might be successive, or which might overlap, or of which one or more might not occur. The first phase was the non-concussive propagation; the second the ampler vibratory motion; this passed into the third or detonating stage, when the amplitude of vibration was a maximum; and the fourth stage was an extinction of the flame before complete combustion of the whole mixture had been effected either by a dissociation phenomenon or from some other cause.

A most interesting practical investigation at Columbia University has shown that these phenomena can easily be manifested in explosive engines. The presence in the combustion-chamber volume of pockets in which explosive mixtures can be imprisoned with narrow channels connecting them may easily cause the vibratory movement referred to above. The first ignition pressure receives a distinct secondary impulse, and the whole mass becomes subject to pressure waves in motion. This can be made to show itself on the indicator-card of the engine, first in pressure much above the normal, and secondly in a wave effect on the expansion line far above any record which can be attributed to inertia of the reciprocating weights of the indicator itself. This can be proved by the simple test of adding to a normal engine an extra volume to its clearance, connected thereto by a narrow neck of pipe, and connecting the indicator to this second or supplementary volume. The indicator which gave normal cards before the addition of the supplementary chamber gave strong vibratory lines after the connection of the chamber was made, all other conditions being the same. The piston compressions may produce similar vibrations, and, in the case of advanced sparking adjustment, superpose the ignition pressure upon the vibration effect, making the engine thump badly, and perhaps even stopping it before the end of the stroke is reached.

It is apparent, therefore, that the cylinder casting must be strong enough to withstand not only the computed or normal ignition pressure, but must be able to resist the very much higher stress which the explosive wave may cause. The clearance must also be free from subdivisions, from which vibratory effect can be started. The results of the explosive wave action may be compared to those due to water-hammer in steam-pipes, and the presence of excess of water in steam-engine cylinders which are not fitted with relief-valves.

229. Concluding Comment.—It will be apparent from the foregoing review of work already begun and carried forward in this field that it is at this point that the student and laboratory investigator touch most vitally upon the problems of the actual designer of the gas-engine. It is the knowledge of the formation of these explosive mixtures, their behavior in the cylinder under the conditions there prevailing, and the constants which are to be introduced as coefficients in formulæ and computations, which make this subject and its possibilities particularly inviting. It is to be hoped that additions to the stock of knowledge in existence and on record along these lines may be made both rapidly and to great extent.

CHAPTER XX.

230. Introductory. — There are difficulties inseparable at present from an effort to give satisfactory discussion of the economic aspect of the internal combustion motor. Among these are:

1. The meagerness of reliable data in unprejudiced hands. In America, while the building of small motors has been in progress for some time, the operation of large units is quite new. The builders have much accumulated information, but if it is favorable to their product they prefer to retain the facts in their possession with a view to making use of them in furthering their sales: and if unfavorable to them in competition, they prefer to retain their acquired knowledge lest it be used against them. In Europe, the industry is older, and the legislative requirements frequently compel publication of costs as features of municipal ownership or control of public service power plants. Hence there are data available there. But two difficulties appear here and should be guarded against. The one is the difference in social and labor and fuel conditions on the two sides of the Atlantic, with their effects upon all economic questions; and the other is the fact that municipal control and management (whether complicated or not with municipal ownership) has always up-to-date managed its technical and engineering matters without proper attention to upkeep and renewal. In the early days of the plant too little is set aside for proper maintenance, in order to make a good showing; and as the result of this the cost of producing the power unit becomes too great in the later years after the plant has run down. Both causes throw

529

suspicion upon the figures in the reports, and lessen their value as guides.

2. The figures of cost, which serve as basis for comparison, are changing continually. The internal combustion engine enters the power producing field as a rival or a supplanter of the steam engine. The elements of cost of the latter are, therefore, the usual elements of comparison. But the cost of steam power is not a stationary standard: new designs or new processes continually advance beyond the older standards and compel new values to be considered. Fuel consumptions and fuel costs are varying bases, and facts and figures of any year are superseded and become out of date.

3. Fuel economy and fuel costs are not the only basis upon which selection of a motor can be or should be made. In the motor car for example, when used for pleasure purposes by untrained operators, fuel expense may be entirely secondary to other considerations, such as flexibility, silence, absence of shock or jar and simplicity of detail.

In the figures which follow, therefore, the changeable character of the values used must not be lost sight of: nor the possible inapplicability of the foreign figures to American conditions of private ownership control.

4. It should also be remembered that in many industrial plants, the heat of the exhaust steam from the engines is used either in processes or in heating the buildings in winter. Costs of power in this case should not be charged with the price of such fuel or heat units as are applied to heating. If gas engines were used, steam would have to be supplied in addition; or it may happen that in effect the power fuel costs nothing, and the engine is a sort of reducing valve between the boilers and the heating coils.

5. As against this may be offset the advantage offered by the internal combustion engine as respects subdivided units of power in different buildings or parts of a plant. The small steam engine is not so economical or efficient as the large unit: the gas engine does not suffer from this limitation. Hence the

gas may be made by a central producer, and distributed by mains to a large number of small gas-engine units for use when and where required. To distribute steam from such a central station is wasteful of heat in transmission and the small steam unit is wasteful. In gas distribution there is little waste, and the small unit nearly as economical as the large one. Electrical subdivision is expensive in first cost, and the loss of efficiency in the double transformation from mechanical into electrical energy and then back again entails more loss than that of the combined producer piping and gas-engine. The gas can also be taken from the producer pipes and burned in suitable stoves for heating. In some climates the jacket water can be used in hot water circulating coils as heat radiators for warming the building, and the economy of the gas-engine be greatly enhanced.

Great caution should, therefore, be exercised in generalizing from the limited number of particular instances referred to in this chapter. Many more will be required before general conclusions should be enunciated, and the question in each case should be studied from all of its aspects.

231. The Elements of Cost. — The elements which go to make up the cost of power in gas-engines are essentially the same as in all power generating stations. The first and obvious division is into the first cost and the operating cost; but as the money to pay for the first cost of the installation in buildings and machinery will either be borrowed directly or will be withdrawn from other investments in which it is earning the usual interest, it is easier to put this yearly interest as an element of the operating cost. It is generally true that money intelligently spent in good engineering in the first cost reduces the operating expense more than the increased interest on such first cost, by diminishing repairs and renewals.

The fuel energy may be bought in the form of gas from outside makers, or it may be made from the raw material in the plant itself. In the former case the buyer pays to the seller a price for fuel which covers the latter's expenditure in manufacturing, and his interest on the plant used in distributing together with the usual profit in his business. If he makes the gas himself he pays for these same elements of cost, but usually less than he pays to the outsider, and the profits inure to himself. If the gas generating or producing plant within his own walls is large enough to derive the advantages of large scale manufacture, it will be cheaper to make his own gas from intramural producers. If on the other hand the demand for gas is very small, so that it can be met by carbureters, here also the process will be cheaper to make the gas as the engine requires and not to depend on outside concerns. While on general principles the large scale manufacture will be the cheaper, yet the losses in leakage and the expense of distribution of the gas give the producer and especially the suction producer a very great opportunity to work success-fully in competition. The elements of cost in the general case will therefore be:

1. **Interest on the First Cost.** — If the cost of the producer is to be included, the gas plant is about the same as the steam plant, less the chimney. If this is costly in the competitive steam plant, or is equated by the cost of a mechanical draft outfit of engine and fans, the gas plant comes out ahead. In large installations the individual gas-engine unit is not so large as yet as the steam-driven unit. More units are, therefore, required to attain the same maximum power. Engineers do not seem inclined to re-commend single electric units la:ger than 900 kilowatts capacity per cylinder-end with gas-engine power so that the capital invest-ment during the continuance of this condition will average about 25 per cent greater for the gas plant than the equivalent steam plant. If, on the other hand, the station is of a capacity making comparison of units possible by their output, the space occupied by a steam-engine and steam generating plant being called 100, the space for the gas-generators and engines will be 80. The cost of buildings will be as 70 to 60. If the gas is delivered from mains from without, the space will be 30 on the basis of 100 for the steam plant, and the cost as 50 to the ratios of 70 and 60 above.

2. Depreciation and Sinking Fund. — The moment that the machinery and buildings have been paid for, they are second-hand articles and are not worth what was paid for them. They also are running down in quality by use and wear and by the process of becoming obsolete by time. Sound financing demands, therefore, that each year their book value in the accounting be diminished, until after a period of *n* years (to be fixed by the owners and investors) their value shall disappear as an asset on the books. This period may be the same or shorter than that at which an investment of new machinery will have to be made if the plant is to continue in business; and at the end of this time money or available capital should be in bank or investment to pay for this renewal without impairing the capital again or bringing in new subscribers. The depreciation of gas-engines is greater or their life period shorter than that of the equivalent steam-engine, by reason of the sudden character of the stresses, the high temperature at which they work, the difficulties with dirty gas and defective lubrication and wear of the valves. Gas-engines at present also are changing type and undergoing such rapid improvement that they become old in a shorter time. The depreciation and sinking fund factor will be greater in a gas-driven than in the equivalent steam-driven plant. In the steam plant it is usually 5–7 per cent or is based on a presumption of life of 15–20 years. In the gas-engine plant the presumption of life is nearer 10 years or even as low as eight in small engines. In larger, it will be 12–15. In the motor-car for pleasure it is rarely over three or even two years, by reason of the hard service, in unskilled hands for the most part, and also from the rapid changes in design. The demand for lightness of weight induces close economy in computing bearing or wearing areas, with consequent shortened life. The sinking fund for both machinery and plant may be fixed arbitrarily by a determination to pay for the plant in so many years.

3. Renewals.—The distinction between a renewal and a repair is that a renewal makes that part as good as when first put in, and should, therefore, have its full presumption of life counted from the date of such renewal. A repair on the other hand makes good only to the degree of excellence which existed after the previous wear and tear had taken place. Renewal by installment may be the same in effect as expenditure from the sinking fund, giving a substantially new machine when these have been completed.

4. Repairs and Maintenances. — The plant may be allowed to deteriorate in quality, and usually, therefore, also in economy, by neglect in its upkeep. Leaky piston-rings and valves, lost motion in the bearings, due to cutting and wear, produce a greater invasion of operating economy in the gas-engine than in the steam-engine, particularly the compound or multiple expansion type of the latter. Dirty gas and deformations by heat, by reason of defective cooling systems, make the contact surfaces lose their shape and pressure and fuel leaks away. Repairs, properly so-called as distinguished from renewals, ought not to exceed 5 per cent in a new plant. The labor of the employees otherwise required can usually be relied on for simple repairs.

5. Fuel, Oil and Supplies. — The probable fuel expense will be treated in a following paragraph. It will increase with the age of the plant, or carelessness in maintenance, whereby pistons and valves grow leaky. More fuel is required under these conditions to furnish the same power. The use of poor oil or that ill chosen for the conditions prevents the piston-rings from performing their functions, and by its presence as gum causes execssive wear and leakage. It is almost a generalization from the facts of public or municipal control of gas-driven plants, that under their conditions of management where the responsibility for the capital investment is remote or is a community matter, the fuel or operating expense is greater before long than in the private plant of the same age. These conditions seem due to carelessness or mistaken economy, whose consequence is a lowered physical condition of the plant, and, therefore, a greater fuel waste or demand.

6. Labor. — With mechanical handling of the fuel to the producers, the labor requirement in the boiler-room of the parallel steam plant will be about the same. With hand-firing, the producer plant is the cheaper to operate. The engine-room price will be the same for the two cases.

7. Water. — The cost of the cooling water will be determined by the source from which it comes. If city water from the mains is used, the water tax in a large plant will be a considerable sum. Near a flowing river or large pond the cost will be only that of the interest and operation cost of the circulating apparatus less the return from any other uses to which the water may be put. Where water is scarce, and impounding reservoirs and cooling towers or other apparatus must be installed, the cost of these adjuncts may appear either in the elements of first cost in the accounting or may be charged to the water cost. Where air-cooling can give satisfactory results, the water expense is eliminated, but much more massive apparatus in fans and ducts is required for efficient cooling, and in large plants, where the quantity of heat in the cylinder per ignition becomes large, the air-cooling system has not as yet been found satisfactory.

8. Insurance. — The charge for insurance is often treated as an overhead or office account, detached from the engineering factors; but the design and conception of the plant may materially affect the rate. If the fuel used is gas from the city mains no effect is produced upon the rate, but the requirement is made that the exhaust pipe be carried to the open air in full metallic strength, and not simply led into a brick or masonry flue. This is to diminish fire-hazard from possible flaming explosions in such flue. If the fuel is liquid gasoline or kerosene, the fire-laws may limit the quantity on storage at one time, or may compel storage in an outside building or in a vault underground with special construction or safe guards and consequent installation expenses. If the fuel is to be made from producers, the insurance requirements may be made practically prohibitory in some cities, and their presence vitiate all insurance on the rest of the plant.

If the producer can be isolated, it may be operated without insurance. The normal view would make the producer insurance the same as the boiler-plant rate. The combustible character of the gas made in the producer should offset the higher tem-perature of the boiler furnace and its gases, and the danger from the rupture of the steam-reservoir under pressure.

9. Overhead Charges. — These are the elements of non-productive expense of the plant which do not appear directly as elements of the operating or as factors of the output. Salaries not chargeable to power production, the office expenses, taxes, rental, interest, royalties, fees and any other elements not group-able under one of the preceding headings, will be included under this heading.

In the following tables, the capital investment and the operating expense from various British and continental reports are brought together, the British pound being equated to five dollars, and the English penny to two cents in the American currency. The English Board of Trade Unit, is the kilowatt per hour, equivalent to $\frac{1000}{746} = 1\frac{1}{3}$ horse power per hour.

COST OF PRODUCER AND KILOWATT.

Capacity of Station in Kilowatts.	Cost of Gas Producers Dollars.	Capacity of P.roducers Cu ft. Per Hour.	Cost per Kilo-watt for Plant Dollars.	Cost per Kilo-watt Producers and Engines Dollars.	Source or Authority.
					Dawson
220	14,400	13,320	...	65*	
600	24,000	24,200	...	40*	Dawson
300	...	...	364	126	Lausanne
235	...	...	404	122	Zurich
170	3,725	...	180	135†	Krone

* Producers only.   † Includes generators.

COST OF 400 H. P. GERMAN ENGINES (KÖRTINO).

Item.	Cost Dollars.	
	Street Gas.	Producer Gas.
Gas Plant and Erection..........................	...	4125
Engine and Bolts	12,725	12,800
Piping...	750	625
Starter	500	250
Pressure Regulator..............................	300	...
Erection in Place................................	300	300
Foundations.....................................	400	600
Total..	14,970	18,700
Cost per Kilowatt Exclusive of Buildings.........	100	120

OPERATING COSTS (CENTS PER KILOWATT HOUR).

Item.	Location, Source or Authority.						
	Laus-anne 1899.	Rugby 1898.	Leyton 1898.	Orleans* St. Ry. 1899.	Linden 1899.*	Zurich 1899.*	Zurich 1899.*
Fuel	0.938	1.420	1.0228	.500	1.080	.588	.968
Oil and Supplies	.074	.450	.1866	.094	.312	.044	.150
Wages and Salaries.....	.370	1.710	2.4642	.600	2.664	.268	4.50
Maintenance and Repairs	.184	.288	.0718	.400	.360	.230	.230
Water..................	.066	...	...	...	...	.052	.108
Sundries...............	.024	...	...	...	.642	...	...
General Expense........	.642	2.084	1.0892	...	...	...	.154
Cost per Kilowatt......	2.373	5.952	4.8346	1.640	5.040	...	2.060
Cost of Fuel per Ton...	$6–$7.	...	...	$5.40	...	...	$6.–$6.50

* Does not include interest or sinking fund.

If these be reduced to percentages of the fuel cost as 100 per cent, the following table will result, although its applicability to American conditions is greatly lessened by the exceptionally high fuel cost.

RELATIVE COST OF ITEMS TO FUEL COST IN PERCENTAGES.

Item.	Laus-anne.	Rugby	Leyton.	Orleans.	Linden.	Zurich.	Zurich.
Fuel	100	100	100	100	100	100	100
Supplies	7	31	18	18	30	7	15
Wages	35	120	223	120	270	45	44
Maintenance	19	20	7	80	37	37	22
Water	6	...	...	...	...	6	10
Sundries	2	...	...	...	...	...	..
General Expense	68	146	104	...	...	...	15

The differing methods of accounting must also be recognized as making a general comparison or generalization impossible and erroneous.

COST OF KILOWATT PER HOUR IN CENTS.

Item.	Cost of Kilowatt per hour in cents.	
	Gas-Engine Plant.	Steam-Engine Plant.
Fuel	From 0.500 to 1.080	From 0.18 to 4.40
Oil and Supplies	" .120 " .258	" .01 " 0.76
Wages and Salaries...........	" .268 " 2.664	" .06 " 3.20
Maintenance	" .072 " 0.400	" .005 " 1.20
Total	" 0.960 " 4.402	" 0.255 " 9.56

A comparative table of the cost of the kilowatt per hour in steam plants and in gas-driven plants presents the following figures:

FUEL COST OF POWER.

Fuel and type of Plant.	Fuel required per horse-power per hour.	British thermal units required per horse-power hour.	Thermal efficien-cy.	Cost of fuel.	Cost of fuel per kilowatt per hour.
Anthracite coal ;			*Per cent*		*Cents.*
Large steam plant ..	2 pounds.....	25,000	10	$2.50 per ton	0.335
Do............	2 pounds	25,000	10	6.25 per ton	.764
Small steam plant...	7 pounds....	100,000	2½	2.50 per ton	1.34
Do...........	7 pounds.....	100,000	2½	6.25 per ton	2.948
Producer gas plant..	1¼ pounds....	14,000	18	2.50 per ton	.188
Do...........	1¼ pounds ...	14,000	18	6.25 per ton	.415
Do...........	2 pounds.....	25,000	10	2.50 per ton	.335
Do...........	2 pounds.....	25,000	10	6.25 per ton	.764
Illuminating gas	24 cubic feet.	12,000	20	1.00 per 1,000 cubic feet	2.948
Crude oil.............	1.4 pints.....	25,000	10	.04 per gallon..........	.911
Gasoline..............	1.1 pints	13,400	19	.15 per gallon..........	2.278
Do...............	1.1 pints.....	13,400	19	.30 per gallon..........	4.556
Alcohol.............	....	....	a19	.30 per gallon.........	6.700
Do...............	....	....	a19	.40 per gallon..........	8.978

a Efficiency of alcohol is assumed to be the same as that of gasoline for identical conditions of use.

This confirms qualitatively the usual central station figure for steam plants in America, where small consumers pay at a rate of five cents per kilowatt hour, and larger users at a much lower rate, still leaving a margin for overhead charges and profit in the business.

232. The Fuel Cost and Guarantee. — The preceding paragraph has been intended to call attention to the factors to be studied and evaluated when the general question of the use of the gas-engine is under consideration. When this has been settled, and the matter under advisement is one between a buyer and the vendor of a particular engine, the broader elements come less into prominence, and the question of fuel cost becomes the paramount one. The design, the generous bearing and wearing areas, the facility for repair and adjustment are all factors in reducing the operating expense and should be carefully considered in deciding on competing claims. But they are more remote than the fuel cost, and are less definitely predicable. Hence the output per fuel unit is the most usual and frequent cost unit in use. In Chapter II in general, and in paragraphs 29, 40, 46, and 47 in particular, the basis for theoretical and actual fuel-consumption for brake horse-power with a given fuel have been referred to. An English guarantee, based on gas made in a Dawson producer (paragraph 24, 25), calorific power of 12,825 B.T.U. per pound higher value, was a brake horse-power on 85 cubic feet of gas per hour, when it ran 145 B.T.U. per cubic foot. A very common American guarantee in larger plants, developing 1000 H.P. is to give one brake horse-power on one pound of coal gasified, provided the gas runs at 11,000 B.T.U. per pound.

In the last reduction, however, the final arbitrament will be the heat units (or the foot-pounds) secured by the purchaser for dollar expended to buy fuel. If this be conceded, then it will be apparent that the liquid fuel motors are to be preferred to the coal users only where reasons of convenience, manageability, mechanical or automatic supply of fuel or other practical reasons outweigh the purely financial aspect of the cost of the fuel per pound. Repeating the data of Chapter II, it will be apparent from the following table of calorific powers and prices per unit, that the heat units for a dollar, or the horse-power per hour, bear the ratios there stated. The H.P. column is derived from the B.T.U. by dividing the latter by 2545 which is the B. T. U. per H.P. per hour since $\dfrac{33,000}{778} \times 60 = 2545$, assuming efficiency to be 100 per cent.

HEAT UNITS PER DOLLAR, AND HORSE POWER PER HOUR.

Fuel.	Price in Dollars.	Calorific Power per Unit.	Heat Units for a Dollar.	Horse Power per Hour per Dollar E = 100	Actual Efficiency Probable in per cent	Cost of Fuel per Horse Power per Hour in Cents.
Small anthracites	2.50 per ton	12,500 per lb.	10,000,000	3929	10	0.25
Large anthracites	6.25 per ton	14,000 per lb.	4,500,000	1718	10	0.57
City Gas	1.00 per 1000 cu. ft.	500 per cu. ft.	500,000	196	20	2.20
Natural Gas.	.10 per 1000 cu. ft.	1000 per cu. ft.	10,000,000	3929	15	0.16
Crude Oil ..	.04 per gallon	21,000 per lb.	1,200,000	471	10	0.68
Gasoline	.10 per gallon	18,000 per lb.	1,200,000	471	19	1.14
Gasoline	.25 per gallon	do.	480,000	188	19	2.83
Kerosene ...	.10 per gallon	20,000 per lb.	1,200,000	471	19	1.13
Kerosene....	.30 per gallon	20,000 per lb.	400,000	157	19	5.00
Alcohol	.30 per gallon	13,500 per lb.	270,000	106	19	5.00

The fifth column can only be made of practical significance and use by multiplying its results by the corresponding actual value of the efficiency of the heat when transformed into mechanical work in the appropriate motor apparatus. For example, if the large steam plant using poor or cheap coal gives a horse-power on two pounds of coal per hour, or on 25,000 B.T.U. per hour, it attains an efficiency of $\dfrac{2545}{25,000} = .10$ or ten per cent. If on the other hand the more usual figure for small plants is taken, and the consumption per brake horse power approaches seven pounds, then the heat units rise to 100,000 per hour and the efficiency drops to $\dfrac{2545}{100,000}$ = 0.25 or $2\frac{1}{2}$ per cent. With the internal combustion engine, however, the discussion in paragraphs 47 and 203 indicate that the efficiency may rise to 20 per cent. Hence the results in columns 6 and 7 of the table show the application of these principles to a cost comparison of the differing motors using the various fuels. From this the fuel cost is deducible in any given case; regard being paid to any conditions affecting the applicability of the assumed efficiency ratio to the case in hand.

CHAPTER XXI.

CONCLUSION.

240. Historical Summary.—The treatment in the foregoing chapters has been intentionally free from reference in detail to the steps in the transition from the early beginnings to the present state of the art. This was done first because to have done otherwise would have been to turn aside from the main purpose, and secondly because this descriptive work has been so thoroughly done by others in previous treatises. Those interested may be referred particularly to the work of Dugald Clerk, Bryan Donkin, and Wm. Norris in England, and Gardner D. Hiscox in America, referred to in the next section. The dates of important patents may also be found from the very full lists in Clerk and Hiscox. The following summary, however, will perhaps be found useful:

1794. ROBERT STREET designs a pump driven by explosion of turpentine vapor below the motor piston.

1823. SAMUEL BROWN designs a motor to operate by atmospheric pressure; the vacuum under the piston created by an explosive flame to expel the air from a chamber, and a condensation in that chamber by a jet of water.

1833. L. W. WRIGHT. Double-acting motor, supplied with gas and air by separate pumps, and using a water-jacket.

1838. WM. BARNET. Invents compression system of gas-motor. Ignites with flame.

1855. A. V. NEWTON. Ignites charges by contact with hot metal surface.

1857. BARSANTI and MATTEUCCI propose a free-piston engine.

1860. LENOIR of Paris, through M. Hippolyte Marinoni, builds a double-acting gas-engine with electric ignition by jump-spark. It takes mixture by aspiration for half-stroke,

542

explodes it at crank position 90° from dead-centre, and expands during the econd half-stroke. Took 95 cubic feet of gas per H.P. per hour. No compression.

1861. F. MILLION proposes compression and the use of a compression- or combustion-chamber.

1862. ALPHONSE BEAU DE ROCHAS, Paris, in a patent of Jan. 7 and a later pamphlet urges the compression four-stroke cycle now known as the "Otto." C. W. Siemens of England proposes it also.

1865. PIERRE HUGON injects water into the mixed gases in the cylinder. Consumption made 85 cubic feet of gas per H.P. per hour.

1867. N. A. OTTO and EUGEN LANGEN exhibit at Paris their free-piston atmospheric engine, using 44 cubic feet of gas per H.P. per hour.

1872. GEORGE B. BRAYTON of Philadelphia designs the Brayton engine with constant-pressure heating. Called Brayton's "Ready Motor."

1873. JULIUS HOCK of Vienna patents petroleum engine.

1876. DR. OTTO brings out the Otto Silent Gas-engine, applying the Beau de Rochas cycle. Gas consumption cut down to 24 cubic feet per H.P. per hour.

1878. SIMON of Nottingham introduces Brayton cycle in England. Crossley and others begin extensive manufacture of gas-engines in England.

1885. ATKINSON Differential Engine appears with the strokes of the cycle of differing lengths.

1886. ATKINSON Cycle Engine for same purpose but with simplified mechanism.

1886. PRIESTMAN introduces oil-engine.

1892. RUDOLPH DIESEL proposes his Rational Motor.

1892. HORNSBY-AKROYD oil-engine appears.

1895. GOTTLIEB DAIMLER introduces high-speed motor.

1895 to 1907. During this period come the process of carburation to utilize liquid fuels; the utilization of producer-gas for power purposes and the development of power from blast-furnace and coke-oven waste gases; the manufacture of large-size units over 600 H.P. by John Cockerill Co. in Belgium and by Crossley and the Premier Engine in England; the design of the Westinghouse throttling governor, and the Sargent engine with cut-off governing, the rise of the natural gas-engine in large units, and the double-acting gas-engine with compression in America.

BIBLIOGRAPHY.

250. Note.—This list does not include some important sources of information in the transactions of engineering societies and in technical journals, notably such as the *Zeitschrift des Vereins Deutscher Ingenieure,* London *Engineering,* and the special gas-engineering and automobile periodicals. For these references the reader is referred to the standard technical indices of the day. It covers only such reference literature as has appeared in book form.

STRUVE, Paris, 1865. La Machine à Gaz.

JOHN BOURNE, London, 1878. Steam-, Air-, and Gas-engines.

MALLARD et LE CHATELIER, Paris, 1883. Recherches expérimentales et théoriques sur la combustion des mélanges gazeux explosifs.

WM. MACGREGOR, London, 1885. Gas-engines.

THOS. M. GOODEVE, London, 1887. The Gas-engine.

KOHLER, Leipzig, 1887. Theorie der Gas-Motoren.

WM. ROBINSON, London, 1890. Gas- and Petroleum-engines.

R. SCHÖTTLER, Braunschweig, 1890. Die Gas-Maschine. (This has a very full German bibliography.)

WEHRLIN, Paris, 1890. Moteurs à gaz et à pétrole.

GUSTAVE CHAUVEAU, Paris, 1891. Traité théorique et pratique des moteurs à gaz.

GUSTAVE RICHARD, Paris, 1892, 1893, 1894. Moteurs à gaz et à pétrole.

RUDOLPH DIESEL, Berlin, 1893. Theorie und Konstruction eines rationellen Wärmemotors.

PAUL VERMAND, Paris, 1895. Les moteurs à gaz et à pétrole.

AIMÉ WITZ, Paris, 1895. Traité théorique et pratique des moteurs à gaz et à pétrole. 2 vols.

WM. T. BRANNT, Philadelphia, 1896. Petroleum and Natural Gas.

RHYS JENKINS, London, 1896. Index to Literature on Power Locomotion on the Highway.

BOVERTON REDWOOD, London, 1896. Petroleum and its Products.

G. LEICKFELD, London, 1896. Practical Handbook on Care and Management of Gas-engines. Trans. by Geo. Richmond.

WM. NORRIS, London, 1896. Practical Treatise on the Otto Cycle Gas-engine.

BRYAN DONKIN, London, 1896. Gas-, Oil-, and Air-engines.

DUGALD CLERK, London, 1896. The Gas- and Oil-engine. (This has a full list of English patents.)

W. C. POPPLEWELL, Manchester, 1897. Elementary Treatise on Heat and Heat-engines.

B. P. WARWICK, Lynn, 1897. The Gas-engine.

A. J. WALLIS-TAYLOR, London, 1897. Motor Cars.

ELLIOT GRAFFIGNY, New York, 1898. Gas- and Petroleum-engines.

LOUIS LOCKERT, New York, 1898. Petroleum-motor Cars.

INTERNATIONAL TEXT-BOOK Co., Scranton, 1899. A Text-book on the Gas-engine.

J. F. ALLEN, Washington, 1900. Automobile Patent Digest.

E. J. STODDARD, 1900. Gas-engine Design.

GOLDINGHAM, New York, 1900. Design and Construction of Oil-engines.

W. W. BEAUMONT, Philadelphia, 1900. Motors and Motor Vehicles.

GARDNER D. HISCOX, New York, 1900. Horseless Vehicles, Automobiles, and Motor Cycles.

C. C. BRAMWELL, New York, 1901. Construction of Gasoline Motor Vehicles.

GARDNER D. HISCOX, New York, 1901. Gas-, Gasoline-, and Oil-vapor-engines. (Full list of American patents.)

E. W. ROBERTS, Cincinnati, 1901. The Gas-engine Handbook.

E. W. LONGANECKER, Indiana, 1902. The Practical Gas Engineer.

ALFRED C. HARMSWORTH, London, 1902. Motors and Motor Driving.

JAMES E. HOMANS, New York, 1902. Self-propelled Vehicles.

Rud E. Mathot, 1905. Modern Gas-Engines and Producer Gas Plants.

C. E. Lucke, 1905. Gas-Engine Design.

Rankine Kennedy, 1905. Modern Engines and Power Generators.

A. Riedler, 1905. Gross Gasmaschinen.

H. Galdner, 1905. Entwerfen und Berechnumg der Verbrenmungsmotoren.

H. Hæder, 1904. Die Gas Motoren.

S. S. Wyer, 1906. Producer Gas and Gas Producers.

E. Sorel, 1904. Carburation et Combustion dans les Moteurs a alcoöl.

L. Perissé, 1905. Les Carburateurs.

J. G. M'Intosh, 1907. Industrial Alcohol.

J. D. Roots, 1899. The Cycles of Gas- and Oil-Engines.

C. E. Lucke, 1902. The Heat Engine Problem.

S. A. Moss, 1906. Elements of Gas-Engine Design.

L. Marchis, 1905. Moteurs a Essence pour Automobiles.

D. Sidersky, 1903. Les Usages Industriels de l'alcoöl.

Professional Paper No. 48, 1906. Report on Coal Testing Plant of U. S. Geological Survey at St. Louis.

Gas Power, 1903 to date. Pub. at St. Joseph, Mich.

APPENDIX.

LOGARITHMS.

260. In arithmetical computations, the usual base of the system is 10, so that x, the logarithm for a number m, will be the exponent to which 10 is to be raised to give the quantity m, or $x = \log_{10} m$. In analytical mathematical work, the base generally employed is not 10, but is represented by e whose value is 2.71828 +. To convert common or Briggs logarithms into Napierian logarithms, the former are to be multiplied by 2.3026.

The equation of the hyperbola in the form $xy = $ constant leads to the deduction that the area between the hyperbolic curve and its nearest asymptote cut off by two ordinates parallel to the other asymptote and distant respectively from the origin by a and b will be proportional to $\log \dfrac{b}{a}$. Hence it will be true that the integral of $\dfrac{dx}{x}$ will be the hyperbolic logarithm of x. To save trouble of conversion, a table is appended covering the usual ranges required.

HYPERBOLIC LOGARITHMS.

No.	Log.	No.	Log.	No.	Log.	No.	Log.	No.	Log.
1.01	.0099	1.20	.1823	1.39	.3293	1.58	.4574	1.77	.5710
1.02	.0198	1.21	.1906	1.40	.3365	1.59	.4637	1.78	.5766
1.03	.0296	1.22	.1988	1.41	.3436	1.60	.4700	1.79	.5822
1.04	.0392	1·23	.2070	1.42	.3507	1.61	.4762	1.80	.5878
1.05	.0488	1.24	.2151	1.43	.3577	1.62	.4824	1.81	.5933
1.06	.0583	1.25	.2231	1.44	.3646	1.63	.4886	1.82	.5988
1.07	.0677	1.26	.2311	1.45	.3716	1.64	.4947	1.83	.6043
1.08	.0770	1.27	.2390	1.46	.3784	1.65	.5008	1.84	.6098
1.09	.0862	1.28	.2469	1.47	.3853	1.66	.5068	1.85	.6152
1.10	.0953	1.29	.2546	1.48	.3920	1.67	.5128	1.86	.6206
1.11	.1044	1.30	.2624	1.49	.3988	1.68	.5188	1.87	.6259
1.12	.1133	1.31	.2700	1.50	.4055	1.69	.5247	1.88	.6313
1.13	.1222	1.32	.2776	1.51	.4121	1.70	.5306	1.89	.6366
1.14	.1310	1.33	.2852	1.52	.4187	1.71	.5365	1.90	.6419
1.15	.1398	1.34	.2927	1.53	.4253	1.72	.5423	1.91	.6471
1.16	.1484	1.35	.3001	1.54	.4318	1.73	.5481	1.92	.6523
1.17	.1570	1.36	.3075	1.55	.4383	1.74	.5539	1.93	.6575
1.18	.1655	1.37	.3148	1.56	.4447	1.75	.5596	1.94	.6627
1.19	.1740	1.38	.3221	1.57	.4511	1.76	.5653	1.95	.6678

HYPERBOLIC LOGARITHMS.

No.	Log.	No.	Log.	No.	Log.	No.	Log.	No.	Log.
1.96	.6729	2.66	.9783	3.36	1.2119	4.06	1.4012	4.76	1.5602
1.97	.6780	2.67	.9821	3.37	1.2149	4.07	1.4036	4.77	1.5623
1.98	.6831	2.68	.9858	3.38	1.2179	4.08	1.4061	4.78	1.5644
1.99	.6881	2.69	.9895	3.39	1.2208	4.09	1.4085	4.79	1.5665
2.00	.6931	2.70	.9933	3.40	1.2238	4.10	1.4110	4.80	1.5686
2.01	.6981	2.71	.9969	3.41	1.2267	4.11	1.4134	4.81	1.5707
2.02	.7031	2.72	1.0006	3.42	1.2296	4.12	1.4159	4.82	1.5728
2.03	.7080	2.73	1.0043	3.43	1.2326	4.13	1.4183	4.83	1.5748
2.04	.7129	2.74	1.0080	3.44	1.2355	4.14	1.4207	4.84	1.5769
2.05	.7178	2.75	1.0116	3.45	1.2384	4.15	1.4231	4.85	1.5790
2.06	.7227	2.76	1.0152	3.46	1.2413	4.16	1.4255	4.86	1.5810
2.07	.7275	2.77	1.0188	3.47	1.2442	4.17	1.4279	4.87	1.5831
2.08	.7324	2.78	1.0225	3.48	1.2470	4.18	1.4303	4.88	1.5851
2.09	.7372	2.79	1.0260	3.49	1.2499	4.19	1.4327	4.89	1.5872
2.10	.7419	2.80	1.0296	3.50	1.2528	4.20	1.4351	4.90	1.5892
2.11	.7467	2.81	1.0332	3.51	1.2556	4.21	1.4375	4.91	1.5913
2.12	.7514	2.82	1.0367	3.52	1.2585	4.22	1.4398	4.92	1.5933
2.13	.7561	2.83	1.0403	3.53	1.2613	4.23	1.4422	4.93	1.5953
2.14	.7608	2.84	1.0438	3.54	1.2641	4.24	1.4446	4.94	1.5974
2.15	.7655	2.85	1.0473	3.55	1.2669	4.25	1.4469	4.95	1.5994
2.16	.7701	2.86	1.0508	3.56	1.2698	4.26	1.4493	4.96	1.6014
2.17	.7747	2.87	1.0543	3.57	1.2726	4.27	1.4516	4.97	1.6034
2.18	.7793	2.88	1.0578	3.58	1.2754	4.28	1.4540	4.98	1.6054
2.19	.7839	2.89	1.0613	3.59	1.2782	4.29	1.4563	4.99	1.6074
2.20	.7885	2.90	1.0647	3.60	1.2809	4.30	1.4586	5.00	1.6094
2.21	.7930	2.91	1.0682	3.61	1.2837	4.31	1.4609	5.01	1.6114
2.22	.7975	2.92	1.0716	3.62	1.2865	4.32	1.4633	5.02	1.6134
2.23	.8020	2.93	1.0750	3.63	1.2892	4.33	1.4656	5.03	1.6154
2.24	.8065	2.94	1.0784	3.64	1.2920	4.34	1.4679	5.04	1.6174
2.25	.8109	2.95	1.0813	3.65	1.2947	4.35	1.4702	5.05	1.6194
2.26	.8154	2.96	1.0852	3.66	1.2975	4.36	1.4725	5.06	1.6214
2.27	.8198	2.97	1.0886	3.67	1.3002	4.37	1.4748	5.07	1.6233
2.28	.8242	2.98	1.0919	3.68	1.3029	4.38	1.4770	5.08	1.6253
2.29	.8286	2.99	1.0953	3.69	1.3056	4.39	1.4793	5.09	1.6273
2.30	.8329	3.00	1.0986	3.70	1.3083	4.40	1.4816	5.10	1.6292
2.31	.8372	3.01	1.1019	3.71	1.3110	4.41	1.4839	5.11	1.6312
2.32	.8416	3.02	1.1053	3.72	1.3137	4.42	1.4861	5.12	1.6332
2.33	.8458	3.03	1.1086	3.73	1.3164	4.43	1.4884	5.13	1.6351
2.34	.8502	3.04	1.1119	3.74	1.3191	4.44	1.4907	5.14	1.6371
2.35	.8544	3.05	1.1151	3.75	1.3218	4.45	1.4929	5.15	1.6390
2.36	.8587	3.06	1.1184	3.76	1.3244	4.46	1.4951	5.16	1.6409
2.37	.8629	3.07	1.1217	3.77	1.3271	4.47	1.4974	5.17	1.6429
2.38	.8671	3.08	1.1249	3.78	1.3297	4.48	1.4996	5.18	1.6448
2.39	.8713	3.09	1.1282	3.79	1.3324	4.49	1.5019	5.19	1.6467
2.40	.8755	3.10	1.1314	3.80	1.3350	4.50	1.5041	5.20	1.6487
2.41	.8796	3.11	1.1346	3.81	1.3376	4.51	1.5063	5.21	1.6506
2.42	.8838	3.12	1.1378	3.82	1.3403	4.52	1.5085	5.22	1.6525
2.43	.8879	3.13	1.1410	3.83	1.3429	4.53	1.5107	5.23	1.6544
2.44	.8920	3.14	1.1442	3.84	1.3455	4.54	1.5129	5.24	1.6563
2.45	.8961	3.15	1.1474	3.85	1.3481	4.55	1.5151	5.25	1.6582
2.46	.9002	3.16	1.1506	3.86	1.3507	4.56	1.5173	5.26	1.6601
2.47	.9042	3.17	1.1537	3.87	1.3533	4.57	1.5195	5.27	1.6620
2.48	.9083	3.18	1.1569	3.88	1.3558	4.58	1.5217	5.28	1.6639
2.49	.9123	3.19	1.1600	3.89	1.3584	4.59	1.5239	5.29	1.6658
2.50	.9163	3.20	1.1632	3.90	1.3610	4.60	1.5261	5.30	1.6677
2.51	.9203	3.21	1.1663	3.91	1.3635	4.61	1.5282	5.31	1.6696
2.52	.9243	3.22	1.1694	3.92	1.3661	4.62	1.5304	5.32	1.6715
2.53	.9282	3.23	1.1725	3.93	1.3686	4.63	1.5326	5.33	1.6734
2.54	.9322	3.24	1.1756	3.94	1.3712	4.64	1.5347	5.34	1.6752
2.55	.9361	3.25	1.1787	3.95	1.3737	4.65	1.5369	5.35	1.6771
2.56	.9400	3.26	1.1817	3.96	1.3762	4.66	1.5390	5.36	1.6790
2.57	.9439	3.27	1.1848	3.97	1.3788	4.67	1.5412	5.37	1.6808
2.58	.9478	3.28	1.1878	3.98	1.3813	4.68	1.5433	5.38	1.6827
2.59	.9517	3.29	1.1909	3.99	1.3838	4.69	1.5454	5.39	1.6845
2.60	.9555	3.30	1.1939	4.00	1.3863	4.70	1.5476	5.40	1.6864
2.61	.9594	3.31	1.1969	4.01	1.3888	4.71	1.5497	5.41	1.6882
2.62	.9632	3.32	1.1999	4.02	1.3913	4.72	1.5518	5.42	1.6901
2.63	.9670	3.33	1.2030	4.03	1.3938	4.73	1.5539	5.43	1.6919
2.64	.9708	3.34	1.2060	4.04	1.3962	4.74	1.5560	5.44	1.6938
2.65	.9746	3.35	1.2090	4.05	1.3987	4.75	1.5581	5.45	1.6956

HYPERBOLIC LOGARITHMS.

No.	Log.	No.	Log.	No.	Log.	No.	Log.	No.	Log.
5.46	1.6974	6.16	1.8181	6 86	1.9257	7.56	2.0229	8.32	2.1187
5.47	1.6993	6.17	1.8197	6.87	1.9272	7.57	2.0242	8.34	2.1211
5.48	1.7011	6.18	1.8213	6.88	1.9286	7.58	2.0255	8.36	2.1235
5.49	1.7029	6.19	1.8229	6.89	1.9301	7.59	2.0268	8.38	2.1258
5.50	1.7047	6.20	1.8245	6.90	1.9315	7.60	2.0281	8.40	2.1282
5.51	1.7066	6.21	1.8262	6.91	1.9330	7.61	2.0295	8 42	2.1306
5.52	1.7084	6.22	1.8278	6.92	1.9344	7.62	2.0308	8.44	2.1330
5.53	1.7102	6.23	1.8294	6.93	1.9359	7.63	2.0321	8.46	2.1353
5.54	1.7120	6.24	1.8310	6.94	1.9373	7.64	2.0334	8.48	2.1377
5.55	1.7138	6.25	1.8326	6.95	1.9387	7.65	2.0347	8.50	2.1401
5.56	1.7156	6.26	1.8342	6.96	1.9402	7.66	2.0360	8.52	2.1424
5.57	1.7174	6.27	1.8358	6.97	1.9416	7.67	2.0373	8.54	2.1448
5.58	1.7192	6.28	1.8374	6.98	1.9430	7.68	2.0386	8.56	2.1471
5.59	1.7210	6.29	1.8390	6.99	1.9445	7.69	2.0399	8.58	2.1494
5.60	1.7228	6.30	1.8405	7.00	1.9459	7.70	2.0412	8.60	2.1518
5.61	1.7246	6.31	1.8421	7.01	1.9473	7.71	2.0425	8.62	2.1541
5.62	1.7263	6.32	1.8437	7.02	1.9488	7.72	2.0438	8.64	2.1564
5.63	1.7281	6.33	1.8453	7.03	1.9502	7.73	2.0451	8.66	2.1587
5.64	1.7299	6.34	1.8469	7.04	1.9516	7.74	2.0464	8.68	2.1610
5.65	1.7317	6.35	1.8485	7.05	1.9530	7.75	2.0477	8.70	2.1633
5.66	1.7334	6.36	1.8500	7.06	1.9544	7.76	2.0490	8.72	2.1656
5.67	1.7352	6.37	1.8516	7.07	1.9559	7.77	2.0503	8.74	2.1679
5.68	1.7370	6.38	1.8532	7.08	1.9573	7.78	2.0516	8.76	2.1702
5.69	1.7387	6.39	1.8547	7.09	1.9587	7.79	2.0528	8.78	2.1725
5.70	1.7405	6.40	1.8563	7.10	1.9601	7.80	2.0541	8.80	2.1748
5.71	1.7422	6.41	1.8579	7.11	1.9615	7.81	2.0554	8.82	2.1770
5.72	1.7440	6.42	1.8594	7.12	1.9629	7.82	2.0567	8 84	2.1793
5.73	1.7457	6.43	1.8610	7.13	1.9643	7.83	2.0580	8.86	2.1815
5.74	1.7475	6.44	1.8625	7.14	1.9657	7.84	2.0592	8.88	2.1838
5.75	1.7492	6.45	1.8641	7.15	1.9671	7.85	2.0605	8.90	2.1861
5.76	1.7509	6.46	1.8656	7.16	1.9685	7.86	2.0618	8.92	2.1883
5.77	1.7527	6.47	1.8672	7.17	1.9699	7.87	2.0631	8.94	2.1905
5.78	1.7544	6.48	1.8687	7.18	1.9713	7.88	2.0643	8.96	2.1928
5.79	1.7561	6.49	1.8703	7.19	1.9727	7.89	2.0656	8.98	2.1950
5.80	1.7579	6.50	1.8718	7.20	1.9741	7.90	2.0669	9.00	2.1972
5.81	1.7596	6.51	1.8733	7.21	1.9754	7.91	2.0681	9.02	2.1994
5.82	1.7613	6.52	1.8749	7.22	1.9769	7.92	2.0694	9.04	2.2017
5.83	1.7630	6.53	1.8764	7.23	1.9782	7.93	2.0707	9.06	2.2039
5.84	1.7647	6.54	1.8779	7.24	1.9796	7.94	2.0719	9.08	2.2061
5.85	1.7664	6.55	1.8795	7.25	1.9810	7.95	2.0732	9.10	2.2083
5.86	1.7681	6.56	1.8810	7 26	1.9824	7.96	2.0744	9.12	2.2105
5.87	1.7699	6.57	1.8825	7.27	1.9838	7.97	2.0757	9.14	2.2127
5.88	1.7716	6.58	1.8840	7.28	1.9851	7.98	2.0769	9.16	2.2148
5.89	1.7733	6.59	1.8856	7.29	1.9865	7.99	2.0782	9.18	2.2170
5.90	1.7750	6.60	1.8871	7.30	1.9879	8.00	2.0794	9.20	2.2192
5.91	1.7766	6.61	1.8886	7.31	1.9892	8.01	2.0807	9.22	2.2214
5.92	1.7783	6.62	1.8901	7.32	1.9906	8.02	2.0819	9.24	2.2235
5.93	1.7800	6.63	1.8916	7.33	1.9920	8.03	2.0832	9.26	2.2257
5.94	1.7817	6.64	1.8931	7.34	1.9933	8.04	2.0844	9.28	2.2279
5.95	1.7834	6.65	1.8946	7.35	1.9947	8.05	2 0857	9.30	2.2300
5.96	1.7851	6.66	1.8961	7.36	1.9961	8.06	2.0869	9.32	2.2322
5.97	1.7867	6.67	1.8976	7.37	1.9974	8.07	2.0882	9.34	2.2343
5.98	1.7884	6.68	1.8991	7.38	1.9988	8.08	2.0894	9.36	2.2364
5.99	1.7901	6.69	1.9006	7.39	2.0001	8.09	2.0906	9.38	2.2386
6.00	1.7918	6.70	1.9021	7.40	2.0015	8.10	2.0919	9.40	2.2407
6.01	1.7934	6.71	1.9036	7.41	2.0028	8.11	2.0931	9.42	2.2428
6.02	1.7951	6.72	1.9051	7.42	2.0041	8.12	2.0943	9.44	2.2450
6.03	1.7967	6.73	1.9066	7.43	2.0055	8.13	2.0956	9.46	2.2471
6.04	1.7984	6.74	1.9081	7.44	2.0069	8.14	2.0968	9.48	2.2492
6.05	1.8001	6.75	1.9095	7.45	2.0082	8.15	2.0980	9.50	2.2513
6.06	1.8017	6.76	1.9110	7.46	2.0096	8.16	2.0992	9.52	2 2534
6.07	1.8034	6.77	1.9125	7.47	2.0108	8.17	2.1005	9.54	2.2555
6.08	1.8050	6.78	1.9140	7.48	2.0122	8.18	2.1017	9.56	2.2576
6.09	1.8066	6.79	1.9155	7.49	2.0136	8.19	2.1029	9.58	2.2597
6.10	1.8083	6.80	1.9169	7.50	2.0149	8.20	2.1041	9 60	2.2618
6.11	1.8099	6.81	1.9184	7.51	2.0162	8.22	2.1066	9 62	2.2638
6.12	1.8116	6.82	1.9199	7.52	2.0176	8.24	2.1090	9 64	2.2659
6.13	1.8132	6.83	1.9213	7.53	2.0189	8.26	2.1114	9.66	2.2680
6.14	1.8148	6.84	1.9228	7.54	2.0202	8.28	2.1138	9.68	2.2701
6.15	1.8165	6.85	1.9242	7.55	2.0215	8.30	2.1163	9.70	2.2721

APPENDIX.

HYPERBOLIC LOGARITHMS.

No.	Log.	No.	Log.	No.	Log.	No.	Log.	No.	Log.
9.72	2.2742	10.25	2.3279	14.00	2.6391	21	3.0445	36	3.5835
9.74	2.2762	10.50	2.3513	14.25	2.6567	22	3.0910	37	3.6109
9.76	2.2783	10.75	2.3749	14.50	2.6740	23	3.1355	38	3.6376
9.78	2.2803	11.00	2.3979	14.75	2.6913	24	3.1781	39	3.6636
9.80	2.2824	11.25	2.4201	15.00	2.7081	25	3.2189	40	3.6889
9.82	2.2844	11.50	2.4430	15.50	2.7408	26	3.2581	41	3.7136
9.84	2.2865	11.75	2.4636	16.00	2.7726	27	3.2958	42	3.7377
9.86	2.2885	12.00	2.4849	16.50	2.8034	28	3.3322	43	3.7612
9.88	2.2905	12.25	2.5052	17.00	2.8332	29	3.3673	44	3.7842
9.90	2.2925	12.50	2.5262	17.50	2.8621	30	3.4012	45	3.8067
9.92	2.2946	12.75	2.5455	18.00	2.8904	31	3.4340	46	3.8286
9.94	2.2966	13.00	2.5649	18.50	2.9173	32	3.4657	47	3.8501
9.96	2.2986	13.25	2.5840	19.00	2.9444	33	3.4965	48	3.8712
9.98	2.3006	13.50	2.6027	19.50	2.9703	34	3.5263	49	3.8918
10.00	2.3026	13.75	2.6211	20.00	2.9957	35	3.5553	50	3.9120

INDEX.

553

Lightning Source UK Ltd.
Milton Keynes UK
UKHW021551110119
335297UK00008B/588/P